METHODS IN MOLECULAR BIOLOGY

Series Editor
John M. Walker
School of Life Sciences
University of Hertfordshire
Hatfield, Hertfordshire, AL10 9AB, UK

For further volumes:
http://www.springer.com/series/7651

Microbial Biofilms

Methods and Protocols

Edited by

Gianfranco Donelli

Microbial Biofilm Laboratory, Fondazione Santa Lucia IRCCS, Rome, Italy

Humana Press

Editor
Gianfranco Donelli
Microbial Biofilm Laboratory
Fondazione Santa Lucia IRCCS
Rome, Italy

ISSN 1064-3745 ISSN 1940-6029 (electronic)
ISBN 978-1-4939-0466-2 ISBN 978-1-4939-0467-9 (eBook)
DOI 10.1007/978-1-4939-0467-9
Springer New York Heidelberg Dordrecht London

Library of Congress Control Number: 2014934151

Printed on acid-free paper

Humana Press is a brand of Springer
Springer is part of Springer Science+Business Media (www.springer.com)

Preface

Starting from the late 1970s, the biofilm's pioneers Bill Costerton and Niels Hoiby have provided significant information on the ability of microorganisms to stick on biotic and abiotic surfaces and to build communities of cells closely interacting with each other within a self-produced exopolysaccharide matrix. However, only since the early 1990s it has been possible to observe, by a confocal laser scanning microscope, living biofilms of *Pseudomonas aeruginosa*, *Pseudomonas fluorescens*, and *Vibrio parahaemolyticus*, stained with viable fluorescent probes. Biofilms were found to be highly hydrated open structures constituted of 73 to 98 % of extracellular substances and large void spaces allowing the circulation of nutrients and signaling molecules and the removal of microbial catabolites. Thus, the so-called mushroom model was proposed to schematically represent the tridimensional structure of these microbial communities, the dynamics of their sessile growth, and the main interactions among the cells and the surrounding environment.

This novel view of the microbial world has led us in the last decades to the consciousness of the predominance of biofilms not only in natural or engineered ecosystems but also in the human body. As biofilms in the different niches are concerned, a new awareness has been acquired on the pivotal role that these sessile-growing communities of microorganisms play in a number of environmental processes: from the biofouling to the biocorrosion of the pipelines of concrete wastewater pipes, to the clogging of the pipelines in the dairy industry, to the deterioration of stones, frescoes, paintings, books, and other ancient remains. And again, the understanding that most of the chronic infections in humans, including the oral, lung, vaginal, and foreign body-associated infections, are biofilm-based, has prompted the need to design new and properly focused preventive and therapeutic strategies for these diseases. In this framework, the consensus conference organized in 2013 by Niels Hoiby under the umbrella of the Study Group for Biofilms of the European Society of Clinical Microbiology and Infectious Diseases (ESCMID) deserves to be mentioned. The objective of this initiative, made possible by the active contribution of a selected number of scientists working on biofilms of medical interest, has been to draft the "ESCMID guidelines for the diagnosis and treatment of biofilm infections" to be published early in 2014. Of course, the detailed description of most of the better established and validated experimental procedures to investigate microbial biofilms contained in the present book will be of paramount importance for all of those involved in the practical application of the abovementioned guidelines.

In fact, most of the currently available methods and protocols to investigate bacterial and fungal biofilms have been exhaustively illustrated and critically annotated in the 25 chapters by authors well known for their relevant experience in the respective fields. The book has joined together microbiologists and specialists in infectious diseases, hygiene, and public health involved in exploring different aspects of microbial biofilms as well as in designing new methods and/or developing innovative laboratory protocols. Chapters have been subgrouped by dividing the experimental approaches suitable for studying biofilms in health and disease from those more appropriate to assay antibiofilm compounds or evaluate antimicrobial strategies and from those regarding the application of methods to detect biofilms growing in the environment or affecting manufacturing plants. In the whole, readers will have at their disposal a precious working tool to perform experiments focused on both the structural and functional properties of single- and multi-species biofilms as well as their response to matrix-dissolving agents, biocides, sanitizers, and antimicrobial molecules. In this regard, advanced techniques such as the multiplex fluorescence in situ hybridization and the chip calorimetry, and innovative antibiofilm strategies as the photodynamic therapy or the bacteriophage attack, are described. Microbiological methods for in vitro screening of bacterial biofilm inhibitors and antifungal compounds are also detailed. Researchers interested in methods based on in vitro or in vivo biofilm observations, in static or dynamic conditions, by fluorescence, confocal, and scanning electron microscopy, will find in this book all the relative information provided by expert guides, each chapter being rich of useful practical suggestions and warnings. Specific chapters also deal with the most advanced animal models, including the nonmammalian ones, to investigate bacterial and fungal biofilms. Other contributions of particular interest are those related to assay protocols for staphylococcal and enterococcal quorum sensing systems, to study the pharmacokinetics and pharmacodynamics of antibiotics in biofilm-related infections, and to evaluate the efficacy of antibiotic-loaded polymers and polymeric nanoparticles.

I am sure that all the "biofilm's lovers" will enjoy this book.

Rome, Italy *Gianfranco Donelli*

Contents

Part II Investigations on Anti-biofilm Compounds and Strategies

Part III Investigations on Biofilms in the Environment and Manufacturing Plants

Contributors

ALI AL-AHMAD • *Department of Operative Dentistry and Periodontology, Albert-Ludwigs-University, Freiburg, Germany*

MARIOS ARVANITIS • *Infectious Diseases Division, Department of Medicine, Rhode Island Hospital, Warren Alpert Medical School of Brown University, Providence, RI, USA*

JOANA AZEREDO • *Centre of Biological Engineering, University of Minho, Campus de Gualtar Braga, Portugal*

MARC M. BAUM • *Oakcrest Institute of Science, Pasadena, CA, USA*

THOMAS BJARNSHOLT • *Department of International Health, Immunology, and Microbiology, University of Copenhagen, Copenhagen, Denmark; Department of Clinical Microbiology, Copenhagen University Hospital, Copenhagen, Denmark*

ROMAIN BRIANDET • *INRA, UMR 1319 Micalis, Jouy-en-Josas, France*

ARNAUD BRIDIER • *IRSTEA, HBAN, Antony, France*

FRIEDERIKE BUCHHOLZ • *Department of Environmental Microbiology, UFZ-Helmholtz Centre for Environmental Research, Leipzig, Germany*

SARA A. BURGESS • *Institute of Food, Nutrition and Human Health, Massey University and Fonterra Research and Development Centre, Palmerston North, New Zealand*

BETH BURGWYN FUCHS • *Infectious Diseases Division, Department of Medicine, Rhode Island Hospital, Warren Alpert Medical School of Brown University, Providence, RI, USA*

FRANCESCA CAPPITELLI • *Dipartimento di Scienze per gli Alimenti, la Nutrizione e l'Ambiente, Università degli Studi di Milano, Milan, Italy*

JYOTSNA CHANDRA • *Center for Medical Mycology and Department of Dermatology, University Hospitals Case Medical Center and Case Western Reserve University, Cleveland, OH, USA*

WEAN SIN CHEOW • *School of Chemical and Biomedical Engineering, Nanyang Technological University, Singapore, Singapore*

OANA CIOFU • *Department of Clinical Microbiology, Rigshospitalet-Copenhagen University Hospital, Copenhagen, Denmark*

GIANFRANCO DONELLI • *Microbial Biofilm Laboratory, Fondazione Santa Lucia IRCCS, Rome, Italy*

MOHAMMED DWIDAR • *School of Nano-Bioscience and Chemical Engineering, Ulsan National Institute of Science and Technology, Ulsan, Republic of Korea*

STEVE H. FLINT • *Institute of Food, Nutrition and Human Health, Massey University, Palmerston North 4474, New Zealand*

IOLANDA FRANCOLINI • *Department of Chemistry, Sapienza University, Rome, Italy*

MAHMOUD A. GHANNOUM • *Center for Medical Mycology and Department of Dermatology, University Hospitals Case Medical Center and Case Western Reserve University, Cleveland, OH, USA*

MARION GIRARDOT • *Laboratory of ecology and biology of the interactions, Faculty of Medicine Pharmacy, University of Poitiers, Poitiers, France*

FRANCESCO GIULIANI • *Molteni Therapeutics S.r.l., Scandicci, Florence, Italy*

VICENTE GOMEZ-ALVAREZ • *U.S. Environmental Protection Agency, Office of Research and Development, Cincinnati, OH, USA*
MANJULA GUNAWARDANA • *Oakcrest Institute of Science, Pasadena, CA, USA*
KUNN HADINOTO • *School of Chemical and Biomedical Engineering, Nanyang Technological University, Singapore, Singapore*
LUANNE HALL-STOODLEY • *NIHR Wellcome Trust Clinical Research Facility, Southampton, UK; Center for Microbial Interface Biology, College of Medicine, The Ohio State University, Columbus, OH, USA*
ELMAR HELLWIG • *Department of Operative Dentistry and Periodontology, Albert-Ludwigs-University, Freiburg, Germany*
MARGARET C. HENK • *Louisiana State University, Baton Rouge, LA, USA*
WANG HENGZHUANG • *Department of Clinical Microbiology, Rigshospitalet-Copenhagen University Hospital, Copenhagen, Denmark*
NIELS HØIBY • *Department of Clinical Microbiology, Rigshospitalet-Copenhagen University Hospital, Copenhagen, Denmark*
CHRISTINE IMBERT • *Laboratory of ecology and biology of the interactions, Faculty of Medicine Pharmacy, University of Poitiers, Poitiers, France*
JEFFREY B. KAPLAN • *Department of Biology, American University, Washington, DC, USA*
LAMPRINI KARYGIANNI • *Department of Operative Dentistry and Periodontology, Albert-Ludwigs-University, Freiburg, Germany*
PAOLO LANDINI • *Department of Biosciences, Università degli Studi di Milano, Milan, Italy*
JOHN R. LAWRENCE • *Environment Canada, Saskatoon, Saskatchewan, Canada*
LIVIA LEONI • *Department of Sciences, Università "Roma Tre", Rome, Italy*
DENISE LINDSAY • *Fonterra Research and Development Centre, Palmerston North, New Zealand*
FRIDA MARIANA MORAIS • *Department of Environmental Microbiology, UFZ-Helmholtz Centre for Environmental Research, Leipzig, Germany*
THOMAS MASKOW • *Department of Environmental Microbiology, UFZ-Helmholtz Centre for Environmental Research, Leipzig, Germany*
ELEFTHERIOS MYLONAKIS • *Infectious Diseases Division, Department of Medicine, Rhode Island Hospital, Warren Alpert Medical School of Brown University, Providence, RI, USA*
ROBERT J. MITCHELL • *School of Nano-Bioscience and Chemical Engineering, Ulsan National Institute of Science and Technology, Ulsan, Republic of Korea*
CLAUS MOSER • *Department of Clinical Microbiology, Rigshospitalet, Copenhagen University Hospital, Copenhagen, Denmark; Department of International Health, Immunology and Microbiology, Rigshospitalet, Copenhagen University, Copenhagen, Denmark*
JIRO NAKAYAMA • *Department of Bioscience and Biotechnology, Faculty of Agriculture, Kyushu University, Fukuoka, Japan*
THOMAS R. NEU • *Helmholtz Centre for Environmental Research, UFZ, Magdeburg, Germany*
LAURA NISTICO • *Center for Genomic Sciences, Allegheny-Singer Research Institute, Pittsburgh, PA, USA*
ERIC PEARLMAN • *Department of Ophthalmology and Visual Sciences, Case Western Reserve University, Cleveland, OH, USA*
ANTONELLA PIOZZI • *Department of Chemistry, Sapienza University, Rome, Italy*
ANDREA POLO • *Dipartimento di Scienze per gli Alimenti, la Nutrizione e l'Ambiente, Università degli Studi di Milano, Milan, Italy*
MARK SHIRTLIFF • *Department of Microbial Pathogenesis, Dental School and Department of Microbiology and Immunology, School of Medicine, University of Maryland, Baltimore, MD, USA*

Akane Shojima • *Department of Bioscience and Biotechnology, Faculty of Agriculture, Kyushu University, Fukuoka, Japan*
Sanna Sillankorva • *Centre of Biological Engineering, University of Minho, Campus de Gualtar Braga, Portugal*
Claus Sternberg • *Department of Systems Biology, Technical University of Denmark, Lyngby, Denmark*
Paul Stoodley • *Center for Genomic Sciences, Allegheny-Singer Research Institute, Pittsburgh, PA, USA; National Center for Advanced Tribology, University of Southampton, Southampton, UK; Center for Microbial Interface Biology and Department of Orthopeadics, The Ohio State University, Columbus, OH, USA*
Shuichi Takayama • *School of Nano-Bioscience and Chemical Engineering, Ulsan National Institute of Science and Technology, Ulsan, Republic of Korea; Department of Biomedical Engineering and Macromolecular Science and Engineering Program, University of Michigan, Ann Arbor, MI, USA*
Kim Thomsen • *Department of Clinical Microbiology, Rigshospitalet-Copenhagen University Hospital, Copenhagen, Denmark*
Hannah Trøstrup • *Department of Clinical Microbiology, Rigshospitalet-Copenhagen University Hospital, Copenhagen, Denmark*
Federica Villa • *Dipartimento di Scienze per gli Alimenti, la Nutrizione e l'Ambiente, Università degli Studi di Milano, Milan, Italy*
Claudia Vuotto • *Microbial Biofilm Laboratory, Fondazione Santa Lucia IRCCS, Rome, Italy*
Paul Webster • *Oakcrest Institute of Science, Pasadena, CA, USA*

Part I

Investigations on Biofilms in Health and Disease

Chapter 1

Methods for Dynamic Investigations of Surface-Attached In Vitro Bacterial and Fungal Biofilms

Claus Sternberg, Thomas Bjarnsholt, and Mark Shirtliff

Abstract

Three dynamic models for the investigation of in vitro biofilm formation are described in this chapter. In the 6-well plate assay presented here, the placing of the plate on a rotating platform provides shear, thereby making the system dynamic with respect to the static microtiter assay.

The second reported model, especially suitable for harvesting high amounts of cells for transcriptomic or proteomic investigations, is based on numerous glass beads placed in a flask incubated with shaking on a rotating platform, thus increasing the surface area for biofilm formation. Finally, the flow-cell system, that is the driving model for elucidating the biofilm-forming process in vitro as well as the biofilm tolerance towards antibiotics and host defense components, is illustrated here.

Key words 6-Well plate assay, Glass beads, Flow-cell system, Confocal laser scanning microscopy

1 Introduction

Surface-attached biofilms can be studied using many different experimental setups. In this chapter we have chosen to describe three dynamic models for in vitro biofilm formation. All have advantages and disadvantages depending on the experiment and the desired outcome.

The first model is based on 6-well microtiter plates. Microtiter biofilm model was originally a static assay that was developed to study the ability of coagulase-negative staphylococci to adhere to surfaces [1]. At present time two versions of high-throughput screening, static microtiter plate assays exist, one where the biofilm is formed in the wells of the plate [2] and the Calgary Biofilm Device [3], which are 96 (or more) pegs which fits into microtiter plates. These assays can be used to test for biomass buildup, by staining the biomass using crystal violet. Crystal violet staining on the other hand does not discriminate between live and dead bacteria. To test whether the bacteria are being killed in these assays, the bacteria needs to be plated for CFU. These static assays are typically used for screening

Gianfranco Donelli (ed.), *Microbial Biofilms: Methods and Protocols*, Methods in Molecular Biology, vol. 1147,
DOI 10.1007/978-1-4939-0467-9_1,

numerous mutants for biofilm forming capacity [4] or for the first screening of antibiofilm drugs both in terms of killing and dispersal [5]. The 6-well plate assays we present here are still very easy to use and control; however, the embedded cover slip can be removed and examined by microscopy, and placing the plate on a rotating platform provides shear, thereby making the system more dynamic. This is an advantage when investigating more than just biomass buildup. The second model we present is based on numerous glass beads in a flask incubated with shaking on a rotating platform. This increases the surface area for biofilm formation but still has the dynamic shear forces. This is especially important for harvesting high amounts of cells for transcriptomic or proteomic investigations.

To study biofilm development in real time and the different stages of formation and behavior of flow-cell systems [6], colony biofilms [7], drip flow reactors [8], or rotating disk reactors [9] can be used. The flow-cell system which we describe here was developed and described by Christensen et al. [6] and was based on the system described by Wolfaardt et al. [10]. The bacteria grow in small channels with a glass surface through which the biofilm can be monitored noninvasively and continuously using confocal laser scanning microscopy (CLSM). The flow-cell system has been the driving model system for elucidating biofilm-forming process [11–14] in vitro and the biofilm tolerance towards antibiotics [15, 16] and host defense components [15, 17–19]. The flow-cell system is superior for direct and noninvasive biofilm investigations; however, the price is that the number of samples to be tested in each experiment is limited and the method is time consuming compared to the microtiter assay.

The three methods we describe in this chapter can preferably be used in combination. The easier and fairly high-throughput 6-well method can be used to identify mutants, growth conditions, antibiofilm drugs of interest, or lead candidates from large libraries or collections. The bead method serves as a basis for much biofilm and subsequently transcriptomic and proteomic elaborations. The properties of biofilm formation gained in these two models can then be investigated in the flow cells.

2 Materials

2.1 Six-Well Microbial Biofilm Growth with Shear

1. Polystyrene 6-well plates.
2. 22 × 22-1 cover slip.
3. Orbital shaker.
4. Sterile plastic container.
5. Ultrasonic bath.
6. Kinematica Polytron P1200E handheld homogenizer.
7. 70 % ethanol.

8. Phosphate-buffered saline (PBS).
9. *Epi*-fluorescence microscope or confocal laser scanning microscope.
10. Appropriate PNA FISH probes such as Cy3-labeled *C. albicans*/ FITC-labeled *S. aureus* PNA probe cocktail.
11. Vectashield Mounting Media.
12. Clear nail polish.
13. Trypticase soy broth (TSB).
14. Yeast peptone dextrose (YPD).
15. Sabouraud dextrose agar.
16. RPMI 1640 buffered with HEPES and supplemented with L-glutamine.
17. 5 % heat-inactivated fetal bovine serum (RPMI-FBS).
18. YPD containing 5 % FBS medium.

2.2 Glass Bead Biofilms

In order to provide increased surface area for biofilm growth, glass beads are added to 500 ml flasks and then placed on an orbital shaker to provide shear.

1. Media: Brain–heart infusion (BHI) broth.
2. Solid soda lime.
3. Solid borosilicate glass balls.
4. TSB.
5. The protein preservation solution is composed of 10mM Tris•Cl, 1mM EDTA, 0.5 mg/ml PMSF (Phenylmethylsulphonylfluoride), and 10mM sodium azide.
6. RNAprotect reagent (Qiagen, Valencia, CA).
7. Homogenizer and conical tubes (one filled with ethanol, two filled with PBS).
8. Conical tubes (50 ml).
9. Glass 500 ml tissue culture bottles.
10. Plastic container.
11. Ultrasonic bath.
12. Kinematica Polytron P1200E handheld homogenizer.
13. 70 % Ethanol.

2.3 Flow Cells

2.3.1 Components for Assembly of the Flow-Chamber System

1. Bubble traps.
2. Flow chambers.
3. Polycarbonate sheet plastic, 6 mm thick (optional, if flow chambers are to be made locally).
4. Substratum: 50 × 24-mm glass cover slips or other appropriate materials.

5. Marprene® tubing, 3 mm outer diameter, 1 mm inner diameter.
6. Silicone tubing, 3 mm outer diameter, 1 mm inner diameter.
7. Silicone tubing, 4 mm outer diameter, 2 mm inner diameter.
8. Silicone tubing, 7 mm outer diameter, 5 mm inner diameter.
9. Clear polypropylene plastic connectors and T-connectors (Cole Parmer), 1/8 in. (3.175 mm) and 1/16 in. (1.588 mm). Reduction connectors 1/8 to 1/16 in.
10. 2-ml syringe.
11. Injection needles.
12. Medium bottles.
13. Waste container.
14. Silicone glue.
15. 70 and 96 % (v/v) ethanol.
16. 0.5 % (w/v) sodium hypochlorite.
17. H_2O, sterile.
18. 1 % hydrogen peroxide (optional).
19. Medium appropriate for organisms and type of biofilm being grown (e.g., biofilm minimal medium, FeEDTA-AB (FAB) [20]).
20. Peristaltic Pump (Watson-Marlow, 205S).
21. Microscope.
22. Rolling cart for flow systems and pumps (optional).
23. Computer Numerical Control (CNC) tooling machine or a drilling machine mounted on an upright stand and equipped with a milling drill tool (3 mm) (if flow chambers are to be made locally).

2.3.2 Components for Construction of the Bubble Trap (Advanced Version)

1. 35 × 80 × 45-mm polycarbonate block.
2. CNC tooling machine.
3. 5-ml syringes with inner diameter of 12.5 mm.
4. 9 × 2-mm rubber gaskets (M-seals, 221355; http://www.m-seals.dk/cms.ashx).
5. Silicone glue.
6. Stoppers (e.g., http://www.nordson.com/en-us/divisions/efd) or use the leftover needle protective cover from the needles used for inoculating the flow cells (see above).

2.3.3 Components for Construction of the Bubble Trap (Simple Version)

1. A 10 mm thick polycarbonate block, 80 × 35 mm surface area.
2. Drilling machine mounted in a vertical stand.
3. An 8 and a 3 mm drill suitable for drilling in plastic.
4. 2- or 5-ml syringes.
5. Silicone glue and stoppers as above.

2.3.4 Materials for Inoculation and Running of the Flow Cells

1. Inoculum, e.g., fresh overnight culture of the microorganisms under study.
2. 70 and 96 % (v/v) ethanol.
3. Medium (e.g., FAB medium).
4. Silicone glue.
5. Flow-cell system (DTU Systems Biology, Technical University of Denmark, or see below).
6. Syringes and needles (0.4×12 mm, 0.5 ml).
7. Clamps.

2.3.5 Equipment for CLSM of Flow Cell-Grown Biofilms

1. Confocal laser scanning microscope (e.g., Zeiss LSM710).
2. Scalpels.
3. Computer software:
   (a) Imaris (Bitplane; http://www.bitplane.com).
   (b) ImageJ (http://rsb.info.nih.gov/ij).
   (c) Comstat version 2 (DTU Systems Biology, Technical University of Denmark, http://www.comstat.dk).
   (d) Java runtime environment (needed for Comstat v. 2, http://www.java.com).

3 Methods

3.1 Six-Well Microbial Biofilm Growth with Shear

3.1.1 Single-Species Biofilms (See Notes 1 and 2)

1. Starter cultures of bacteria (e.g., *Staphylococcus aureus*, *Staphylococcus epidermidis*, *Pseudomonas aeruginosa*, *Streptococcus pyogenes*, *Bacillus subtilis*, *Escherichia coli*, *Acinetobacter baumannii*, *Klebsiella pneumoniae*, *Salmonella typhimurium*) are derived from frozen glycerol stocks that are streaked onto petri dishes in order to ensure a lack of contamination.
2. A minimum of ten colony-forming units (CFUs) should be picked and added to TSB or another appropriate growth medium and incubated overnight at 37 °C. By choosing ten CFUs, aberrant results due to clonal differences can be avoided.
3. Fresh log-phase bacterial starter cultures are grown by diluting the overnight culture 1:100 in fresh 37 °C pre-warmed TSB for 3 h. Bacterial cultures are then washed twice in sterile PBS.
4. Bacteria are then diluted to an OD of 0.1 at 600 nm, and 50 μl are added to each polystyrene 6-well plate with 5 ml of 37 °C pre-warmed TSB per well and each containing a sterile $22 \times 22 \times 1$ cover slip.
5. Shear is provided by placing plates on a rotating platform during incubation at 100 rpm (ensure that this is consistent for all studies).

6. For growth longer than 24 h, the plate is removed from the incubator and immediately placed at an angle on another unused 6-well plate to allow solution to collect at the well sides.
7. Spent media is removed, and 5 ml of fresh, sterile 37 °C pre-warmed media is added carefully to the side of the well. The 6-well plate is returned to the incubator and allowed to rotate at 100 rpm for an additional 2 min.
8. The plate is removed from the incubator and again immediately placed at a slight angle to allow solution to collect at the well sides. Wash media is removed, and 5 ml of fresh, sterile 37 °C pre-warmed media is added carefully to the side of the well.
9. The 6-well plate is returned to the incubator and allowed to rotate at 100 rpm.
10. If multiple days are required for different biofilm growth stages (early, maturing, and fully mature as described below), repeat every 24 h (*see* **Note 3**).

*3.1.2 Dual-Species Biofilm (Fungi and Bacteria) (See **Notes 1** and **2**)*

1. Bacteria are grown, subcultured to log phase, washed, and diluted as described above.
2. An aliquot of a glycerol stock of *Candida albicans* is grown and maintained on Sabouraud dextrose agar. Cultures are grown overnight in YPD in an orbital shaker (100 rpm) at 37 °C under aerobic conditions. Yeast cells are harvested and washed twice in sterile PBS.
3. *C. albicans* overnight cultures are grown as described above and diluted to an OD of 1.0 at 540 nm. Aliquots of each species suspensions (50 μl) are added to each polystyrene 6-well plate with 5 ml of 37 °C pre-warmed broth (described below) per well and each containing a sterile cover slip [21].
4. Dual-species biofilms (*C. albicans* and bacterial species) are grown in RPMI 1640 buffered with HEPES and supplemented with L-glutamine and 5 % heat-inactivated fetal bovine serum (RPMI-FBS) when hyphal growth by *C. albicans* is needed or YPD containing 5 % FBS medium (YPDFBS) for experiments with yeast cells of *C. albicans.*

3.1.3 Single- or Dual-Species Coverslip Microscopy

1. PNA-FISH hybridization is performed as per the manufacturer's protocol (http://www.advandx.com/Technology/PNA-FISH-Technology.aspx). Use Cy3-labeled *C. albicans*/FITC-labeled *S. aureus* PNA probe cocktail for this cell combination. Non-adherent cells are removed by gently washing with PBS prior to imaging.
2. Add a single drop of Vectashield Mounting Media to the microscope slide, and lay cover slip face down. Seal edges between the cover slip and microscope slide using clear nail polish.

3. For all microbial microscopy counts, a minimum of ten random and blinded fields of view should be evaluated. In addition, each field of view must have a minimum of 20 cells per field of view. If less than 20 cells per field of view, increase the number of fields of view to attain 2,000 cells.

3.1.4 Biofilm Dispersal (See Note 4)

1. Biofilm dispersal method 1 [22, 23]: Each cover slip should be transferred into a sterile plastic container containing 5 ml PBS. The container is sealed and immersed in an ultrasonic bath. Sonication at 30 kHz with a power output of 300 W, as specified by the manufacturer, should be performed at 37 °C for 5 min.
2. Biofilm dispersal method 2 [23, 24]: Using a Kinematica Polytron P1200E handheld homogenizer at maximum speed (30,000 rpm), first disinfect with 70 % ethanol and wash twice in sterile PBS, followed by homogenization for 1 min on ice. Immediately serially dilute and plate for CFU determination.

3.2 Glass Bead Biofilms

3.2.1 Preparation of Glass Beads and Biofilm Culture Bottle

1. Wash beads with a detergent and dry.
2. Add beads to 500 ml glass bottles (approximately half full), add culture media (BHI or TSB) to 2–3 cm above beads so that all beads are submerged (e.g., BHI), and autoclave sterilize.

3.2.2 Protocol for Biofilm Growth

1. A minimum of ten colonies should be picked and added to TSB and incubated overnight at 37 °C. By choosing ten CFUs, aberrant results due to clonal differences can be avoided.
2. Fresh log-phase bacterial starter cultures are grown by diluting the overnight culture 1:100 in fresh 37 °C pre-warmed TSB for 3 h. Bacterial cultures are then washed twice in sterile PBS.
3. Bacteria are then diluted to an OD of 0.1 at 600 nm, and 500 μl are added to each bead and culture-containing 500 ml sterilized bottle.
4. Shear is provided by placing bottle in holding clamp on a rotating platform during incubation at 100 rpm (ensure that this is consistent for all studies) at 37 °C.
5. For growth longer than 24 h, the bottle is removed from the incubator and spent media is immediately removed via pipet. Fresh, sterile 37 °C pre-warmed media is added carefully to the side of the bottle.
6. The bottle is returned to the incubator and allowed to rotate at 100 rpm.
7. If multiple days are required for different biofilm growth stages (early, maturing, and fully mature as described below), repeat every 24 h (*see* **Note 3**).

3.2.3 Harvest

1. The bottle is removed from the incubator, and media containing planktonic bacteria is immediately removed via pipet. Fresh, sterile 37 °C pre-warmed media is added carefully to the side of the bottle, and it is then returned to the incubator and allowed to rotate at 100 rpm for an additional 2 min.
2. The bottle is removed from the incubator, and again wash media is immediately removed to remove all non-adherent bacteria.

*3.2.4 Biofilm Dispersal (See **Note 4**)*

1. Biofilm dispersal method 1 [22, 23]: Beads should be transferred into a sterile plastic container containing 50 ml PBS. The container is sealed and immersed in an ultrasonic bath. Sonication at 30 kHz with a power output of 300 W, as specified by the manufacturer, should be performed at 37 °C for 5 min. Immediately serially dilute and plate for CFU determination.
2. Biofilm dispersal method 2 [23, 24]: Sterile PBS is added to the washed beads in the 500 ml tissue culture bottle to a level just above the level of the beads. Using a Kinematica Polytron P1200E handheld homogenizer at maximum speed (30,000 rpm), first disinfect with 70 % ethanol and wash twice in sterile PBS, followed by homogenization for 1 min on ice. Immediately serially dilute and plate for CFU determination.

3.2.5 Preservation

For RNA or proteomic sample harvest, biofilms must also be dispersed. However, instead of using PBS as the solution to disperse the biofilm into, one should use RNA protect for RNA samples or PBS with protease inhibitor solution for proteomic samples. Following dispersion, transfer solution to centrifuge tubes and centrifuge to concentrate, and pour off supernatant. Freeze at −80 °C until ready for processing.

3.3 Flow Cells

1. Growing biofilms in flow-chamber devices allows for continuous, dynamic observation of living microbial communities and facilitates manipulations which consequences likewise can be monitored online. The requirements for such devices are that they should be mountable on a microscope without interfering with medium supply and that the growing biofilm should be easily observable while on the microscope. Furthermore, manipulations such as challenging the biofilm with antibiotics or changing nutrient availability should be easy to do.
2. The flow chambers used in this chapter are constructed from transparent, non-fluorescent plastic, polycarbonate (PC) which has the additional advantage of being tolerant to autoclaving at 115 °C. The flow chambers are essentially plastic blocks with milled or molded channels (40 × 4 × 1 mm) with an inlet and outlet connector in the ends. On the free side of the

channels a piece of coverslip glass is glued with silicone glue. Nutrient medium is supplied to the flow chamber via silicone tubing, which is very flexible and also permeable to oxygen, ensuring oxygen saturation of the medium at the inlet to the flow chamber. Occasionally small air bubbles will form in the medium lines. These could be catastrophic to the biofilm structure, effectively acting like a razor blade when passing through the flow chamber. To remove such bubbles a device called a bubble trap is inserted in the supply line upstream of the flow chamber. The bubble trap is a small liquid-filled cylinder (a single-use syringe) sitting on top of the passing medium. If an air bubble is in the liquid while passing the trap, the bubble will flow to the top of the cylinder and stay there rather than moving on to the flow chamber. Medium is contained in a flask and led to the bubble traps and flow chambers via peristaltic pump. To prevent the tubing to break because of the wear applied by the pump special tubing is used in the pump. Marprene® tubing has proven to be a suitable material for this purpose—it can be autoclaved and is essentially inverted towards the medium. Marprene® can withstand pump wear for long periods of time and can be reused several times. Waste from the flow chambers is collected in suitable containers downstream of the flow chambers and can subsequently be disposed of by, e.g., autoclaving.

3.3.1 Fabrication of Flow Chambers

Flow chambers can be made by milling and moulding, or if the necessary tools are unavailable several sources for buying ready-made flow chambers exist. To construct flow chambers by milling a milling machine is required. While it is possible to control the milling process manually it is recommended to use a programmable machine (a CNC machine).

1. Use a sheet of polycarbonate, 6×76×26 mm, to mill the flow chamber shown in Fig. 1. This will produce a flow chamber with dimension that fits on a standard microscope slide holder.
2. Start by making a flat flange in each end of the device 1 mm thick by milling away 8 mm into the plastic. The middle part is 6 mm thick and will accommodate the actual flow channels.
3. Mill the flow channels using a 4 mm flat head milling tool. The channels can be of any depth from 0.1 to 5 mm. The standard dimensions are 1 mm depth and 40 mm length. To allow for medium inflow the channels either drill a hole from each end of the flow chamber which allows insertion of a silicone tube or use an advanced milling machine to fabricate a stud in each end of the channels onto which the silicone medium line can be connected. Use a thin drill to open the tubes to connect to the flow channels.

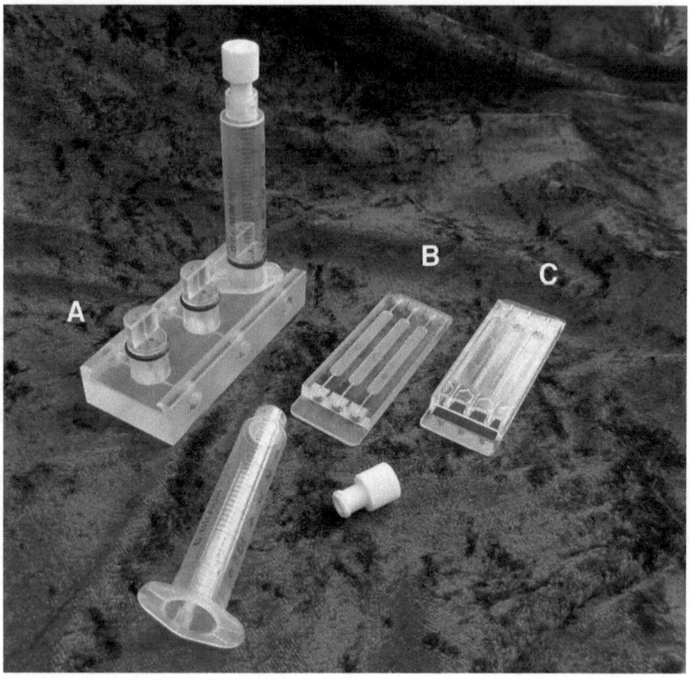

Fig. 1 Bubble traps and flow cells (*A–C*)

4. Attach the selected substratum (usually a glass cover slip, 50 × 24 mm) to the flow chamber. Some silicone glues, in particular types intended for sealing sanitary installations, may contain bacteriostatic or bactericidal compounds and should be avoided. We found 3 M Super Sealant suitable for gluing flow chambers. Place thin strips of silicone glue along the edges of the channels, and take care not to leave gaps in the strips. A single-use 2 ml syringe with a cut pipette tip placed inside the syringe tip is a convenient tool for applying the glue: remove the piston, and insert the thin 1 cm of a 200 μl pipette tip in the syringe. Fill about 1 ml silicone glue in the open end of the syringe, and replace the piston. The filled syringe can now aid in applying the thin strips of glue (*see* **Note 5**).
5. Substrata other than glass can be used: Cut the desired substratum to cover the flow channels; i.e., it should have the same dimensions as a glass cover slip, approximately 50 × 24 mm.
6. Align the substratum to the part of the flow chamber with the applied glue. Gently press the substratum onto the chamber, and ensure that there are no visible leaks, i.e., areas where the glue seem to form gaps. If such gaps can be seen, use either the finger or the piston handle to press the substratum firmly to the chamber.

7. If it is not possible to completely seal the flow chamber, or if silicone glue has entered the channels, remove the substratum and clean the plastic base using 96 % ethanol. Dependent on the type of substratum either dispose and replace or clean using ethanol. Repeat **steps 4–6**.
8. Allow at least several hours for the glue to solidify, preferably overnight before use.
9. If the chambers are fabricated using a machine that has made studs in the in- and outlet ends attach silicone tubing by twisting the tubing onto the studs. Alternatively, place cut tubing ends into the receiver holes and seal with silicone glue (manually fabricated chambers).

3.3.2 Fabrication of Advanced Bubble Trap Using CNC Milling Equipment

The bubble traps in the system are required to remove small air bubbles from reaching the flow chambers, as described above (*see* **Note 6**). The simplest form of a bubble trap can be made from a single-use syringe, a cap, and two needles. A slightly more advanced version utilizes syringes glued to a plastic base with in- and outlet ports. More sophisticated versions must be fabricated using advanced milling tools. These are much easier to handle but also much more expensive. A milled version is shown in Fig. 1.

1. Mill the base of the bubble trap from a single block of polycarbonate (35 × 80 × 45 mm).
2. Use the machine to carve rings to accommodate gaskets to prevent leakage when the syringes are mounted on the base. Also carve a rail along the sides of the base—this can be used to lock the syringes into place by twisting the syringe so that the handles will fit under the rails.
3. Place 5-ml syringes (inner diameter, 12.5 mm) as on each of the columns. Do not twist the syringe to fix it onto the base before the entire system has been tested. In case of a blocked line downstream of the bubble trap, the syringe can then pop off, relieving the built-up pressure. If the syringe is locked damage may occur to the flow chambers in this case.
4. Place a stopper on top of each syringe.

3.3.3 Fabrication of Simple Bubble Traps

1. Remove the piston from a 5 ml syringe, and detach the rubber seal from the plunger.
2. Put the rubber seal back into the end of the syringe.
3. Use silicone glue to attach this onto a flat base.
4. Penetrate two injection needles above the seal into the syringe, one needle 1 cm above the other. The lower needle is now the outlet port and the upper needle the inlet.
5. Place a stopper on top of the syringe.

3.3.4 Fabrication of Slightly More Advanced Bubble Traps

1. Prepare a base plate from a piece of polycarbonate (30 × 110 × 10 mm).
2. Drill three evenly spaced cavities on one side of the plastic (do not drill all the way through the plastic, ca. 8 mm) using an 8 mm drill.
3. Drill in- and outlet channels from each side. Use a 3 mm drill, and place the center 4 mm from the top of the base. Drill far enough that each cavity will have a connection on each side of the plastic.
4. Glue three single-use 2 ml syringes on top of each cavity using silicone glue.
5. Use 3 mm outer diameter silicone tubing in the in- and outlet ports.
6. Place a stopper on top of the syringes.
7. Allow solidifying for at least 24 h.

This construction is quite fragile, and great care should be taken to avoid breaking off the syringes from the base.

3.3.5 Assembly of the Flow-Cell System

Timing is crucial when setting up a flow-chamber experiment. The items that need to be glued should preferably be done so 1 day prior to use. After assembly and checking for leakages the system must be sterilized using hypochlorite for at least 4 h and washed with several passages of sterile water. Only then the system can be filled with medium and inoculated. The system is shown in Fig. 2.

All tubing is of 1 mm inner diameter, except for the medium supply tubing and the waste tubing which have 2 mm inner diameter. All tubing is connected using polypropylene plastic connector. The entire system, except the pump, can be autoclaved before or after assembly.

Start by preparing the flow chambers and bubble traps (*see* Subheadings 3.3.1–3.3.4). Then prepare and autoclave the medium (remember to insert a medium-supply silicone tube into the medium and cap off the end with aluminum foil). The tubing system must then be prepared: a fan-out connector to split the medium supply so that each channel will have its own supply from the pump: this makes it much easier to, e.g., remove channels that are contaminated or broken from the system if needed. The remainders of the flow lines are individual for each channel.

1. Prepare flow chambers and bubble traps according to Subheadings 3.3.1–3.3.4.
2. Calculate the amount of medium you will need. To do this you can calibrate the usage by attaching a piece of Marprene® tubing to the pump and connect it to an inlet and outlet tubing. Set the pump to the desired velocity, and measure how much

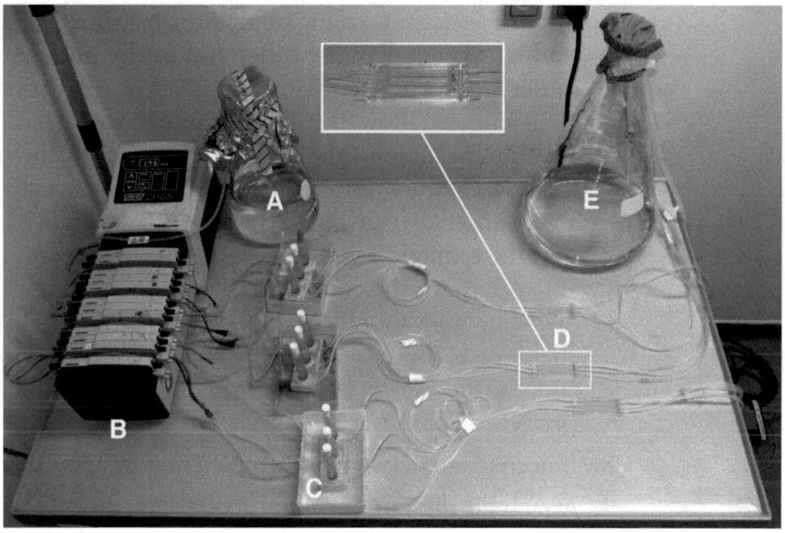

Fig. 2 The assembled flow-cell system (*A–E*)

water is passed through the tubing for a given period, e.g., 1 h. Then you can easily calculate the medium requirement per day. Multiply by the number of channels to find the overall usage per 24 h. It is advisable to prepare more medium than needed to allow for unexpected delays and for having a buffer volume near the bottom of the supply flask.

3. Preparation of medium: Place a sufficient long silicone tube, 2 mm inner diameter, in the bottle. Fix the tubing outside of the bottle using autoclave tape. Attach a straight connector to the end, and protect the end against contamination with aluminum foil. It is essential to clamp off the tubing or siphoning may empty the bottle while autoclaving. Fill the bottle with medium, and cover the bottle opening with aluminum foil. Autoclave immediately.
4. Prepare all the tubing for the rest of the system:
5. Start by making the fan-out component: make one piece of silicone tubing (50 cm) that will attach to the silicone tubing from the medium bottle. Cover the end with aluminum foil. To the other end attach a T-connector (1/16 in.). To each end attach 5 cm of silicone tubing (1 mm inner diameter). To the two new ends add additional T-connector and 5 cm pieces of tubing. Continue until there is the same number of free ends as channels in the system.
6. Prepare short pieces of Marprene® tubing, just long enough to pass through the pump. In each end place a 1/16 in. straight connector. To prevent the pump from pulling off the connectors, wind four winds of autoclave tape around the Marprene® next to the connector on the inlet side of the pump.

7. On the outlet side of the pump attach a short (35 cm) piece of tubing and connect the other end to the bubble trap. On the outlet side of the bubble trap attach a long piece of tubing (150 cm, to allow the flow chamber to be taken from the experiment table to the microscope). Attach the other end to the flow chamber. At the outlet end of the flow chamber attach a very short (10 cm) piece of tubing and then a 1/16 to 1/8 straight reduction connector. Finally, attach the waste (effluent) tubing to the connector. The waste tubing is 2 mm inner diameter and should be sufficiently long to allow the flow chamber to be moved to the microscope while keeping the end in the waste container (120 cm is usually sufficient) (*see* **Note 7**).
8. If needed, autoclave the assembled flow system (recommended).
9. Attach all the pump tubing to the pump. It is highly recommended to keep all the tubing in order, such that the first flow channel in the first flow chamber is connected to the first channel in the first bubble trap and the first channel in the pump. This will make it much easier to find and fix problems such as leakages in the system later on.
10. Optionally place the entire setup on a rolling cart to make transport to the microscope and incubation room (if required) easy.
11. Place the waste lines into a suitable reception container.

3.3.6 Sterilization of the Flow-Cell System

1. Always use sterile water when operating the system, even before sterilization.
2. Remove all stoppers from the bubble traps, and store them in 70 % ethanol. Place all components on a flat surface, and make sure that the waste container is at the same level as the flow chambers.
3. Place the free end of the fan-out connector in 1 l of water. Fill the system by setting the pump to maximum speed. When the bubble traps are filled with water wait for about 15 s and then place the stopper on the syringe. Then observe that the rest of the system is filled. At this point it is important to detect and seal any leaks in the system.
4. Remove air bubbles (even tiny ones) in the flow chambers by tapping them firmly on the ends on a hard surface. This will release the bubbles, and they can flow away due to the high velocity of the liquid.
5. Lift the inlet tube into the air, and allow the system to be emptied completely.
6. Remove all the stoppers and fill the system as described above, with a solution of 0.5 % sodium hypochloride in water. When the system is completely filled and all air bubbles removed con-

tinue flow at maximum speed for another 5–10 min. Reduce flow rate to the minimum of the pump and sterilize for at least 1 h. It is not recommended to leave the sterilizing agent in the system for extended periods and never more than 24 h.

7. The system is now sterile, and handling should be done to prevent contamination.
8. Empty and fill the system three consecutive times with sterile water to remove all traces of the hypochloride. Optionally the second pass can be water with 1 % hydrogen peroxide.
9. If the system is to be used immediately, empty the system and fill with medium. The system is now ready for inoculation.
10. If the system is not used at once it is recommended to leave it with water, running at a low rate until use.

3.3.7 Inoculation of the Flow Cells (See **Note 8**)

The amount of cells to inoculate is very dependent on both the strain and the medium. As a starting point cells from an overnight culture must be diluted before inoculation. The exact dilution must be empirically determined. For *Escherichia coli* a dilution of 1:100 is suitable, while *Acinetobacter* spp. should only be diluted ten times. *Pseudomonas aeruginosa* is usually diluted 1:1,000.

1. Withdraw about 250 μl of the diluted culture into a syringe equipped with a thin needle, e.g., 27 G. Wipe the tubing on the inlet side of the flow chamber with 70 % ethanol. Stop the pump, and penetrate the silicone tubing as near the inlet of the channel as possible. Inject the inoculum into the channel, and observe that no air bubbles are injected. If that happens remove the air bubble using the syringe. Important: Never place fingers behind the tubing while inserting the needle—the risk of being stung by the needle is significant.
2. Seal the injection hole with silicone glue.
3. Leave the flow chamber with the substratum facing the table for 1 h without flow to allow initial adhesion of the cells to the substratum.
4. Reverse the flow chambers, and resume flow. In this system a flow rate of 1–3 ml per channel per hour is suitable for *E. coli* and *P. aeruginosa*. For other organisms other flow rates may be needed.

3.3.8 Running of the Flow-Cell System (See **Note 9**)

The system can run for several days or even weeks without interruption. However, check frequently that medium is available and the waste is not overflowing. Deal with any leaks. A broken flow chamber cannot be fixed and must be isolated from the system.

Take care to ensure uninterrupted flow as some strains react by detachment for even short periods of no-flow (*see* **Note 10**).

Dispose of waste according to local regulations.

*3.3.9 Microscopic Inspection and Imaging of Flow Cell: Grown Biofilms (See **Note 11**)*

The flow cells with the transparent glass cover slip which also serve as the substratum for the biofilm enable easy and non-invention visualization. All microscopes can basically be used; however, the optimal method of choice for visualization of flow chamber-grown biofilms is to use CLSM.

Practical information when using CLSM and flow cells:

1. The flow cell system should be close to the confocal microscope to avoid interruption of growth if the system has to be transported.
2. Make sure that the tubing in both ends on the flow cells are long enough for the flow cell to be placed under the microscope without interfering with the setup.
3. Make sure that the flow cell is firmly attached in the specimen holder. Adhesive tape to assist mounting the sample onto the microscope is a possibility.
4. WARNING: Before starting the microscopy be careful if the microscope is adjusted to automatically set parfocality as for normal microscope slides. The flow chambers are much thicker; as a consequence the microscope may position the lens wrongly and destroy the sample. An empty flow cell can be used to calibrate the microscope adjustments before mounting a real experiment.

3.3.10 End of Experiment

Many of the flow-cell components can be reused and should be thoroughly cleaned after each experiment. Other components such as the upstream silicone tubes should not be reused and discarded after each experiment.

1. Disassemble the inlet tube to allow emptying the remaining liquid from the still assembled system by filling with air. All waste should be collected and disposed of following local biohazard regulations.
2. Detach the flow chambers. The pump tubing and the downstream waste tubing can be reused if autoclaved.
3. The flow chambers are disassembled using a scalpel to remove the substratum. If the substratum is made of fragile material such as glass it will inevitably break in this process. Remember to use plastic gloves and protective eye wear.
4. Remove remains of silicone glue from the flow chamber using mechanical abrasion and 96 % ethanol.
5. The reusable tubing and the clean flow cells can be wrapped in metal foil, autoclaved, and stored for subsequent experiments.

3.3.11 Image Analysis

The CLSM imaging can be either used as qualitatively descriptive images or for quantitative measurements. In both instances the CLSM images obtained of the biofilm experiment by CLSM should most often be processed since most CLSM files are in a specific format.

For qualitative and descriptive analysis of the images, several software programs are available. We recommend the commercial package Imaris® software suite (http://www.bitplane.ch) which can create two- and three-dimensional visualizations with simple measurements, time-lapse movies, as well as animations. A freeware alternative is the program ImageJ [25] (http://rsb.info.nhi.gov/ij) which can be supplemented by user-written plug-ins to perform several graphical visualizations of CLSM images and extensive qualitative measurements. Using nonspecialized programs such as ImageJ does require more from the user to get full benefit of its capabilities than do commercial dedicated packages.

For quantitative analysis several programs have been developed, such as the ISA3D [26], Comstat [27] (http://www.comstat.dk), and Daime [28] (http://www.microbial-ecology.net/daime). Using these programs basic parameters from CLSM image stacks, such as biomass, roughness, and average thickness, can be calculated. The quantitative measurements are numbers, rather than images, and provide a way to directly evaluate both reproducibility of experiments and statistically compare different biofilms which qualitatively seem similar.

4 Notes

1. All experiments should be performed in triplicate (i.e., three true replicates).
2. Several strains of each species should be used including recent representative clinical isolates.
3. Each stage of biofilm growth is as different as biofilms are to planktonic cultures [29]. If multiple growth stages of biofilms are required, perform biofilm growth curves for each strain in which CFU/cm^2 is determined at multiple time points following inoculation. The various stages include the following: (a) early-stage biofilms are those with monolayers soon after microbial attachment, (b) fully mature biofilms are those where the CFU/cm^2 reaches a maximum and static level, and (c) maturing biofilms are those with CFU/cm^2 at 50 % of fully mature biofilms.
4. In order to get representative CFU counts from biofilms, it is important to break up the biofilm conglomerates by either sonication or homogenization.
5. As with any liquid, even a small leak will render the entire flow cell useless and should be sealed before the experiment is initiated. Leaks are easy to spot when the setup is filled with water.

Use normal water before the system is sterilized. If leaks are observed, either change the damaged part or seal with silicone glue; remember to allow for hardening. It is usually very difficult to seal leaks once the system and experiment are initiated since the flow should be turned off to allow for the silicone to dry. The glass cover slips on the flow cells are very fragile, so please be careful. If the flow cell leaks while experiment with bacteria is running the biohazard risk is great. Especially if the flow cell is mounted on the microscope, it can become harmful to the equipment and difficult to decontaminate. Either immediately attempt to reseal the leak or remove the flow cell from the microscope. While running an experiment, it is not possible to drain the system or even stop the flow temporarily. If leaks occur during an experiment clean and dry the area containing the leak, and keep a piece of paper towel at hand to repeatedly soak away leaking fluid. Apply excess of silicone glue to the leak, and continue to remove leaking liquid until it stops.

6. As with any liquid-changing temperature or flow rates, gas or air bubbles can form. These air bubbles can introduce artifacts in the developing biofilm eventually destroying or altering the three-dimensional structure of biofilms. Due to this it is recommended to use bubble traps to catch the gas bubbles before they reach the flow cells. Since the change in temperature, for example, when cooling the media from autoclaving to room temperature to heat it up to the experimental temperature, increases the bubble formation, it is recommended not to cool the medium after autoclaving, and to place it immediately at the correct temperature for the experiment. Furthermore, if running of the flow-cell system above room (e.g., 37 °C) the change of bubble formation increases. A recently described [30] setup with a modified medium container may be employed.
7. Place the end of the waste tubes downstream of the flow cell above the surface of the waste reservoir to avoid reverse siphoning from the waste container.
8. As of any other experimental setup, microbial growth in flow cells may be influenced by biological variation, such as selection of mutants. Due to this we recommend always to run at least two independent flow channels of each bacterium or strain, and as with any other biological setup experiments should always be performed in replicates.
9. Since many bacteria are motile and tend to migrate towards chemical gradients, such as the fresh medium supply, bacterial growth might occur in the tubing upstream of the flow cell also known as "backgrowth." If such fouling occurs we recommend removing the contaminated part of the tubing since biofilm in the upstream tubing will use substrate and release waste products that may affect the biofilm in the flow cell.

The contaminated tubing is removed by first clamping the tubing on the effluent side of the flow cells, so that waste medium does not flow backwards through the chamber, destroying the sample. Hence cut off the contaminated part, reconnect, and slowly remove clamping. It is in this case critical not to open clamps too abruptly as the concurrent liquid movement can disrupt the biofilm. The effluent tubes can also be changed if they become heavily fouled, but here clamping is not critical, but the biohazard of the contaminated waste should be avoided.

10. The biofilms formed by some microbial species will disperse in response to even short periods of change in flow rate. If the bacteria of a given experiment are expected to easily disperse it may be important to ensure constant running of the peristaltic pump.
11. Always perform the microscopy in the same area of the flow cell. The biofilm formation in the flow cell is very different in the inlet part compared to the outlet part, because the bacteria use substrate and release waste products. We recommend the microscopic inspection, and imaging of the biofilm should be done near the medium inlet to avoid uncontrolled conditions.

References

1. Christensen GD, Simpson WA, Younger JJ et al (1985) Adherence of coagulase-negative staphylococci to plastic tissue culture plates: a quantitative model for the adherence of staphylococci to medical devices. J Clin Microbiol 22:996–1006
2. O'Toole GA, Kolter R (1998) Initiation of biofilm formation in *Pseudomonas fluorescens* WCS365 proceeds via multiple, convergent signalling pathways: a genetic analysis. Mol Microbiol 28:449–461
3. Ceri H, Olson ME, Stremick C et al (1999) The Calgary biofilm device: new technology for rapid determination of antibiotic susceptibilities of bacterial biofilms. J Clin Microbiol 37:1771–1776
4. Lee B, Haagensen JA, Ciofu O et al (2005) Heterogeneity of biofilms formed by nonmucoid *Pseudomonas aeruginosa* isolates from patients with cystic fibrosis. J Clin Microbiol 43:5247–5255
5. Hengzhuang W, Wu H, Ciofu O et al (2011) Pharmacokinetics/pharmacodynamics of colistin and imipenem on mucoid and nonmucoid *Pseudomonas aeruginosa* biofilms. Antimicrob Agents Chemother 55:4469–4474
6. Christensen BB, Sternberg C, Andersen JB et al (1999) Molecular tools for study of biofilm physiology. Methods Enzymol 310: 20–42
7. Anderl JN, Franklin MJ, Stewart PS (2000) Role of antibiotic penetration limitation in *Klebsiella pneumoniae* biofilm resistance to ampicillin and ciprofloxacin. Antimicrob Agents Chemother 44:1818–1824
8. Goeres DM, Hamilton MA, Beck NA et al (2009) A method for growing a biofilm under low shear at the air-liquid interface using the drip flow biofilm reactor. Nat Protoc 4: 783–788
9. Zelver N, Hamilton M, Pitts B et al (1999) Measuring antimicrobial effects on biofilm bacteria: from laboratory to field. Methods Enzymol 310:608–628
10. Wolfaardt GM, Lawrence JR, Robarts RD et al (1994) Multicellular organization in a degradative biofilm community. Appl Environ Microbiol 60:434–446
11. Klausen M, Heydorn A, Ragas P et al (2003) Biofilm formation by *Pseudomonas aeruginosa* wild type, flagella and type IV pili mutants. Mol Microbiol 48:1511–1524
12. Klausen M, Aaes-Jorgensen A, Molin S, Tolker-Nielsen T (2003) Involvement of bacterial migration in the development of com-

plex multicellular structures in *Pseudomonas aeruginosa* biofilms. Mol Microbiol 50:61–68
13. Sauer K, Camper AK, Ehrlich GD et al (2002) *Pseudomonas aeruginosa* displays multiple phenotypes during development as a biofilm. J Bacteriol 184:1140–1154
14. Stoodley P, Sauer K, Davies DG, Costerton JW (2002) Biofilms as complex differentiated communities. Annu Rev Microbiol 56: 187–209
15. Bjarnsholt T, Jensen PØ, Burmølle M et al (2005) *Pseudomonas aeruginosa* tolerance to tobramycin, hydrogen peroxide and polymorphonuclear leukocytes is quorum-sensing dependent. Microbiology 151:373–383
16. Pamp SJ, Gjermansen M, Johansen HK, Tolker-Nielsen T (2008) Tolerance to the antimicrobial peptide colistin in *Pseudomonas aeruginosa* biofilms is linked to metabolically active cells, and depends on the pmr and mexAB-oprM genes. Mol Microbiol 68: 223–240
17. Alhede M, Bjarnsholt T, Jensen PØ et al (2009) *Pseudomonas aeruginosa* recognizes and responds aggressively to the presence of polymorphonuclear leukocytes. Microbiology 155:3500–3508
18. Jensen PO, Bjarnsholt T, Phipps R et al (2007) Rapid necrotic killing of polymorphonuclear leukocytes is caused by quorum-sensing-controlled production of rhamnolipid by *Pseudomonas aeruginosa*. Microbiology 153: 1329–1338
19. van Gennip M, Christensen LD, Alhede M et al (2009) Inactivation of the rhlA gene in *Pseudomonas aeruginosa* prevents rhamnolipid production, disabling the protection against polymorphonuclear leukocytes. APMIS 117: 537–546
20. Pamp SJ, Tolker-Nielsen T (2007) Multiple roles of biosurfactants in structural biofilm development by *Pseudomonas aeruginosa*. J Bacteriol 189:2531–2539
21. Peters BM, Ovchinnikova ES, Krom BP et al (2012) *Staphylococcus aureus* adherence to *Candida albicans* hyphae is mediated by the hyphal adhesin Als3p. Microbiology 158: 2975–2986
22. Bjerkan G, Witso E, Bergh K (2009) Sonication is superior to scraping for retrieval of bacteria in biofilm on titanium and steel surfaces in vitro. Acta Orthop 80:245–250
23. Fux CA, Wilson S, Stoodley P (2004) Detachment characteristics and oxacillin resistance of *Staphylococcus aureus* biofilm emboli in an in vitro catheter infection model. J Bacteriol 186:4486–4491
24. Brady RA, O'May GA, Leid JG et al (2011) Resolution of *Staphylococcus aureus* biofilm infection using vaccination and antibiotic treatment. Infect Immun 79:1797–1803
25. Abrámoff MD, Magalhâes PJ, Ram SJ (2004) Image processing with ImageJ. Biophoton Int 11:36–41
26. Beyenal H, Donovan C, Lewandowski Z, Harkin G (2004) Three-dimensional biofilm structure quantification. J Microbiol Methods 59:395–413
27. Heydorn A, Nielsen AT, Hentzer M et al (2000) Quantification of biofilm structures by the novel computer program COMSTAT. Microbiology 146:2395–2407
28. Daims H, Lucker S, Wagner M (2006) Daime, a novel image analysis program for microbial ecology and biofilm research. Environ Microbiol 8:200–213
29. Southey-Pillig CJ, Davies DG, Sauer K (2005) Characterization of temporal protein production in *Pseudomonas aeruginosa* biofilms. J Bacteriol 187:8114–8126
30. Crusz SA, Popat R, Rybtke MT et al (2012) Bursting the bubble on bacterial biofilms: a flow cell methodology. Biofouling 28: 835–842

Chapter 2

Aqueous Two-Phase System Technology for Patterning Bacterial Communities and Biofilms

Mohammed Dwidar, Shuichi Takayama, and Robert J. Mitchell

Abstract

We describe a novel method which makes use of polymer-based aqueous two-phase systems to pattern bacterial communities inside Petri dishes. This method allows us to culture submillimeter-sized bacterial communities in spatially distinct spots while maintaining a degree of chemical connectedness to each other through the aqueous phase. Given sufficient time, these bacterial cultures develop biofilms, each corresponding to the footprint of the droplet spot. This method can be used to study the interactions between bacterial communities and biofilms spotted adjacent to each other. Furthermore, it can be extended to study the interactions between different bacterial communities and an underlying epithelial cell layer.

Key words Aqueous two-phase system (ATPS), Biofilm, Micro-patterning, Localized culturing, Bacterial–bacterial interaction, Bacterial–epithelial interaction, Cell printing

1 Introduction

Bacterial biofilms found in the environment or even those present inside the body are usually a complex consortia of bacteria [1]. Despite being enclosed inside the biofilm matrix [2], these bacteria are not isolated from the surrounding environment. They communicate with each other inside the biofilm as well as with surrounding biofilms and microcolonies formed by the same or different species [3]. They also affect and are affected by the underlying surface, which in certain circumstances can be living structures, such as epithelial cell layers. These interactions are carried out by a wide range of effectors, both known and yet unknown, and include various proteins, nutrients, hormones, and quorum sensing molecules [4–7]. Such interactions, however, are usually difficult to study systematically in the lab using common biofilm construction protocols, such as in 96-well plates and flow cells. To mimic the interactions between biofilms and microcolonies present in nature, it would benefit scientists to be able to pattern small bacterial communities and biofilms adjacent to each other

Gianfranco Donelli (ed.), *Microbial Biofilms: Methods and Protocols*, Methods in Molecular Biology, vol. 1147,
DOI 10.1007/978-1-4939-0467-9_2,

so that they are chemically communicating but physically segregated. Currently, this has been approached by different groups using microfluidic devices, including the Weibel's group devices to pattern bacteria and biofilms [8, 9], the microfluidic device developed by Ismagilov's group to study the chemical interactions between spatially distinct bacterial communities [10], and the system developed by Jayaraman's group to study host–pathogen interactions [11]. While these microdevice-based systems are very useful, they require specialized equipment or fabrication facilities that are not available to many labs.

Here, using aqueous two-phase system (ATPS) techniques [12, 13], we describe how to pattern bacterial communities of submillimeter diameters beside each other using materials commonly accessible to labs [14–16]. The fully aqueous nature of the ATPSs allows for the free diffusion of most small molecules and ions between these bacterial communities. However, due to bacteria partitioning preferences and interfacial tension between the polyethylene glycol (PEG) and dextran (DEX) phases, many bacterial species can be confined within the DEX-rich microdroplets. This method has been used to successfully pattern diverse bacterial strains, including *Escherichia coli* and the human pathogens *Staphylococcus aureus* and *Pseudomonas aeruginosa*. In principle, it should also work with other bacterial strains as long as the cells partition into the DEX phase. In addition, when the ATPS-patterned bacterial suspensions are given sufficient time to grow, they develop into distinct biofilms with sizes corresponding to the footprints of the DEX drops that were patterned.

Using the same technique discussed above, we were also able to pattern populations of bacteria on epithelial cells in Petri dishes. Although not rigorously confirmed as being biofilms, after 24 h of DEX droplet patterning, the bacterial communities were attached to the underlying epithelial cells. This was true even after removal of the medium and gentle washing of the plate.

This protocol can be used, for instance, to compare two mutants or different strains for their effects on the underlying epithelial layer within the same Petri dish. Also, it can be used to study the influence some bacterial strains have on the virulence of other bacteria. For example, we recently used this technique to study the effects the bacterial predatory strain, *Bdellovibrio bacteriovorus*, had on human pathogens [17].

2 Materials

2.1 ATPS for Patterning Bacterial Communities in Petri Dishes

1. Overnight bacterial culture (1 ml): Culture your bacterial strain (e.g.: *E. coli*) in a suitable culturing medium such as Luria Broth (LB) or tryptic soy broth (TSB) overnight. Antibiotics can be used for this culturing as needed.

2. ATPS components (*see* **Notes 1** and **2**):
   (a) PEG (M.W. 200,000).
   (b) DEX (M.W. 20,000).
3. 35 mm Petri dishes.
4. Micropipette capable of dispensing volumes down to 0.1 μm (*see* **Note 3**).
5. Pure deionized water.
6. Balance.
7. Centrifuge.
8. Clean bench.
9. Non-shaking incubator with 100 % humidity.
10. Water bath adjusted to 37 °C.

2.2 ATPS for Patterning Bacterial Communities on Epithelial Monolayers

1. ATPS components (*see* **Note 4**):
   (a) PEG (M.W. 35,000).
   (b) DEX (M.W. 500,000).
2. Petri dishes coated with a viable confluent epithelial cell monolayer or other cell type of interest (in our case, we used human mammary epithelial cells, MCF 10a).
3. Suitable culturing medium for your epithelial cell line (we used DMEM/F12 medium).

3 Methods

3.1 ATPS for Patterning Bacterial Communities in Petri Dishes

3.1.1 Preparation of the Aqueous Two-Phase System Solutions

1. Dissolve PEG in the required non-sterile culturing medium (such as LB) (*see* **Note 5**) to a final concentration of 14 % (w/w).
2. Similarly dissolve DEX separately in a second container containing non-sterile culturing medium, also to a final concentration of 14 % (w/w) (*see* **Note 6**).
3. Independently stir the two solutions vigorously, and then autoclave each solution to sterilize them.
4. After cooling to room temperature, centrifuge both solutions briefly to remove any insoluble components.
5. Aseptically mix equal volumes of the two solutions.
6. Centrifuge the mixture at 3,000 × *g* for 30 min at 4 °C.
7. The mixture should separate into two phases—the upper PEG-rich phase and lower DEX-rich phase. Note that the volume of the DEX-rich phase will be less than that of the PEG-rich phase.
8. Separate the two phases from each other, and discard the interface.
9. Both solutions can be stored aseptically at 4 °C until needed.

3.1.2 Suspending the Bacteria in the DEX-Rich Phase

1. Pre-warm the DEX-rich phase in a water bath to the appropriate temperature.
2. Centrifuge 1 ml of an overnight bacterial culture at 3,000 × *g* for 5 min (RT).
3. Discard the supernatant (*see* **Note 7**).
4. Wash the bacterial pellet with 0.5 ml of the DEX-rich solution.
5. Resuspend the pellet in the DEX-rich solution to give a final OD of 0.05 (*see* **Notes 8** and **9**).

3.1.3 Bacterial Patterning and Biofilm Formation

1. Pre-warm the PEG-rich phase in a water bath to the appropriate temperature.
2. Aseptically fill the 35 mm Petri dish with 2.5 ml of the PEG-rich solution.
3. Using a conventional manual micropipette, dispense (*see* **Note 10**) 0.3 μl droplets of the DEX-rich bacterial suspension (*see* **Note 11**) at the desired positions.
4. Incubate the plate under 100 % humidity and at a temperature suitable for growth/biofilm formation according to your strain (e.g., 37 °C in case of *Staphylococcus aureus*) (*see* **Notes 12** and **13**).
5. After 24 h of incubation, the bacterial spot should be visible to the naked eye (*see* Fig. 1).

3.1.4 Biofilm Washing and Analysis

1. Submerge the plate gently in a water bath adjusted to 37 °C, invert it, and incubate it upside down for 10 min.
2. This will allow the ATPS solutions and the planktonic cells to drain away by gravity while the attached biofilm will not.
3. The resulting biofilms can be seen by the naked eye after staining with 0.1 % crystal violet (CV) for 20 min. Otherwise, it can be analyzed under the microscope, particularly if the bacteria are expressing a fluorescent protein.

3.2 ATPS for Patterning Bacterial Communities on Epithelial Monolayers

3.2.1 Preparation of ATPS Solutions

1. Add both PEG (M.W. 35,000) and DEX (M.W. 500,000) to a suitable epithelial cell culturing medium (such as DMEM/F12) (*see* **Notes 4** and **14**) at final concentrations of 2.5 and 3.2 % (w/w), respectively.
2. Stir the mixture thoroughly to dissolve the two polymers completely.
3. Centrifuge the mixture at 3,000 × *g* and 4 °C for 30 min.
4. This will yield an upper PEG-rich phase and a lower DEX-rich phase.
5. Separate the two phases, and filter-sterilize them individually using a 0.22 μm syringe filter.
6. Store both solutions at 4 °C until needed.

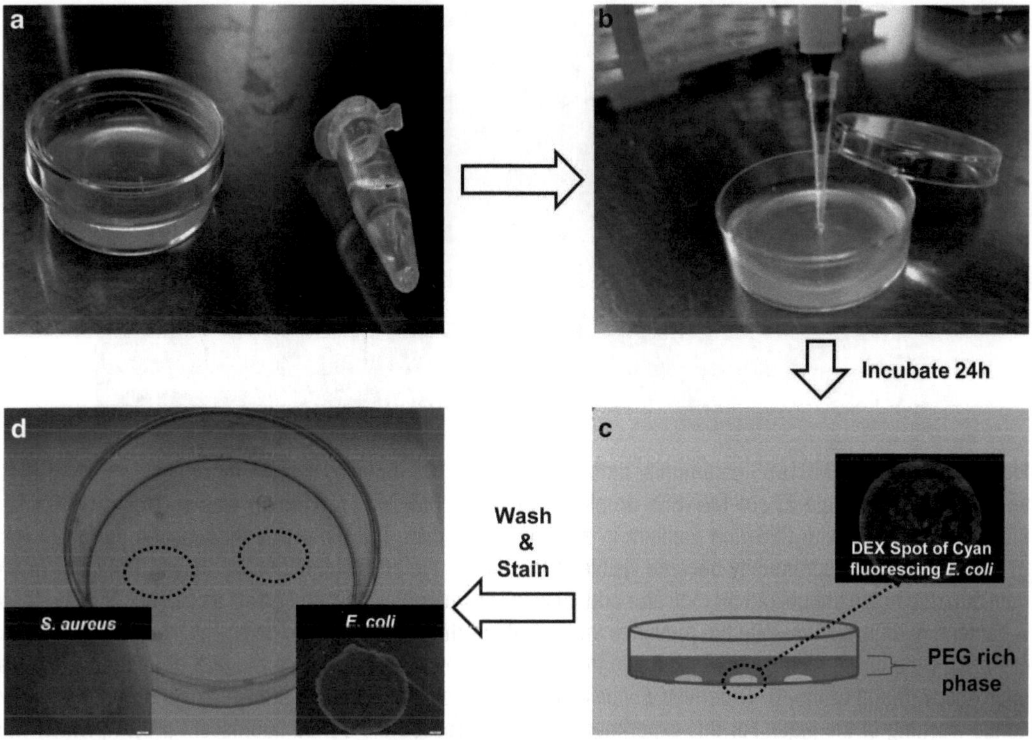

Fig. 1 Schematic diagram describing the procedure used to pattern bacterial biofilms on polystyrene surfaces in 35 mm Petri dish. (**a**) Prepare the DEX- and PEG-rich solutions in an appropriate culturing medium (e.g., TSB). Resuspend the overnight bacterial culture in the DEX-rich phase at an O.D. of 0.05. Fill the Petri dish with 2.5 ml of the PEG-rich phase. (**b**) Dispense the DEX-rich droplet(s) containing your bacterial strain(s) at the desired locations inside the PEG-rich phase. (**c**) After 24 h of incubation, the patterned bacterial communities should be visible to the naked eye and can be analyzed under a microscope. The image shown on the *right* represents a typical bacterial spot formed by *E. coli* MG1655 expressing the cyan fluorescent protein. This image was taken using an inverted epifluorescence microscope (Olympus IX71) operated by Metamorph software and connected to an Andor Luca 658 camera. (**d**) After washing the Petri dish to remove the ATPS components and the planktonic cells, the plate was stained with crystal violet and observed. The *central photo* shows a Petri dish with six patterned biofilms. The three biofilms on the *right side* were formed by *E. coli* MG1655, while the three on the *left side* were formed by *S. aureus* KACC 10768. The two *inset* photos on the *right* and *left corners* show the CV-stained bright-field images for *E. coli* and *S. aureus* biofilms, respectively. These images were taken using SZX16 stereoscope connected to a DP72 CCD 126 camera and operated by DP2-BSW imaging software (Olympus). Scale bar: 200 μm

3.2.2 Suspending the Bacteria in the DEX-Rich Phase

1. Pre-warm the DEX-rich phase in a 37 °C water bath.
2. Centrifuge 1 ml of overnight bacterial culture at 3,000 × *g* for 5 min.
3. Discard the supernatant.
4. Wash the bacterial pellet with 0.5 ml of the DEX-rich solution.
5. Resuspend the pellet in DEX-rich solution to give a final OD of 0.05.

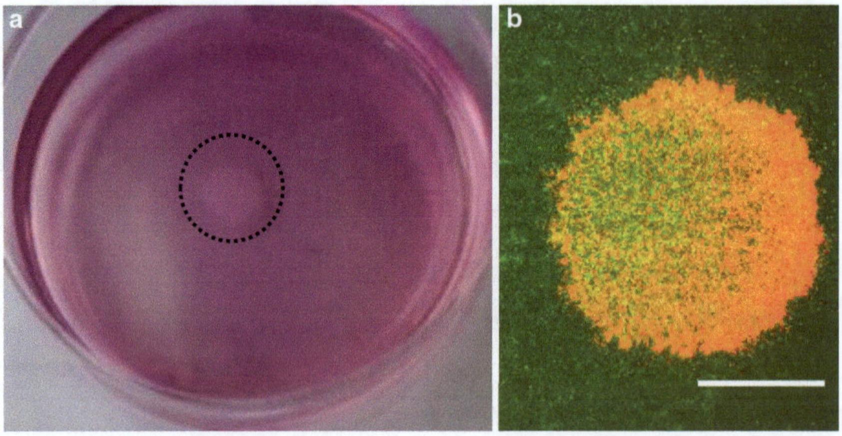

Fig. 2 Typical *E. coli* MG1655 community patterned on MCF 10a monolayer in 35 mm Petri dish using ATPS technology. (**a**) A single *E. coli* MG1655 droplet in DEX-rich DMEM/F12 medium was spotted on MCF 10a monolayer covered with PEG-rich medium in a 35 mm Petri dish. The plate was then incubated for 24 h at 37 °C. The bacterial community became visible to the naked eye after incubation as shown. The color of the medium (due to the phenol red pH indicator added) was still red (not yellow and acidic as can occur when there is bacteria overgrowth) despite the presence and growth of the bacterial community. This is due to the nature of the ATPS culturing technique which limits the bacterial growth to only within the DEX droplet. (**b**) Confocal microscopy image of the ATPS-derived *E. coli* community shown in Panel (**a**) after removal of the media and gentle washing of the plate. For this experiment, *E. coli* expressed the DsRed fluorescent protein, which was encoded on plasmid pHKT3 [19] The epithelial cells underneath were stained with calcein AM, which confers green fluorescence. This image was taken using an LSM700 confocal microscope (Carl Zeiss) operated by ZEN 2009 software. Scale bar: 2 mm

3.2.3 Bacterial Patterning and Biofilm Formation

1. Culture your epithelial cells as usual in 35 mm Petri dishes.
2. After the epithelial cells form a homogenous, nearly confluent layer (approximately 80 % confluent layer when observed under optical microscope), aspirate the medium.
3. Pre-warm the PEG-rich phase in a 37 °C water bath.
4. Wash the epithelial monolayer with phosphate-buffered saline (PBS) to remove any traces of antibiotics which could be present in the culturing medium.
5. Fill the plate with 2.5 ml of the PEG-rich solution.
6. Using the conventional manual micropipette, dispense 0.3 μl droplets in the desired positions inside the plate.
7. Incubate the plate at 37 °C for 24 h under 100 % humidity.
8. After 24 h of incubation, the bacterial spots (*see* **Note 15**) should be visible to the naked eye (*see* Fig. 2).

3.2.4 Biofilm Washing and Analysis

1. Aspirate the medium from the plate gently without disturbing the formed bacterial communities.
2. Wash the plate gently with PBS.
3. The resulting biofilms and the underlying epithelial layer can be analyzed under a microscope after staining with the appropriate stains (*see* **Notes 16** and **17**).

4 Notes

1. The ATPS mixture described here was successful and convenient in patterning bacterial biofilms on inanimate surfaces such as polystyrene because the composition is far from the critical concentration where the two phases become one. Thus, it provides stability and is suitable for long-term bacterial culturing (>24 h). Whereas other ATPS mixtures can also be used, it should be noted that mixtures employing high polymer concentrations are more reliable for long-term cultures. On the other hand, the viscous nature of these mixtures can affect the convection and diffusion of the chemicals and ions inside the dish. Furthermore, for ATPS patterning using manual pipettes, the lower viscosity associated with lower concentrations is preferred as the dispensing occurs quickly and accommodates a "less steady" hand during patterning.
2. The ATPS mixture described here was not that successful upon trying it on viable epithelial cell monolayers as it had a slight but significantly deleterious effect on the viability of the epithelial monolayer. Thus we had to use a different formula as described below.
3. The need for a micropipette capable of dispensing submicroliter volumes is to be able to pattern submillimeter-sized bacterial communities. Otherwise, you can scale up the technique and dispense larger volumes, which will generate larger bacterial communities.
4. The ATPS mixture used here is much less viscous than the previous one described. The reason for using this mixture here is that the one listed in the first protocol slightly but significantly affected the viability of the underlying epithelial layer in a negative manner. In contrast, the mixture described here had no adverse effect on epithelial cell viability.
5. In the case where the media is supplied as a dehydrated powder, such as LB or TSB, simply weigh the ATPS polymer, add the required amount of the medium components, add the appropriate amount of water, mix thoroughly, and then autoclave.
6. The concentrations of the ATPS components are given in (w/w) to ensure an accurate composition of the final ATPS phases.
7. Discard as much of the culture supernatant as possible, and take only the bacterial pellet to ensure that the composition of the final DEX-rich solution is consistent and well defined.
8. The initial OD does not have to be strict for many experiments as it does not have a significant effect on the final biofilm derived after 24 h of incubation. An initial OD of 0.05 was chosen because it is commonly used in biofilm studies.

9. Handle sterilized ATPS mixtures, and perform all the bacterial patterning and incubations under aseptic conditions to avoid contamination.
10. The final area of the biofilm can be changed by altering the droplet volume. In our case, we tried volumes ranging from 0.2 up to 0.8 μl, which gave biofilm areas of about 0.5–2.4 mm^2, respectively. Also note that the area of spreading for a given droplet volume will change with different ATPS formulations and underlying surfaces.
11. Upon dispensing the bacterial droplets in the PEG phase, hold the pipette vertically and try not to touch the bottom of the plate. In fact, it is better to dispense the droplet just below the PEG surface and allow it to settle down slowly by gravity. Also, to ensure accuracy and the absence of air bubbles, press the pipette plunger down to its lowest level (the expulsion volume level) and take up the maximum amount of the bacterial suspension from the DEX tube. Afterwards, upon delivering it to the PEG phase, press down to the first level and then withdraw the pipette. A portion of the bacterial suspension should still remain within the tip and can be discarded.
12. Try not to move the Petri dishes after patterning, and, if necessary, keep this to minimum and do it with caution to avoid disturbance of the bacterial droplets.
13. Make sure that you incubate your Petri dishes under humid conditions to avoid evaporation from the plate, which can distort your ATPS composition and the resulting biofilms.
14. Make sure that the medium used for culturing the epithelial cells is not supplemented with any antibiotic that can affect the bacterial growth.
15. Note that the diameters of the final bacterial spots derived here are bigger than those obtained in the previous protocol when using the same droplet volume. For example, a 0.6 μl droplet will generate a spot area of 27 mm^2, compared to 1.3 mm^2 upon using the previous protocol. This is attributed to both the differences in the surface chemistry between the surfaces, i.e., polystyrene and epithelial cells, as well as the different ATPS formulation being used.
16. We recommend viewing of a video protocol showing the ATPS patterning technique used when patterning mammalian cell cultures as many of the procedures for patterning of bacteria are analogous [18].
17. Note that you cannot pattern a large number of bacterial spots within the same plate. Attempts to do so will cause the pH of the medium to drop dramatically and the underlying epithelial cells to die and detach from the surface. Using this protocol,

we successfully patterned three *E. coli* droplets (at an initial O.D. of 0.05 and volume of 0.3 μl) on MCF10a monolayer inside a 35 mm Petri dish for 24 h without a dramatic change in the medium pH or a loss in epithelial cell viability.

Acknowledgments

We would like to thank Dr. Toshiyuki Yaguchi at Nagoya Institute of Technology, Japan, and Dr. Brendan M. Leung at the University of Michigan, USA, for their contributions in establishing the protocols discussed in this chapter.

This work was supported by a grant from the Creativity and Innovation Project funded by Ulsan National Institute of Science and Technology (UNIST) (Grant #1.120051.01) and by the World Class University (WCU) program (No. R322008000200540) through the National Research Foundation of Korea (NRF) as funded by the Ministry of Education, Science and Technology (MEST). The authors appreciate the support.

References

1. McEldowney S, Fletcher M (1987) Adhesion of bacteria from mixed cell suspension to solid surfaces. Arch Microbiol 148:57–62
2. Flemming HC, Wingender J (2010) The biofilm matrix. Nat Rev Microbiol 8:623–633
3. Rao D, Webb JS, Kjelleberg S (2005) Competitive interactions in mixed-species biofilms containing the marine bacterium *Pseudoalteromonas tunicata*. Appl Environ Microbiol 71:1729–1736
4. Lee JH, Lee J (2010) Indole as an intercellular signal in microbial communities. FEMS Microbiol Rev 34:426–444
5. Bansal T, Englert D, Lee J et al (2007) Differential effects of epinephrine, norepinephrine, and indole on *Escherichia coli* O157: H7 chemotaxis, colonization, and gene expression. Infect Immun 75:4597–4607
6. Sperandio V, Torres AG, Jarvis B et al (2003) Bacteria-host communication: the language of hormones. Proc Natl Acad Sci U S A 100:8951–8956
7. Richardson AE, Barea JM, McNeill AM et al (2009) Acquisition of phosphorus and nitrogen in the rhizosphere and plant growth promotion by microorganisms. Plant Soil 321:305–339
8. Weibel DB, Lee A, Mayer M et al (2005) Bacterial printing press that regenerates its ink: contact-printing bacteria using hydrogel stamps. Langmuir 21:6436–6442
9. Eun YJ, Weibel DB (2009) Fabrication of microbial biofilm arrays by geometric control of cell adhesion. Langmuir 25:4643–4654
10. Kim HJ, Boedicker JQ, Choi JW et al (2008) Defined spatial structure stabilizes a synthetic multispecies bacterial community. Proc Natl Acad Sci U S A 105:18188–18193
11. Kim J, Hegde M, Jayaraman A (2010) Co-culture of epithelial cells and bacteria for investigating host-pathogen interactions. Lab Chip 10:43–50
12. Tavana H, Jovic A, Mosadegh B et al (2009) Nanolitre liquid patterning in aqueous environments for spatially defined reagent delivery to mammalian cells. Nat Mater 8:736–741
13. Tavana H, Mosadegh B, Takayama S (2010) Polymeric aqueous biphasic systems for non-contact cell printing on cells: engineering heterocellular embryonic stem cell niches. Adv Mater 22:2628–2631
14. Yaguchi T, Lee S, Choi WS et al (2010) Micropatterning bacterial suspensions using aqueous two phase systems. Analyst 135:2848–2852
15. Yaguchi T, Dwidar M, Byun CK et al (2012) Aqueous two-phase system-derived biofilms for bacterial interaction studies. Biomacromolecules 13:2655–2661
16. Byun CK, Hwang H, Choi WS et al (2013) Productive chemical interaction between a bacte-

rial microcolony couple is enhanced by periodic relocation. J Am Chem Soc 135:2242–2247

17. Dwidar M, Leung BM, Yaguchi T et al (2013) Patterning bacterial communities on epithelial cells. PLoS One 8:e67165

18. Frampton JP, White JB, Abraham AT et al (2013) Cell co-culture patterning using aqueous two-phase systems. J Vis Exp. 73: e50304 doi:10.3791/50304

19. Tomlin KL, Clark SRD, Ceri H (2004) Green and red fluorescent protein vectors for use in biofilm studies of the intrinsically resistant *Burkholderia cepacia* complex. J Microbiol Meth 57:95–106

Chapter 3

Quorum Sensing in Gram-Positive Bacteria: Assay Protocols for Staphylococcal *agr* and Enterococcal *fsr* Systems

Akane Shojima and Jiro Nakayama

Abstract

A thiolactone/lactone peptide-mediated quorum sensing (QS) system is commonly employed in gram-positive bacteria to control the expression of a variety of phenotypes, including the production of virulence factors and biofilm formation. Here, we describe assay protocols for the well-studied QS systems (*agr* and *fsr*) of two representative gram-positive pathogens, *Staphylococcus aureus* and *Enterococcus faecalis*. These convenient assay systems are useful for the screening of QS inhibitors as well as for basic research to address the mechanism of these QS systems.

Key words Quorum sensing, *Staphylococcus aureus*, *Enterococcus faecalis*, Gelatinase

1 Introduction

Quorum sensing (QS) is a common regulatory system in unicellular microorganisms that is used to control the expression of a certain phenotype in response to cell density. This regulatory system is often mediated by signal molecules, leading to cell-to-cell communication networks. Through QS, bacteria orchestrate phenotypic shifts, such as from planktonic to biofilm states or non-virulent to virulent forms [1]. The signal molecules for QS can differ between gram-positive and gram-negative bacteria, such as *N*-acyl homoserine lactones in gram-negative bacteria and peptides in gram-positive bacteria, or they can be of a common type, such as AI-2 [1]. Among the diversified-structure peptides employed as QS signal molecules, the Agr-type autoinducing peptide (AIP) with a thiolactone/lactone structure has been found commonly among the low-GC gram-positive phylum Firmicutes [1, 2]. The *agr* QS system was originally found in *Staphylococcus aureus* [3], and a homologous QS system termed the *fsr* system was consequently found in *Enterococcus faecalis* [4]. Furthermore, the recent

Gianfranco Donelli (ed.), *Microbial Biofilms: Methods and Protocols*, Methods in Molecular Biology, vol. 1147,
DOI 10.1007/978-1-4939-0467-9_3, © Springer Science+Business Media New York 2014

increasing availability of bacterial genome sequence data has revealed the widespread presence of *agr*-type QS in Firmicutes, including a number of clostridia and listeria, in addition to staphylococci and enterococci [5].

S. aureus controls the expression of a series of virulence-associated proteins under the *agr* QS system [6]. The *agr* system is encoded by an *agrBDCA* operon. AgrD is processed by a cysteine protease-like function of AgrB, and a cyclic peptide with thiolactone is generated as an AIP. The AIP triggers the AgrC–AgrA two-component regulatory system and activates the *agr*P3 promoter, leading to the transcription of RNAIII that encodes δ-hemolysin. RNAIII also acts as a regulatory RNA and eventually controls the expression of a series of virulence genes by either transcription or translation. *E. faecalis* employs the *fsr* system to control the expression of two pathogenicity-related extracellular proteases, gelatinase (GelE) and serine protease (SprE) [4, 7]. Like the *agr* system, the *fsr* system consists of an AIP propeptide (FsrD), a processing enzyme (FsrB′), and the two-component sensor-regulator proteins (FsrC–FsrA) [4, 7, 8]. However, the *fsr* gene cluster is divided into two dependent transcriptional units (*fsrA* and *fsrBDC* operon), and a cyclic peptide with lactone instead of thiolactone, named gelatinase biosynthesis-activating pheromone (GBAP), functions as the AIP [9]. The *gelE–sprE* promoter is directly controlled by the FsrC–FsrA two-component regulatory system, instead of via the RNAIII-like regulatory molecule.

The emergence of drug-resistant staphylococci and enterococci, notably methicillin-resistant *S. aureus* (MRSA) and vancomycin-resistant enterococci (VRE), has complicated the treatment of these bacteria because of the limited choice of effective antibiotics [10, 11]. Under this situation, anti-pathogenic therapy targeting *agr*-type QS has attracted interest in terms of its mode of action, whereby only the expression of virulence is suppressed and without bactericidal action, leading to the selection of surviving resistant strains. Thus far, a number of researches have been conducted to develop QS inhibitors that target staphylococcal *agr* and enterococcal *fsr* systems [12–18]. While drug designs of inhibitors of AIP signaling or AIP biosynthesis have been successful [14–16], compound screening has successfully explored a number of QS inhibitors [12, 13, 17, 18].

In this chapter, assay protocols for quantifying the QS activity in *S. aureus* and *E. faecalis* will be noted in detail. The staphylococcal *agr* QS assay employs a dual-reporter strain, *S. aureus* 8325-4 (pSB2035), which develops luminescence and fluorescence in response to an AIP stimulus [19]. This reporter strain allows for a one-step assay in a multiwell plate, enabling the high-throughput screening of QS inhibitors [18]. On the other hand, the enterococcal QS activity is measured by the enzymatic activity of the chromosomally encoded gelatinase that is naturally regulated by the *fsr* QS system [9]. This assay is somewhat laborious because of its two-step procedure, in which an enzyme assay is performed following culturing of the indicator strain. However, with this assay system, a number of QS inhibitors have been screened while the detailed molecular mechanism in the *fsr* system has been investigated [12, 13].

2 Materials

2.1 Staphylococcal Agr System

1. Bacterial strains: *S. aureus* 8325-4 (pSB2035) and *S. aureus* 12600^T (*see* **Note 1**).
2. Culture media: Luria-Bertani (LB) broth (10 g of tryptone, 5 g of yeast extract, and 10 g of NaCl per liter; autoclaved at 121 °C for 15 min). Chloramphenicol (7 μg/mL) is added to the medium for *S. aureus* 8325-4 (pSB2035).
3. Cell washing buffer: PBS (8 g of NaCl, 0.2 g of KCl, 2.9 g of Na_2HPO_4, and 0.2 g of KH_2PO_4 per liter; pH 7.4; autoclaved at 121 °C for 15 min).
4. Multi-mode microplate reader: Infinite F200 Pro (Tecan) (*see* **Note 2**).
5. Microplate (*see* **Note 3**).

2.2 Enterococcal fsr System

1. Bacterial strain: *E. faecalis* OG1RF or *E. faecalis* OU510.
2. Culture medium: Dissolve 36.4 g of Todd Hewitt Broth (THB, Oxoid) per liter of water, and then autoclave it at 121 °C for 15 min.
3. Azocoll (*see* **Note 4**): (1) Suspend 250 mg of azocoll substrate (Azocoll™ substrate, 50 or 100 mesh: Calbiochem) in 50 mL of 50 mM Tris–HCl (pH 7.8) containing 1 mM $CaCl_2$ (*see* **Note 5**). (2) Leave to stand for 90 min at 37 °C. (3) Filter the suspension through an Advantec No.2 filter paper (discard the filtrate). (4) Wash the residue on the filter paper with 50 mL of 50 mM Tris–HCl containing 1 mM $CaCl_2$ (*see* **Note 6**).
4. Microtube mixer: Shaker SN-30B (Nissin Scientific Corp., Tokyo, Japan) (*see* **Note 7**).
5. Microplate absorbance reader: Sunrise (Tecan).

3 Methods

3.1 Staphylococcal Agr System

S. aureus 8325-4 (*agr* group I) was transformed with pSB2035, which encodes the *Photorhabdus luminescens* luciferase gene cluster (*luxABCDE*) and the enhanced green fluorescence gene (*gfp*) under the *agr*P3 promoter. The transformant, 8325-4 (pSB2035), autoinduces luciferase and GFP in a QS fashion [19]. By culturing this strain with the sample, both QS-enhancing and QS-inhibitory activities can be measured (Fig. 1). Since the *Photorhabdus* luciferase demands flavin mononucleotide as a cofactor, the luminescence depends on metabolic activity in addition to QS activity. Therefore, the other index, GFP, should also be measured for the screening study and so forth.

1. Preculture: *S. aureus* 8325-4 (pSB2035) and *S. aureus* 12600^T are cultured overnight in LB broth, at 37 °C with gentle agitation.
2. Samples to be tested for QS induction or suppression are dispensed into wells of a 96-well microplate (*see* **Note 8**). For the dose–response experiment, a serial dilution is prepared, as shown in Fig. 1.2.
3. Overnight cultures of *S. aureus* 8325-4 (pSB2035) and *S. aureus* 12600^T are diluted 1:100 in fresh LB broth, and then 200 μL is dispensed into each well of a microplate.
4. The microplate is set in the plate reader and incubated at 37 °C with shaking (*see* **Note 9**).
5. The OD_{620} and luminescence are measured at certain intervals (*see* **Note 10** and Fig. 1.3 for data example).
6. For measurement of fluorescence, the culture is transferred into a 1.5-mL microtube, and cells are harvested by centrifugation at 13,000 × *g* for 2 min. The supernatant is removed, and the cells are washed with 200 μL of PBS. After repeating this washing process two times, the cells are suspended in PBS and dispensed into each well of a microplate (black wall). Then, the fluorescence is measured at the excitation wavelength of 485 nm and emission wavelength of 535 nm (*see* **Note 11**).

3.2 Enterococcal *fsr* System

E. faecalis OG1RF is a gelatinase-positive strain, in which gelatinase production is autoinduced by endogenous GBAP. Since this QS system starts at the mid-log phase and continues to the late-log phase, the gelatinase activity in the late-log-phase culture supernatant represents its QS activity [9]. *E. faecalis* OU510 is a GBAP-negative, but GBAP-responsive, strain. By culturing this strain with the sample and measuring the induced gelatinase activity, the GBAP activity of the sample can be titrated [13] (Fig. 2).

1. *E. faecalis* OG1RF or OU510 is cultured overnight in THB medium, at 37 °C with gentle agitation.
2. Samples to be tested for Fsr QS are dispensed into 1.5-mL microtubes (*see* **Note 12**).
3. The overnight culture of *E. faecalis* is diluted 1:100 in fresh THB medium.

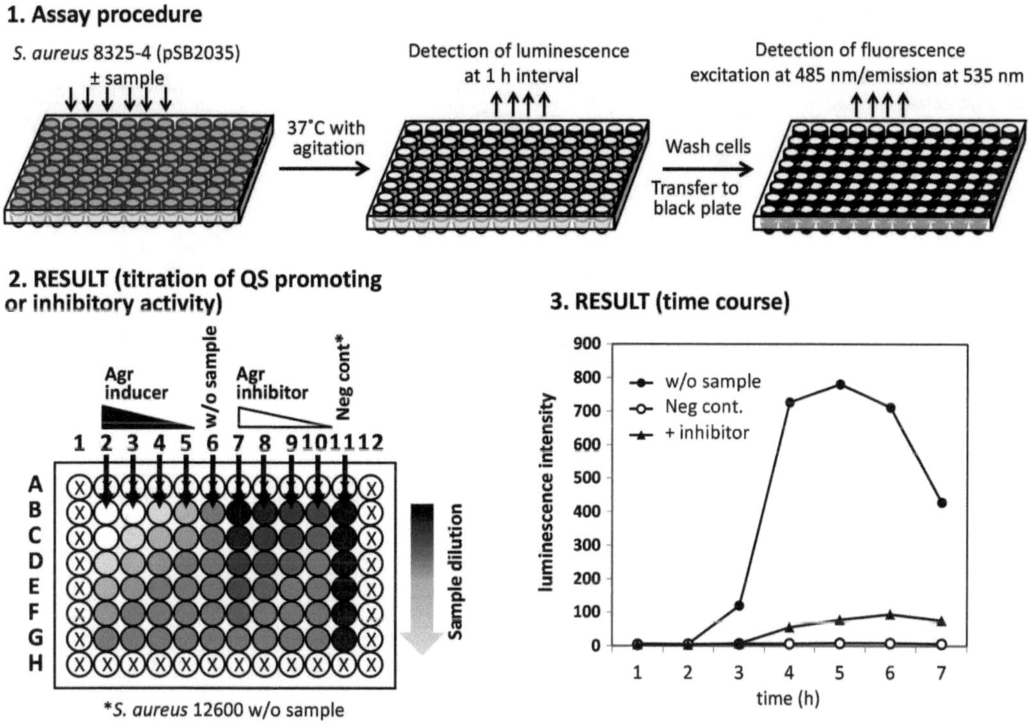

Fig. 1 Assay for the staphylococcal *agr* QS system. (*1*) Assay procedure. *S. aureus* 8325-4 (pSB2035) is dispensed into each well of a microplate containing tested sample and cultured. Cell growth (OD_{620}) and luminescence of luciferase are monitored in a plate reader. When cell growth reaches appropriate phase, cells are harvested and washed. Then, the fluorescence of cells is measured in the plate reader. (*2*) Examples of *agr* QS promoter or inhibitor titers. In the *upper rows* in *columns 2–5*, more luminescence (*whiter* in this illustration) is observed compared with the control without sample in *column 6*, which is indicative of the *agr*-inducing activity of samples applied in *columns 2–5*. On the other hand, in the *upper rows* in *columns 7–10*, less luminescence (*darker* in this illustration) is observed compared with the control in *column 6*, which is indicative of *agr*-inhibitory activity. (*3*) Example of time-course data. The *closed circle* represents the luminescence of the positive control (*S. aureus* 8325-4 (pSB2035) without sample). This datum indicates that the QS of this strain is at its maximum at 5 h after inoculation. The *closed triangle* represents the luminescence of the inhibitor. The *open circle* represents the luminescence of *S. aureus* 12600^T as a background of *S. aureus* cells

4. Five hundred microliters of the culture is dispensed into the 1.5-mL microtube containing the sample.
5. The tubes are cultured for 5 h at 37 °C, with shaking at 120 rpm.
6. After 5 h, 200 μL of the culture is dispensed into each well of a 96 well flat-bottom microtiter plate and the cell growth is measured at OD_{620}.
7. The rest of the culture in the microtube is centrifuged at 9,100 × *g* for 5 min.
8. Eight hundred microliters of azocoll solution is dispensed into a new 1.5-mL microtube and preincubated by shaking at 170 rpm and 37 °C for 15 min.

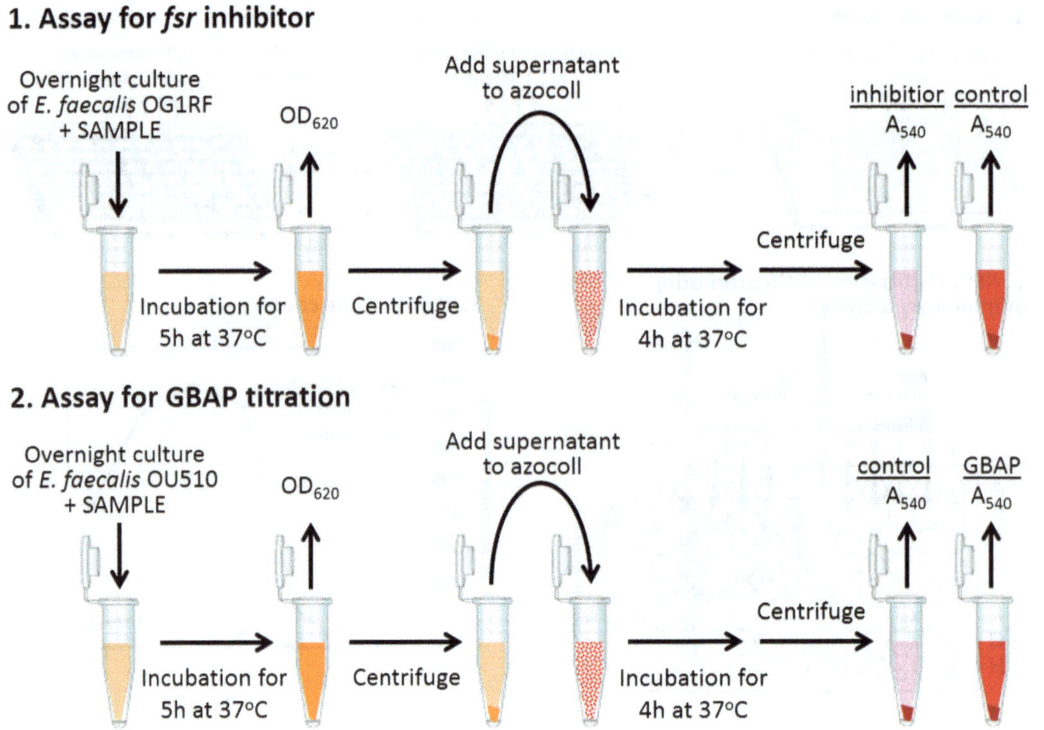

Fig. 2 Procedures for assaying the *fsr* inhibitor (*1*) and the GBAP titer (*2*). An overnight culture of *E. faecalis* OG1RF or OU510 is cultured with samples for 5 h at 37 °C. After 5 h, the culture supernatant is added to azocoll solution and incubated for 4 h at 37 °C. Then, the supernatant is measured at OD_{540} for gelatinase activity

9. Forty microliters of the culture supernatant is added to the azocoll suspension, and the mixture is shaken at 170 rpm and 37 °C for 4 h.
10. After 4 h, the azocoll suspension is centrifuged at 20,400 × *g* for 5 min.
11. Two hundred microliters of the supernatant is dispensed into the wells of a 96-well microtiter plate, and the absorbance is measured at OD_{540} (*see* **Note 13**).

4 Notes

1. *S. aureus* 12600^{T} is used to obtain the background values of luminescence and fluorescence of *S. aureus* cells.
2. This microplate reader is equipped with multiple detection modes of luminescence, fluorescence, and absorbance and a plate shaking and heating system. With this system, the luminescence and fluorescence from bacterial cells are measured in real time while culturing the bacteria at 37 °C under a certain aerobic condition.

3. The plate with a transparent bottom must be used when cell growth is being monitored by absorbance in real time. A white wall plate should be used for luminescence, whereas a black wall plate should be used for fluorescence.
4. Azocoll, which is an Azo dye-impregnated collagen, is used as a gelatinase substrate [20]. The peptide fragments that have been solubilized upon proteolysis are quantified by measuring the A_{540} of the supernatant.
5. The 500 mM $CaCl_2$ solution is sterilized by filtration and is then mixed with Tris–HCl buffer immediately before use. (For long storage periods, the $CaCl_2$ solution should be stored in a plastic tube instead of glassware.)
6. The azocoll suspension can be stored at 4 °C for up to 2 weeks.
7. This is like a cocktail shaker for microtube to keep azocoll resin suspended in the microtube.
8. The outermost wells in the plate should not be used for samples because the liquid is easily evaporated. It is best to fill medium into the outermost wells in order to suppress the evaporation in the inner wells. A lipophilic substance can be dissolved in methanol before application. Up to 5 % of methanol at final concentration can be applied. If the methanol in the sample is higher than 5 %, the excess methanol can be evaporated in a speed-vac concentrator before adding the sample to the reporter strain culture.
9. Shaking conditions, such as rotation rate, rotation mode (reciprocal or rotational), and swing amplitude, should be optimized for your experiment.
10. Since QS expression depends on the culture condition and the luminescence of luciferase is transient, the expression of QS should be monitored at a certain interval, such as every 1 h.
11. The luciferase luminescence in the cell depends on flavin mononucleotide. Therefore, the luminescence intensity depends on the metabolic activity in the cells. In order to check for false-positive QS inhibition due to the suppression of primary metabolism, the GFP fluorescence, which directly reflects the *agr*P3 promoter activity, should be measured simultaneously. Because LB broth shows a high background of fluorescence, it is better to remove and wash the medium for accurate and high-sensitivity assays. Since GFP fluorescence is sustainable in living cells for at least a few hours, a time lag is allowed after the luminescence has reached its maximum.
12. A lipophilic substance can be dissolved in methanol before application. Up to 1 % of methanol at final concentration can be applied. If the methanol is higher than 1 %, the excess

methanol can be evaporated in a speed-vac concentrator before adding the sample to the reporter strain culture.

13. For data example, *see* refs. 12, 13.

Acknowledgments

We would like to thank Prof. Paul Williams for providing the reporter strain *S. aureus* 8325-4(p2035). This research was supported in part by grants in aid for Scientific Research (B) Nos. 2138006 and 24380050 from the Japan Society for the Promotion of Science and from the Adaptable and Seamless Technology Transfer Program through Target-Driven R&D (A-step) (AS232Z02064G) from the Japan Science and Technology Agency.

References

1. Miller MB, Bassler BL (2001) Quorum sensing in bacteria. Annu Rev Microbiol 55: 165–199
2. Horinouchi S, Ueda K, Nakayama J, Ikeda T (2010) Cell-to-cell communications among microorganisms. In: Lew Mander L, Liu H-W (eds) Comprehensive natural products II, vol 4. Elsevier Science, Amsterdam, pp 283–337
3. Recsei P, Kreiswirth B, O'Reilly M et al (1985) Regulation of exoprotein gene expression by *agr*. Mol Gen Genet 202:58–61
4. Qin X, Singh KV, Weinstock GM, Murray BE (2000) Effects of *Enterococcus faecalis fsr* genes on production of gelatinase and a serine protease and virulence. Infect Immun 68: 2579–2586
5. Wuster A, Babu MM (2008) Conservation and evolutionary dynamics of the *agr* cell-to-cell communication system across firmicutes. J Bacteriol 190:743–746
6. Novick RP (2003) Autoinduction and signal transduction in the regulation of staphylococcal virulence. Mol Microbiol 48:1429–1449
7. Qin X, Singh KV, Weinstock GM, Murray BE (2001) Characterization of fsr, a regulator controlling expression of gelatinase and serine protease in *Enterococcus faecalis* OG1RF. J Bacteriol 183:3372–3382
8. Nakayama J, Chen S, Oyama N et al (2006) Revised model for *Enterococcus faecalis fsr* quorum-sensing system: the small open reading frame *fsrD* encodes the gelatinase biosynthesis-activating pheromone propeptide corresponding to staphylococcal *agrD*. J Bacteriol 188:8321–8326
9. Nakayama J, Cao Y, Horii T et al (2001) Gelatinase biosynthesis-activating pheromone: a peptide lactone that mediates a quorum sensing in *Enterococcus faecalis*. Mol Microbiol 41:145–154
10. DeLeo FR, Otto M, Kreiswirth BN, Chambers HF (2010) Community-associated meticillin-resistant *Staphylococcus aureus*. Lancet 375: 1557–1568
11. Murray BE (2000) Vancomycin-resistant enterococcal infections. N Engl J Med 342:710–721
12. Nakayama J, Uemura Y, Nishiguchi K et al (2009) Ambuic acid inhibits the biosynthesis of cyclic peptide quormones in gram-positive bacteria. Antimicrob Agents Chemother 53:580–586
13. Nakayama J, Tanaka E, Kariyama R et al (2007) Siamycin attenuates fsr quorum sensing mediated by a gelatinase biosynthesis-activating pheromone in *Enterococcus faecalis*. J Bacteriol 189:1358–1365
14. Lyon GJ, Mayville P, Muir TW, Novick RP (2000) Rational design of a global inhibitor of the virulence response in *Staphylococcus aureus*, based in part on localization of the site of inhibition to the receptor-histidine kinase, AgrC. Proc Natl Acad Sci U S A 97:13330–13335
15. Nakayama J, Yokohata R, Sato M et al (2013) Development of a peptide antagonist against fsr quorum sensing of *Enterococcus faecalis*. ACS Chem Biol 8:804–811
16. Chan WC, Coyle BJ, Williams P (2004) Virulence regulation and quorum sensing in staphylococcal infections: competitive AgrC antagonists as quorum sensing inhibitors. J Med Chem 47:4633–4641

17. Zhao ZG, Yan SS, Yu YM et al (2013) An aqueous extract of Yunnan Baiyao inhibits the quorum-sensing-related virulence of *Pseudomonas aeruginosa*. J Microbiol 51: 207–212
18. Desouky SE, Nishiguchi K, Zendo T et al (2013) High-throughput screening of inhibitors targeting *agr/fsr* quorum sensing in *Staphylococcus aureus* and *Enterococcus faecalis*. Biosci Biotechnol Biochem 77:923–927
19. Qazi SN, Counil E, Morrissey J et al (2001) *agr* expression precedes escape of internalized *Staphylococcus aureus* from the host endosome. Infect Immun 69:7074–7082
20. Mäkinen PL, Clewell DB, An F, Mäkinen KK (1989) Purification and substrate specificity of a strongly hydrophobic extracellular metalloendopeptidase ("gelatinase") from *Streptococcus faecalis* (strain 0G1-10). J Biol Chem 264:3325–3334

Chapter 4

Advanced Techniques for In Situ Analysis of the Biofilm Matrix (Structure, Composition, Dynamics) by Means of Laser Scanning Microscopy

Thomas R. Neu and John R. Lawrence

Abstract

The extracellular constituents in bioaggregates and biofilms can be imaged four dimensionally by using laser scanning microscopy. In this protocol we provide guidance on how to examine the various extracellular compartments in between microbial cells and communities associated with interfaces. The current options for fluorescence staining of matrix compounds and extracellular microhabitats are presented. Furthermore, practical aspects are discussed and useful notes are added. The chapter ends with a brief introduction to other approaches for EPS analysis and an outlook for future needs.

Key words Biofilm, Bioaggregate, Biofilm matrix, Extracellular polymeric substances (EPS), Confocal laser scanning microscopy (CLSM), Laser scanning microscopy (LSM), Fluorescence, Fluorescence lectin-binding analysis (FLBA)

1 Introduction

Today laser scanning microscopy (LSM) represents a well-established technique for structure–function studies of microbial aggregates and films. The main reason for applying LSM is its three-dimensional sectioning capability of fully hydrated, living, thick microbial communities. The LSM approach allows multichannel imaging of cellular and extracellular constituents. In addition there are a variety of probes for the microenvironment. Due to the specificity of fluorochromes and fluor-conjugated probes analytical details can be extensive. With modern instruments up to five different parameters can be recorded simultaneously or sequentially. The two-dimensional digital image series recorded may be examined by multidimensional visualization or semiquantitatively analyzed. Additional information on basic aspects of LSM and how to apply LSM for examination of microbiological communities has been presented elsewhere. The reader is referred to a number of

Gianfranco Donelli (ed.), *Microbial Biofilms: Methods and Protocols*, Methods in Molecular Biology, vol. 1147,
DOI 10.1007/978-1-4939-0467-9_4, © Springer Science+Business Media New York 2014

Table 1
Review articles on the application of laser scanning microscopy in different areas of microbiology and biofilm research

Topics	Reference
First overview on CLSM applications	[88]
Medical biofilms	[89]
Medical biofilms	[90]
Environmental applications	[91]
Comprehensive review of CLSM	[92]
CLSM methodology	[93]
Short CLSM overview	[94]
Medical biofilms	[95]
Structured CLSM approach	[96]
Environmental applications	[97]
CLSM in soil microbiology	[98]
CLSM immuno/molecular techniques	[99]
One-photon LSM versus two-photon LSM	[100]
Spatiotemporal approaches	[101]
Environmental applications	[102]
CLSM techniques and protocols	[103]
CLSM of aggregates	[104]
CLSM of microbes on hydrocarbons	[105]
CLSM–MRI–STXM	[106]

other reviews discussing the applicability of LSM for analysis of microbial communities at interfaces (Table 1).

Microbial biofilm systems are composed of microbial cells and their polymeric compounds which they produce and release into extracellular space. The topic of extracellular polymeric substances (EPS) was first compiled in a comprehensive book discussing the state of the art at the time [1]. More recently the EPS topic was comprehensively discussed by reviewing the major publications in the field [2]. The theme was picked up again in an article elaborating on different aspects of the biofilm matrix [3].

To start with we first have to define the various constituents present in between microbial cells in bioaggregates and biofilms. These are generally known as EPS although the acronym EPS has been defined in several different ways [2, 4]. Further the view of what makes up EPS has changed over several decades of biofilm research. Initially the idea was that only polysaccharides and proteins are present. In the meantime it is acknowledged that many different constituents are present in the extracellular space. These include various types of polysaccharides, a range of different proteins, extracellular nucleic acids, amphiphilic compounds, extracellular membrane vesicles, as well as bacteria-derived refractory compounds [2, 3, 5]. All of these constituents are involved in the functionality of the matrix in between bacteria. The issue of matrix function has been summarized in a still-evolving concept regarding the roles of EPS compounds [2]. Due to the biochemical diversity of EPS constituents it is impossible to use a single-staining approach. EPS imaging and analysis require targeting each specific group of biochemical compounds separately. Consequently we have to distinguish different staining procedures for each type of EPS. The main targets for fluorescence staining comprise *glycoconjugates*, *different proteins*, *nucleic acids*, *lipophilic compounds*, as well as *microhabitats*. At this point it should be mentioned that a major challenge of fluorescence staining will be the separation of extracellular signals from signals derived from microbial cells or cell surfaces.

In order to properly apply fluorochromes in epifluorescence or laser scanning microscopy, their absorption and emission spectrum should be established. In many cases only the excitation and emission maxima are published. This is however not sufficient for optimal staining and imaging especially if combinations of fluorochromes are used. Issues may arise due to the laser lines available (excitation) or potential cross talk between two channels (emission). On the other hand with new super-continuum light sources (white lasers) the maximum of the absorption peak can be excited. Despite these remarks, in the protocols below the excitation and emission maxima are given for simplicity. In any case each instrument is equipped with particular combinations of lasers on the excitation side. Similarly on the emission side different filter cubes, a spectral photometer, or an array of PMTs might be available. As a result the settings which have to be selected or defined are different at each individual instrument for the same application. A useful tool in this respect is a spectra viewer offered by some institutes and companies:

http://probes.invitrogen.com/servlets/spectraviewer

http://www.bdbiosciences.com/research/multicolor/spectrum_viewer/index.jsp

http://www.ebioscience.com/resources/fluorplan-spectra-viewer.htm

https://www.micro-shop.zeiss.com/?s=20002666494270&l=en&p=de&f=f&a=i

https://www.mcb.arizona.edu/ipc/fret/index.html

In addition, each instrument has the option for running a lambda scan. This option records the emission signal at a defined interval across the whole emission range. Instruments with traditional laser sources allow a lambda emission scan at a selected laser line, whereas instruments with a super-continuum light source can do both a lambda emission and a lambda excitation scan.

Clearly there is a desire to have a single probe for the overall biofilm matrix. In fact there was one attempt to image the complete matrix indirectly via the reflection of potassium permanganate (US patent 7,871,791 B2 from 18.01.2011). However, to the best of our knowledge, the technique is not widely used in the field of biofilm research. Rather than a general probe, and due to the complexity of the biofilm matrix, currently it might be more appropriate to aim at probes for the individual matrix constituents and the matrix microhabitats. This calls for a correlative approach using multiple fluorescent probes which may finally allow the simultaneous or the sequential imaging of several matrix constituents of interest. Despite this need, there is not even a single fluorescent probe for staining, e.g., all the polysaccharides within a biofilm matrix. Consequently there is still a requirement for new fluorochromes and probes which will target different matrix constituents and matrix microhabitats.

From the technical side, LSM will remain the first method of choice as the samples can be imaged in the hydrated state and multiple channels. In addition sample mounting allows many options for examination of bioaggregates and biofilms (coverwell chamber, coverslip chamber, Petri dish, flow cell, etc.). In terms of high-resolution new imaging, the so-called nanoscopy techniques are available. Compared to LSM with about 250 nm resolution, they allow imaging at a resolution of 120 nm (structured illumination microscopy), 80 nm (stimulated emission depletion microscopy), and 30 nm (localization/stochastic/blink microscopy). Although these techniques allow high-resolution imaging, the sample has to adapt to the technical requirements, e.g., for mounting, and there are a number of other trade-offs [6]. In addition, other new emerging techniques will offer analytical imaging in a new dimension. By chemical imaging using confocal Raman microscopy/spectroscopy, nanoSIMS, and synchrotron imaging (STXM) new insights into the complex polymer matrix of microbial communities may be gained [7–10]. Again, due to sample and mounting limitations these advanced chemical imaging techniques are ideally combined with traditional LSM. By applying a combined approach using sophisticated probes for EPS constituents combined with LSM, nanoscopy, and chemical imaging, this complex scaffold of extracellular space, its variable biochemistry, as well as its function and dynamics should become more accessible and understandable.

2 Materials

2.1 Polysaccharides

There is no general fluorescent stain for all polysaccharides with only two exceptions. However, it should be clear that these two options are working with specific types of polysaccharides only. For example calcofluor white (CFW) can be used for staining polysaccharides such as β-D-glucans having a (1→3) or (1→4) linkage (*see* **Notes 1** and **2**). From experience, in rare cases, this UV fluorochrome is sometimes useful for staining biofilms but only if exactly this type of polysaccharide is present. The same binding specificity is true for Congo red which may be used instead of CFW.

2.1.1 Calcofluor White M2R

1. Add 35 mg of CFW to 7 ml water.
2. Add a few drops of 10 N NaOH in order to dissolve CFW; the pH should be at 10 or 11.
3. Add water to have a final stock solution of 10 ml.
4. Prepare aliquots and store at −20 °C in the dark [11].
5. If the high pH is an issue, CFW may be directly dissolved in water and remaining crystals removed by centrifugation.

2.2 Glycoconjugates

In biofilms glycoconjugates are present at the bacterial cell surface, e.g., in the form of lipopolysaccharides or lipoteichoic acids, more extended as capsule or sometimes as adhesive material detectable as microbial "footprints" [12, 13]. In addition these glycoconjugates act to link bacterial cells, microcolonies, biofilms, bioaggregates, as well as forming microbial mats. In biofilm systems glycoconjugates may also appear as cloud-like structures devoid of cells as a result of bacterial detachment and dispersion. These glycoconjugates can be probed using lectins which are non-immunogenic proteins having a specificity for carbohydrates [14]. In most cases lectin specificity is not for a monosaccharide but for a di- or rather an oligosaccharide with a defined three-dimensional conformation. Fluor-labelled lectins currently represent the only probe which can be used for in situ staining of glycoconjugates in microbiological samples from the environment (Fig. 1). Only if pure culture or defined mixed culture studies are examined, fluorescently labelled antibodies may be applied for specific staining of cell surface antigens or extracellular carbohydrate antigens. This approach was heavily used by the research groups of Hartman (roots) and Kolenbrander (oral). Sample references of interest may be the following [15–23]. The general methodology as well as a critical examination of the lectin technique have been reported previously [24, 25]. The best approach for an unknown sample is to run a lectin screening with all the commercially available lectins for identification of a useful panel of probes [26–29] (Fig. 2). The suitability of lectins for selectively targeting and differentiating glycoconjugates was demonstrated using three different lectins for

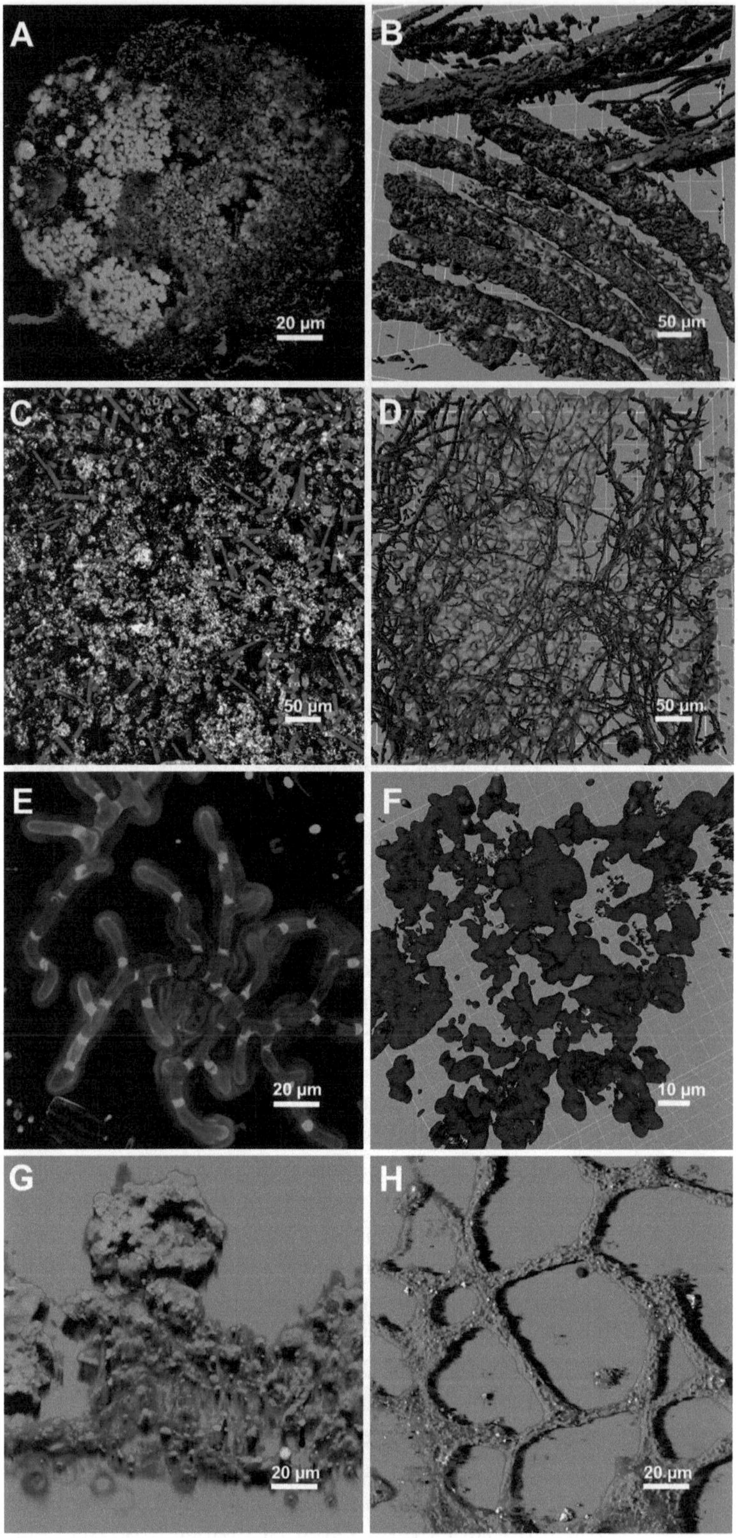

Fig. 1 Fluorescence lectin-binding analysis (FLBA) of biofilm systems from a variety of different habitats (some of the lectins were identified during a screening with all commercially available lectins as being useful for a particular biofilm): (**a**) Biofilm on pumice from a fluidized bed reactor showing two types of cell clusters,

detection of microhabitats within bacterial microcolonies [30]. Similarly lectins were used to identify the various adhesive glycoconjugates of *Deinococcus geothermalis* microcolonies and biofilms [31, 32]. The lectin approach is now termed fluorescence lectin-binding analysis (FLBA) which usually involves an initial lectin screening and subsequent staining with a panel of suitable lectins (*see*, e.g., ref. 27).

2.2.1 Fluorescence Lectin-Binding Analysis

1. FLBA requires the user to obtain commercial lectins labelled with fluorochromes.
2. The lectins are either labelled with traditional fluorochromes (e.g., FITC, TRITC) or with newly developed fluorochromes (e.g., different types of Alexa).
3. Lectins are supplied by Sigma-Aldrich, Vector Laboratories, EY Laboratories, Invitrogen, and some other biochemical companies.
4. If the lectin is only available without a fluorescence label, it may be conjugated with fluorochromes using commercial labelling kits, e.g., from Invitrogen, Pierce, Dyomics, AttoTec, and Abberior.
5. The fluorescence labelling is straightforward. Just follow the protocol supplied with the kit.
6. Using this approach, a fluorochrome with a particular property (excitation/emission) can be attached to the lectin matching a possible autofluorescence of the sample or for use with other fluorochromes (e.g., nucleic acid specific) being applied.
7. Lectins may also be labelled using quantum dots if an extremely stable fluorescence is needed.

Fig. 1 (continued) with and without producing lectin-specific glycoconjugates. The granule was stained with AAL-Alexa488 lectin (*green*) and Syto 60 (*red*). (**b**) Biofilm from a rotating annular reactor shown as 3D isosurface projection. The biofilm produced strands of polymer (streamers) which are colonized by other bacteria. The biofilm was stained by AAL-Alexa488 lectin (*green*) and Syto 60 (*red*). (**c**) Biofilm from a creek showing reflection of minerals (*white*), autofluorescence of cyanobacteria (*pink*), and WGA-FITC lectin staining (*green*). The autofluorescence of cyanobacteria (phycobilins/chlorophyll A) results in an overlay of the two signals. Please take notice of the lectin-stained sheath of cyanobacteria, many of which are empty. (**d**) Biofilm from a tube reactor shown as 3D isosurface projection. The base biofilm was embedded in glycoconjugates (semi-transparent), whereas the filamentous overlaying bacteria did not bind this particular lectin. The biofilm was stained by AAL-Alexa488 lectin (*green*) and Syto 60 (*red*). (**e**) Algal microcolony of an aggregate shown as maximum intensity projection. Please take notice of the differential staining of three glycoconjugate types by the AAL-Alexa488 lectin. The lectin stained the outer sheath of algae, the direct cell surface of algae, as well as the connections between individual algal cells. (**f**) Aggregate from the river Danube shown as 3D isosurface projection. The aggregate was stained with PhaE-Alexa633 lectin (*blue*) and SybGreen (*green*). (**g**) Cryosection of a rotating annular reactor biofilm shown as shadow projection. The sample was stained before sectioning with AAL-Alexa-488 lectin and Syto 60. (**h**) Phototrophic biofilm from a flow lane setup shown as shadow projection. The bacteria developed into a netlike structure stained with ACL-Alexa488 lectin

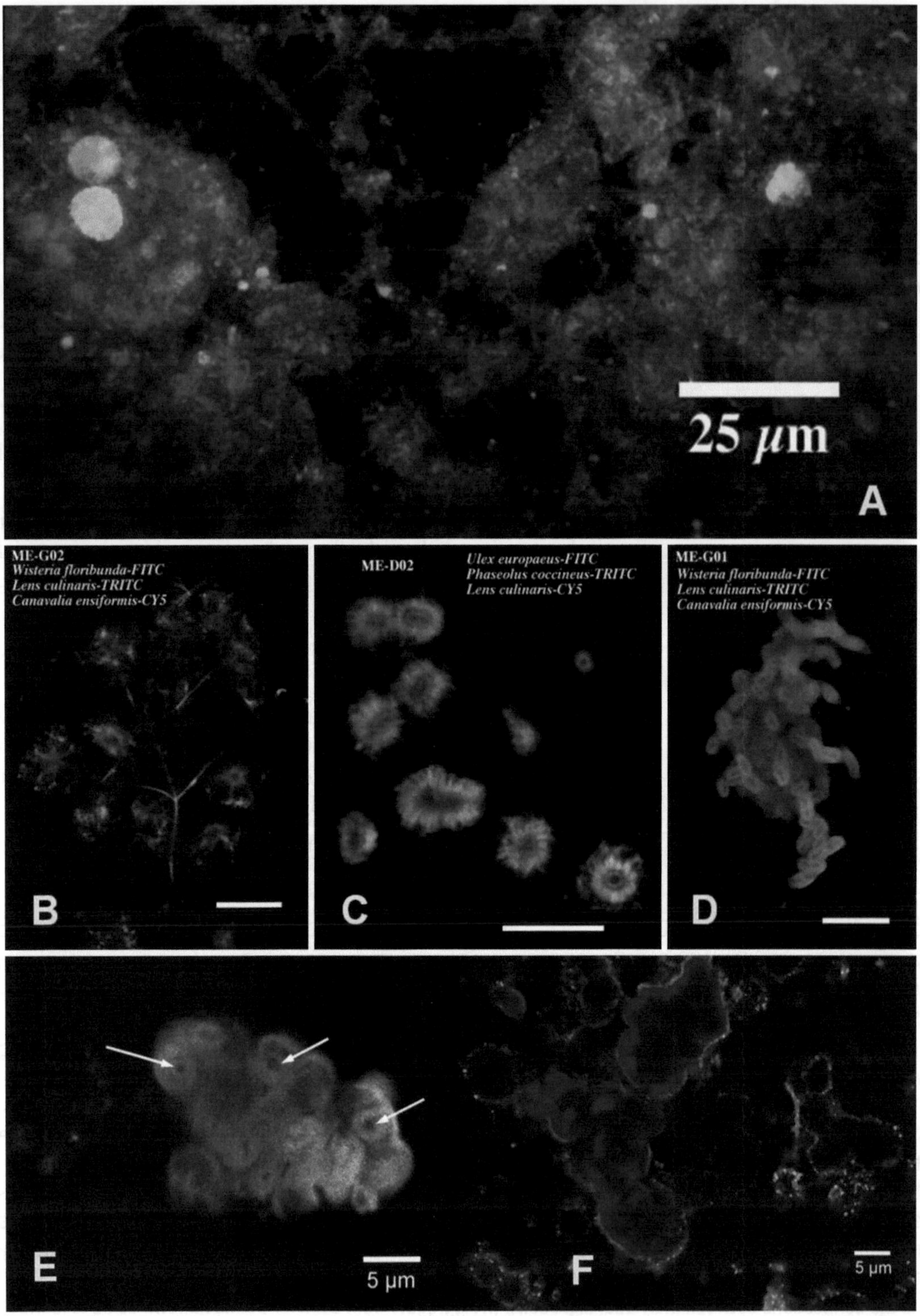

Fig. 2 Fluorescence lectin-binding analysis (FLBA) applied to river biofilms shown as maximum intensity projection: (**a**) Result of the sequential imaging of a triple-labelled microbial sample using the nucleic acid stain Syto9 (*green*), the hydrophobic lipid-sensitive stain Nile red (*red*), and polysaccharide using the lectin of *Ulex europaeus*-CY5 (*blue*). (**b**) Bacterial microcolony stained with the lectins *Wisteria floribunda*-FITC (*green*), *Lens culinaris*-TRITC (*red*), and *Canavalia ensiformis*-CY5 (*blue/not detected*). (**c**) Bacterial microcolonies stained

2.3 Proteins

For highly sensitive detection of proteins in electrophoresis gels different fluorescent stains are available. These stains may also be used on microbiological samples. The Sypro Red stain has been demonstrated to be suitable for measuring protein content of bacterial samples by means of flow cell cytometry [33]. We have found Sypro fluorochromes to be useful for staining detailed structural features in microbial communities in combination with CLSM [7, 34, 35]. Basically Sypro stains will bind to extracellular proteins but also to proteins at the cell surface. As a result the bacterial cells show an outer boundary staining. If cell surface and extracellular protein must be distinguished, counterstaining using permeable nucleic acid-specific fluorochromes may be necessary.

2.3.1 Sypro

1. Sypro protein stain is available as Sypro Orange, Red, Ruby, Tangerine (Invitrogen), all of which have different excitation/emission properties.
2. We have had good experience with Sypro Orange and Red on various microbiological sample types.
3. Other protein-specific fluorochromes are supplied by, e.g., Bio-Rad and GE-Healthcare.

2.4 Amyloids

There is an emerging awareness that amyloids may represent the major proteinaceous component of the EPS matrix. They show a characteristic structure with β-sheet-rich fibers. These types of proteins have been described using other terms as well, for example curli, tafi, chaplins, harpins, hydrophobins, or repellents. Amyloids have been suggested to function as a structural compound, toxin, surface-active fiber, genetic material, adhesin, and host mimetics [36, 37]. Amyloids in biofilms can be imaged by using different staining strategies, either directly with fluorochromes or via β-sheet-specific antibodies [38]. Amyloids are also discussed as functional amyloids not only in bacteria but also in yeasts [39–43].

1. Thioflavin T is usually applied for direct staining [44].
2. Conformation-specific antibodies, such as WO1 and WO2 [45], are available for antibody staining.

Fig. 2 (continued) with lectins *Ulex europaeus*-FITC (*green*), *Phaseolus coccineus-TRITC* (*red*), and *Lens culinaris*-CY5 (*blue*). (**d**) Bacterial microcolony stained with *Wisteria floribunda*-FITC, *Lens culinaris-TRITC*, and *Canavalia ensiformis*-CY5. (**e**) Bacterial microcolony after staining with ELF97-phosphatase (*green*), *Phaseolus vulgaris*-TRITC (*red*), and *Arachis hypogaea*-CY5 (*blue*); note green fluorescence in the surrounding EPS; arrows indicate the location of blue cells. (**f**) Biofilm stained with 100 nm sulfonated (*red*) and 100 nm carboxylated (*green*) latex beads as well as *Arachis hypogaea*-CY5 lectin (*blue*). Scale bars = 5 μm unless otherwise specified

2.5 Enzymes

There are many enzyme assays which are fluorescence based. With respect to laser microscopy the problem is diffusion of the fluorescent enzyme product to be recorded. However, there is one particular assay perfectly suited to examine the sample by means of LSM. The ELF 97 phosphatase assay results in an enzyme product which will precipitate at the point of enzyme activity. One of the first applications in microbiology was reported for activated sludge flocs [46]. Later the assay was used to measure phosphatase production of bacteria in photosynthetic biofilms [47]. Other studies have applied the ELF 97 phosphatase assay to marine [48] and freshwater [49–51] plankton samples. In addition, the ELF approach was employed to localize phosphatase activity in arbuscular mycorrhizal fungi [52].

1. ELF 97 phosphatase kit (Invitrogen).

2.6 Extracellular Nucleic Acids

Extracellular nucleic acids have been known for a long time to be part of microbial communities (e.g., refs. 53, 54). However, a short publication in science [55] triggered a rush into studies of extracellular DNA (eDNA) in microbial communities. In the meantime it is accepted that eDNA has a crucial role in biofilm and bioaggregate development and dynamics. In order to image eDNA principally all cell-impermeable nucleic acid-specific fluorochromes can be used. However, in most reports 7-hydroxy-9H-(1,3-dichloro-9,9-dimethylacridin-2-one) (DDAO) has been selected as the preferred fluorochrome for staining eDNA [56, 57].

1. DDAO from Invitrogen.

2.7 Amphiphilic Compounds/Outer Membrane Vesicles

Microorganisms release a range of different amphiphilic compounds which may be involved in microbial interaction with interfaces and each other [58]. These surface-active and lipophilic compounds may have low molecular weight or they can be polymeric as for example emulsan produced by Acinetobacter [59]. At a certain concentration they may aggregate and form lipophilic compartments. In addition, it has been known for a long time that bacteria release lipid vesicles from the cell surface. These are called outer membrane vesicles (OMVs) or just membrane vesicles (MVs). However, only a recent publication has brought this aspect into the context of EPS and their functionality [60]. In any case, most of the literature published to date has only used electron microscopy for imaging MVs. Only in one report FITC was used to visualize MVs although at low resolution [61]. Nevertheless, for both amphiphiles and MVs, the same lipophilic fluorescent stains may be applied, for example those binding to lipid vesicles (e.g., Nile red), fluorochromes intercalating with membranes (e.g., FM dyes) or other lipophilic fluorochromes:

1. Nile red (e.g., from Sigma Aldrich).
2. FM 1-43 and FM 4-64 (Invitrogen).

3 Methods

3.1 Polysaccharides

1. Apply a few drops of CFW to the sample and incubate for 10 min.
2. Rinse sample once to reduce background signal and obtain higher contrast.
3. Imaging can be done with a UV laser at 365 nm (strong CFW emission) or with a 405 nm laser diode (weaker CFW emission).

3.2 Glycoconjugates

3.2.1 Fluorescence Lectin-Binding Analysis (See Notes 3–6)

Lectins are usually delivered freeze-dried in portions of 1 mg. This is also the preferred form, since it eliminates unnecessary handling of these toxic materials (see below). Add 1 ml of water to dissolve the original sample and dilute 1:10 with the appropriate buffer or water (*see* **Note 7**). For some lectins a special lectin buffer is required [62]. Cover the samples with a few droplets of the lectin solution and incubate for 20 min in the dark at room temperature. The sample then has to be destained by carefully removing the unbound lectin. Washing of the sample has to be repeated four times. For lectin staining and removal of unbound lectins different procedures can be used:

1. With delicate biofilms the lectin solution can be carefully added to the biofilm surface, later the lectin solution is removed by drawing it off using a filter paper, and then new buffer or water is added and removed again (*see* **Note 8**).
2. Biofilms in flow cells may be stained by pumping the lectin solution into the flow cell, and then the lectin can be flushed out again with buffer or water by using the pump at the same speed.
3. If the biofilms are solid, developed at high shear force, the lectin solution can be added and flushed off using a Pasteur pipette.
4. Biofilms on solid particles (e.g., sand or aquifer material) can be stained in tubes where they will settle at the bottom; to remove the unbound lectin the supernatant is carefully drawn off using a pipette (glass, Pasteur, or μl pipette), and then buffer or water is added.
5. Aggregates can be stained in tubes (e.g., Eppendorf tubes); usually the aggregates will settle at the bottom of the tube, then the supernatant can be carefully removed, and new buffer or water added; if the aggregates do not settle they can be spun down using a small tabletop centrifuge at low speed.
6. Very delicate, non-washable samples may be stained with a lower lectin concentration, e.g., at a dilution of 1:100 or 1:1,000. Incubation time should then be extended up to 60 min or longer; this will avoid destaining although some background signal has to be accepted.

3.3 Proteins

3.3.1 Sypro

1. Take 5 μl of Sypro, and dilute it in 5 ml buffer or water (*see* **Note 9**).
2. Add a few droplets to the sample and incubate for 15 min.
3. The sample can be directly examined using the appropriate settings at the epifluorescence microscope (inspection by eye) or laser microscope (data recording) (*see* **Note 10**).
4. Excitation/emission:

Sypro Orange	470/570 nm
Sypro Red	550/630 nm
Sypro Ruby	450/610 nm
Sypro Tangerine	490/640 nm

3.4 Amyloids

The staining of amyloids in biofilms as described originates from the first publication on amyloids in biofilm systems [38].

3.4.1 Thioflavin T

1. In biofilms a thioflavin T concentration of 1.5 μM (*E. coli*) and 7.5 μM (natural biofilms) has been used.
2. The thioflavin solution is added to the cells or the biofilms and incubated for 5–10 min.
3. Samples do not require washing since the emission of bound stain is shifted from that of unbound (385 nm) (*see* **Note 11**).
4. The excitation maximum of thioflavin T upon binding to the fibrils is at 450 nm, and the signals from the sample can be recorded from 460 to 550 nm.

3.4.2 W01 W02 Antibodies (See **Note 12**)

1. First the sample has to be treated with a blocking agent. 1 % gelatine can be used as BSA interferes with fluorescence in situ hybridization.
2. The monoclonal antibodies WO1 and WO2 are developed in mouse spleen. They are added at a concentration of 10 nM with Tween 20 (0.05 %), and samples are incubated for 2 h.
3. The primary antibody is washed off with PBS/gelatine (with 0.1 % Triton) and resuspended in the same solution containing 1:256 working dilution of a fluor-conjugated second IgM antibody with 0.025 % Tween 20.
4. Then the sample is incubated for 1 h and washed again three times with PBS containing 0.1 % Triton X-100.

3.5 Enzymes

The ELF 97 phosphatase kit may be applied as suggested by the company's protocol. However, other protocols have been proposed for detection of extracellular phosphatase in plankton samples [50, 51].

1. The ELF 97 phosphatase substrate is diluted 20× and filtered to remove precipitates (*see* **Note 13**).

2. Then 100–200 μl is added to the sample (*see* **Note 14**).
3. After an appropriate time interval (*see* **Note 15**) the reaction (*see* **Note 16**) is stopped using PBS/levamisole as described in the protocol.
4. Excitation/emission: 360/550 nm.

3.6 Extracellular Nucleic Acids

1. DDAO may be applied at a final concentration of 1 μM for staining of eDNA (*see* **Notes 17** and **18**).
2. A 10-min incubation followed by careful washing 4× is usually sufficient; however, testing is recommended.
3. Excitation/emission: 640/660 nm.

3.7 Amphiphilic Compounds/Outer Membrane Vesicles (See Notes 19 and 20)

1. Prepare stock solution of FM dyes at 1 mg/ml in DMSO, and make aliquots of 10 μl and store at −20 °C.
2. Dilute 1:100 in buffer or water, add few drops to sample, and incubate for 10 min.
3. Samples can be immediately used for observation; no destaining is required.
4. Excitation/emissions:

Nile red	550/650 nm
FM 1-43	472/580 nm
FM 4-64	510/734 nm

3.8 Extracellular Microhabitat

The extracellular microhabitat in biofilms and bioaggregates can be characterized by different types of probes. For example fluorescent beads can be applied to determine the porosity of the biofilm matrix. Similarly fluorescently labelled dextrans and ficols of different molecular size can be employed [63]. The beads are available with anionic or cationic groups. For dextrans a range of molecular sizes may be employed with molecular weights of 4,000, 40,000, 70,000, 500,000, and 2,000,000. In order to get optimal results the incubation time has to be tested using time scales from 1 to 24 h.

Core shell silica nanoparticles (C-dots) have been described as "lab on a particle" which allows measurement of the pH in biofilms [64]. C-dots contain two fluorochromes, one for localization and the other for measuring. A similar approach was suggested for monitoring oxygen in biofilms [65]. In fact this method based on a sensing particle approach seems to be promising for measuring other dissolved compounds and gradients within biofilms. The advantage of this approach versus point measurements using microelectrodes is clearly on the aerial and volumetric information received.

Recently amphiphilic quantum dots were reported to image hydrophobic microhabitats within the biofilm matrix [66, 67]. In fact surface-engineered quantum dots may be useful for targeting different features of EPS. Related to this approach, it has been demonstrated that a hydrophobic layer is present at the biofilm surface of *Bacillus subtilis* [68, 69].

3.9 Counterstaining of Bacteria

In most cases EPS staining is combined with fluorescence staining of bacterial cells. For this purpose a wide range of nucleic acid-specific fluorochromes are available. These include the classical nucleic acid stains such as AO, DAPI, and PI. Usually cell-permeant fluorochromes from the Syto series are applied as they are available in a wide range of different colors. We frequently used Syto9, formerly part of the BacLight staining kit but now also available as a single compound. In addition other dyes such as SybrGreen and PicoGreen have been employed. Some studies applied other cell-impermeant fluorochromes for fixed samples, e.g., from the Sytox or POPO and PO-PRO series. Detailed information on the individual fluorochromes is available at the Invitrogen website. Despite their dominance, other companies also offer nucleic acid-specific fluorochromes, e.g., the company Biotium.

3.10 Other Approaches for EPS Analysis

Apart from in situ staining of EPS in many cases extraction and chemical analysis of EPS compounds have been *used* [70]. Several manuscripts report a number of procedures for extraction of EPS from activated sludge or microbial biofilms. Often they were done in parallel in order to assess the efficiency of the extraction. In any case potential cell lysis represents a critical issue in all the extraction protocols suggested (Table 2).

In the age of -omics EPS has also been examined by molecular techniques for example using proteomics. The results are very limited as only a few reports have been published so far. In a study with *Haemophilus influenzae* biofilms 265 proteins were identified in the EPS [71]. The matrix of *Myxococcus xanthus* contains polysaccharides but also proteins which have been found to have novel functions [72]. A comparison of the intra- and extracellular proteome was reported for *Listeria monocytogenes* grown in liquid culture [73]. In a study of dental plaque biofilms and their EPS composition it was shown that sucrose had a strong effect on EPS protein composition [74]. The EPS of Shewanella biofilms was examined using infrared spectroscopy in combination with proteomics showing 58 extracellular proteins associated with biofilm-specific functions such as redox activity of the matrix [2, 75].

Another approach takes advantage of chemical imaging for characterization of the biofilm matrix. In two publications Raman microscopy was used to examine matrix locations inside biofilms. In a first study the applicability of Raman microscopy to record EPS-specific signals was tested. Within the matrix, several Raman bands could be assigned to carbohydrates, proteins, nucleic acids,

Table 2
Extraction and chemical analysis of EPS from biofilms and activated sludge

Focus	Reference
Comparison of methods/sludge	[107]
Cation-exchange resin/biofilms	[108]
Comparison of methods/biofilms	[109]
Comparison of methods/sludge	[110]
Comparison of methods/sludge	[111]
EPS from cyanobacteria	[112]
Protein extraction/sludge	[113]
EPS from extreme acidic biofilms	[114]
eDNA/biofilm	[115]
Comparison of methods/sludge	[116]
Comparison of methods/sludge	[117]
Comparison of methods/assessment	[118]
Comparison of methods/sludge	[119]

Selected publications comparing different methods, isolating specific EPS types, or focusing on specific groups of bacteria

and carotenoids [76]. In the second study a combined approach taking advantage of confocal laser scanning and Raman microscopy was used. For this purpose biofilms were developed under different growth conditions. Examination of the biofilms matrix was done by FLBA and by analysis of single-spot Raman spectra [77]. Finally the Raman approach may be of advantage if Raman mapping of extended areas can be analyzed and assigned to specific groups of bacteria or microcolonies.

Very recently advanced chemical imaging was suggested as a tool to analyze EPS. For example the combination of Raman spectroscopy, STXM, and nanoSIMS was used to examine isolated EPS produced by *Bacillus subtilis* before and after adsorption to goethite [8]. However, it should be accepted that for this type of analysis the biofilm sample is restricted in terms of size/thickness, mounting, hydration, and depth of analysis.

4 Notes

1. CFW will also stain chitin, cellulose, keratin, collagen, and elastin in tissue.
2. CFW is also used to stain fungal cell surfaces. Therefore it is useful for fungal biofilms and aquatic fungi.

3. A positive lectin staining is clearly visible by eye using an epi-fluoresence microscope with the appropriate filter cube. If green lectins are used (e.g., Alexa488) the bound lectin should show a crisp green signal which usually is associated with a biological structure (e.g., bacterial cells).
4. If combinations of lectins are applied, tests should be made in advance as some lectins will bind to each other and form precipitates. The precipitates are easily visible under the microscope due to their appearance and the combined color of the two fluorochromes conjugated to the lectins.
5. Controls should be made. But be careful using monosaccharides for inhibition of lectin binding. The expected effect of the sugar is a lowering of the emission signal. However, you may increase the emission signal due to creation of a new specificity by saturating only one part of the mostly complex binding sites.
6. The targets of lectins are glycoconjugates of any kind. As a result glycoproteins and glycolipids may also give a positive result.
7. Some lectins are extremely toxic. *Be careful*! *Do not handle freeze-dried lectins but only lectins in aqueous solution*! *Do not open the vial*! The vials usually have a rubber cap covered with an aluminum seal. Open the aluminum seal in the center. Use a syringe with needle to add water, and dissolve the lectin. In order to take out a portion of the lectin to prepare the stock solution also use a syringe.
8. For drawing off lectins using some sort of sorbent many options are available. In the laboratory several types of sorptive papers are usually found (paper tissues, lens tissues, household tissue rolls, toilet paper, filtration papers). For this particular purpose the cut-in-half triangular absorbent swabs (Kettenbach, Germany), which have a strong suction capacity, are quite helpful.
9. It is a good idea to prepare aliquots from the stain as delivered in order to avoid damaging freeze–thaw cycles.
10. For observation of the far-red Sypro stain it is helpful to have dark-adapted eyes and work nearly in the dark with dimmed light.
11. Depending on the sample type and additional staining procedures different blocking agents may be used.
12. The most specific detection can be achieved using the WO1 and WO2 antibodies.
13. For the ELF 97 phosphatase substrate it is necessary to try a range of different dilutions.

14. Obviously for measuring extracellular phosphatase, fixation has to be omitted.
15. It is also important to test different time intervals of incubation.
16. In order to stop the reaction various reagents may be used [51].
17. Other fluorochromes used for eDNA staining include:
    (a) Syto9 [78, 79].
    (b) DAPI [80].
    (c) Propidium iodide [81].
    (d) Toto3 [82].
    (e) SytoxOrange [83].
    (f) BOBO3 [84].
18. In many studies staining of eDNA with DDAO was combined with either bacteria expressing fluorescent proteins (e.g., [56, 57, 85, 86, 87]).
19. FM fluorochromes are excellent stains for bacterial membranes. Therefore they should also be suitable for staining bacterial membrane vesicles.
20. Due to the MV's size of 50–250 nm they are at the edge of the resolution of a CLSM.

References

1. Wingender J, Neu TR, Flemming H-C (1999) Microbial extracellular polymeric substances. Springer, Heidelberg
2. Neu TR, Lawrence JR (2009) Extracellular polymeric substances in microbial biofilms. In: Moran A, Brenan P, Holst O, von Itzstein M (eds) Microbial glycobiology: structures, relevance and applications. Elsevier, San Diego, pp 735–758
3. Flemming HC, Wingender J (2010) The biofilm matrix. Nat Rev Microbiol 8:623–633
4. Allison DG, Sutherland IW, Neu TR (2003) EPS: what's in an acronym? In: Mc Bain A, Allison DG, Brading M, Rickard A, Verran J, Walker J (eds) Biofilm communities: order from chaos? BioLine, Cardiff, pp 381–387
5. Miltner A, Bombach P, Schmidt-Brücken B, Kästner M (2012) SOM genesis: microbial biomass as a significant source. Biogeochemistry 111:41–55
6. Schermelleh L, Heintzmann R, Leonhardt H (2010) A guide to super-resolution fluorescence microscopy. J Cell Biol 190:165–175
7. Lawrence JR, Swerhone GDW, Leppard GG et al (2003) Scanning transmission x-ray, laser scanning, and transmission electron microscopy mapping of the exopolymeric matrix of microbial biofilms. Appl Environ Microbiol 69:5543–5554
8. Liu X, Eusterhues K, Thieme J et al (2013) STXM and NanoSIMS investigations on EPS fractions before and after adsorption to goethite. Environ Sci Technol 47:3158–3166
9. Behrens S, Kappler A, Obst M (2012) Linking environmental processes to the in situ functioning of microorganisms by high-resolution secondary ion mass spectrometry (NanoSIMS) and scanning transmission X-ray microscopy (STXM). Environ Microbiol 14: 2851–2869
10. Remusat L, Hatton PJ, Nico PS et al (2012) NanoSIMS study of organic matter associated with soil aggregates: advantages, limitations, and combination with STXM. Environ Sci Technol 46:3943–3949
11. Rasconi S, Jobard M, Jouve L, Sime-Ngando T (2009) Use of calcofluor white for detection, identification, and quantification of phytoplanktonic fungal parasites. Appl Environ Microbiol 75:2545–2553

12. Neu TR, Marshall KC (1991) Microbial "footprints": a new approach to adhesive polymers. Biofouling 3:101–112
13. Neu TR (1992) Microbial "footprints" and the general ability of microorganisms to label interfaces. Can J Microbiol 38:1005–1008
14. Sharon N, Lis H (2003) Lectins. Kluwer Academic Publishers, Dordrecht
15. Schloter M, Borlinghaus R, Bode W, Hartmann A (1993) Direct identification, and localization of Azospirillum in the rhizosphere of wheat using fluorescence-labelled monoclonal antibodies and confocal scanning laser microscopy. J Microsc 171:173–177
16. Assmus B, Schloter M, Kirchhof G et al (1997) Improved in situ tracking of rhizosphere bacteria using dual staining with fluorescence-labeled antibodies and rRNA-targeted oligonucleotides. Microb Ecol 33:32–40
17. Schloter M, Wiehe W, Assmus B et al (1997) Root colonization of different plants by plant-growth-promoting *Rhizobium leguminosarum* bv. trifolii R39 studied with monospecific polyclonal antisera. Appl Environ Microbiol 63:2038–2046
18. Gilbert B, Assmus B, Hartmann A, Frenzel P (1998) In situ localization of two methanotrophic strains in the rhizosphere of rice plants. FEMS Microbiol Ecol 25:117–128
19. Palmer RJ Jr, Gordon SM, Cisar JO, Kolenbrander PE (2003) Coaggregation-mediated interactions of Streptococci and Actinomyces detected in initial human dental plague. J Bacteriol 185:3400–3409
20. Chalmers NI, Palmer RJ Jr, Du-Thumm L et al (2007) Use of quantum dot luminescent probes to achieve single: cell resolution of human oral bacteria in biofilms. Appl Environ Microbiol 73:630–636
21. Periasamy S, Chalmers NI, Du-Thumm L, Kolenbrander PE (2009) *Fusobacterium nucleatum* ATCC 10953 requires *Actinomyces naeslundii* ATCC 43146 for growth on saliva in a three-species community that includes *Streptococcus oralis* 34. Appl Environ Microbiol 75:3250–3257
22. Periasamy S, Kolenbrander PE (2009) Mutualistic biofilm communities develop with *Porphyromonas gingivalis* and initial, early and late colonizers of enamel. J Bacteriol 191:6804–6811
23. Periasamy S, Kolenbrander PE (2010) Central role of the early colonizer *Veillonella sp.* in establishing multispecies biofilm communities with initial, middle, and late colonizers of enamel. J Bacteriol 192:2965–2972
24. Neu TR, Lawrence JR (1999) Lectin-binding-analysis in biofilm systems. Methods Enzymol 310:145–152
25. Neu TR, Swerhone GDW, Lawrence JR (2001) Assessment of lectin-binding analysis for in situ detection of glycoconjugates in biofilm systems. Microbiology 147:299–313
26. Staudt C, Horn H, Hempel DC, Neu TR (2003) Screening of lectins for staining lectin-specific glycoconjugates in the EPS of biofilms. In: Lens P, Moran AP, Mahony T, Stoodley P, O'Flaherty V (eds) Biofilms in medicine, industry and environmental technology. IWA Publishing, UK, pp 308–327
27. Zippel B, Neu TR (2011) Characterization of glycoconjugates of extracellular polymeric substances in tufa-associated biofilms by using fluorescence lectin-binding analysis. Appl Environ Microbiol 77:505–516
28. Lu S, Chourey K, Reiche M et al (2013) Insights into the structure and metabolic function of microbes that shape pelagic iron-rich aggregates ("iron snow"). Appl Environ Microbiol 79:4272–4281
29. Bennke CM, Neu TR, Fuchs BM, Amann R (2013) Mapping glycoconjugate-mediated interactions of marine Bacteroidetes with diatoms. Syst Appl Microbiol 36:417–425
30. Lawrence JR, Swerhone GDW, Kuhlicke U, Neu TR (2007) In situ evidence for microdomains in the polymer matrix of bacterial microcolonies. Can J Microbiol 53: 450–458
31. Saarima C, Peltola M, Raulio M et al (2006) Characterisation of adhesion threads of *Deinococcus geothermalis* as Type IV pili. J Bacteriol 188:7016–7021
32. Peltola M, Neu TR, Kanto-Oqvist L et al (2008) Architecture of *Deinococcus geothermalis* biofilms on glass and steel: a lectin study. Environ Microbiol 10:1752–1759
33. Zubkov MV, Fuchs BM, Eilers H et al (1999) Determination of total protein content of bacterial cells by SYPRO staining and flow cytometry. Appl Environ Microbiol 65: 3251–3257
34. Hinck S, Mußmann M, Salman V et al (2011) Vacuolated *Beggiatoa*-like filaments from different hypersaline environments form a novel genus. Environ Microbiol 13: 3194–3205
35. Rusznyak A, Akob DM, Nietzsche S et al (2012) Calcite biomineralization by bacterial isolates from the recently discovered pristine karstic Herrenberg cave. Appl Environ Microbiol 78:1157 1167

36. DePas WH, Chapman MR (2012) Microbial manipulation of the amyloid fold. Res Microbiol 163:592–606
37. Schwartz K, Boles BR (2013) Microbial amyloids – functions and interactions within the host. Curr Opin Microbiol 16:93–99
38. Larsen P, Nielsen JL, Dueholm MS et al (2007) Amyloid adhesins are abundant in natural biofilms. Environ Microbiol 9: 3077–3090
39. Gebbink MFBG, Claessen D, Bouma B et al (2005) Amyloids: a functional coat for microorganisms. Nat Rev Microbiol 3:333–341
40. Jordal PB, Dueholm MS, Larsen P et al (2009) Widespread abundance of functional bacterial amyloid in mycolata and other gram-positive bacteria. Appl Environ Microbiol 75:4101–4110
41. Dueholm MS, Petersen SV, Soenderkaer M et al (2010) Functional amyloid in Pseudomonas. Mol Microbiol 77:1009–1020
42. Blanco LP, Evans ML, Smith DR et al (2012) Diversity, biogenesis and function of microbial amyloids. Trends Microbiol 20:66–73
43. Oli MW, Otoo HN, Crowley PJ et al (2012) Functional amyloid formation by *Streptococcus mutans*. Microbiology 158:2903–2916
44. Krebs MRH, Bromley EHC, Donald AM (2005) The binding of thioflavin-T to amyloid fibrils: localisation and implications. J Struct Biol 149:30–37
45. O'Nuallain B, Wetzel R (2002) Conformational Abs recognizing a generic amyloid fibril epitope. Proc Natl Acad Sci U S A 99: 1485–1490
46. Kloecke FO, Geesey GG (1999) Localization and identification of populations of phosphatase: active bacterial cells associated with activated sludge flocs. Microb Ecol 38:201–214
47. Espeland EM, Wetzel RG (2001) Effects of photosynthesis on bacterial phosphatase production in biofilms. Microb Ecol 42: 328–337
48. Gonzalez-Gil S, Kaefer BA, Jovine RVM et al (1998) Detection and quantification of alkaline phosphatase in single cells of phosphorous-starved marine phytoplankton. Mar Ecol Prog Ser 164:21–35
49. Rengefors K, Pettersson K, Blenckner T, Anderson DM (2001) Species-specific alkaline phosphatase activity in freshwater spring phytoplankton: application of a novel method. J Plankton Res 23:435–443
50. Nedoma J, Strojsova A, Vrba J et al (2003) Extracellular phosphatase activity of natural plankton studied with ELF97 phosphate: fluorescence quantification and labelling kits. Environ Microbiol 5:462–472
51. Strojsova A, Vrba J, Nedoma J, Komarkova J, Znachor P (2003) Seasonal study of extracellular phosphatase expression in the phytoplankton of a eutrophic reservoir. Eur J Phycol 38:295–306
52. van Aarle IM, Olsson PA, Söderström B (2001) Microscopic detection of phosphatase activity of saprophytic and arbuscular mycorrhizal fungi using a fluorogenic substrate. Mycologia 93:17–24
53. Murakawa T (1973) Slime production by *Pseudomonas aeruginosa* IV. Chemical analysis of two varieties of slime produced by *Pseudomonas aeruginosa*. Jpn J Microbiol 17:513–520
54. Palmgren R, Nielsen PH (1996) Accumulation of DNA in the extracellular matrix of activated sludge and bacterial cultures. Water Sci Technol 34:233–240
55. Whitchurch CB, Tolker-Nielsen T, Ragas P, Mattick JS (2002) Extracellular DNA required for bacterial biofilm formation. Science 295:1487
56. Allesen-Holm M, Bundvik Barken K, Yang L et al (2006) A characterisation of DNA release in *Pseudomonas aeruginosa* cultures and biofilms. Mol Microbiol 59:1114–1128
57. Yang L, Barken KB, Skindersoe ME et al (2007) Effects of iron on DNA release and biofilm development by *Pseudomonas aeruginosa*. Microbiology 153:1318–1328
58. Neu TR (1996) Significance of bacterial surface-active compounds in interaction of bacteria with interfaces. Microbiol Rev 60. 151–166
59. Rosenberg E, Ron EZ (1999) High- and low-molecular-mass microbial surfactants. Appl Microbiol Biotechnol 52:154–162
60. Schooling SR, Beveridge TJ (2006) Membrane vesicles: an overlooked component of the matrices of biofilms. J Bacteriol 188:5945–5957
61. Tashiro Y, Ichikawa S, Shimizu M et al (2010) Variation of physiochemical properties and cell association activity of membrane vesicles with growth phase in *Pseudomonas aeruginosa*. Appl Environ Microbiol 76:3732–3739
62. Brooks SA, Leathem AJC, Schumacher U (1997) Lectin histochemistry. A concise practical handbook. Bios Scientific Publishers, Oxford
63. Lawrence JR, Wolfaardt GM, Korber DR (1994) Determination of diffusion coefficients in biofilms by confocal laser microscopy. Appl Environ Microbiol 60:1166–1173

64. Hidalgo G, Burns A, Herz E et al (2009) Functional tomographic fluorescence imaging of pH microenvironments in microbial biofilms by use of silica nanoparticle sensors. Appl Environ Microbiol 75:7426–7435
65. Acosta MA, Velasquez M, Williams K et al (2012) Fluorescent silica particles for monitoring oxygen levels in three-dimensional heterogeneous cellular structures. Biotechnol Bioeng 109:2663–2670
66. Aldeek F, Mustin C, Balan L et al (2011) Surface-engineered quantum dots for the labeling of hydrophobic microdomains in bacterial biofilms. Biomaterials 32:5459–5470
67. Aldeek F, Schneider R, Fontaine-Aupart MP et al (2013) Patterned hydrophobic domains in the exopolymer matrix of *Shewanella oneidensis* MR-1 biofilms. Appl Environ Microbiol 79:1400–1402
68. Epstein AK, Pokroy B, Seminara A, Aizenberg J (2011) Bacterial biofilm shows persistent resistance to liquid wetting and gas penetration. Proc Natl Acad Sci U S A 108: 995–1000
69. Kobayashi K, Iwano M (2012) BslA(YuaB) forms a hydrophobic layer on the surface of *Bacillus subtilis* biofilms. Mol Microbiol 85:51–66
70. Nielsen PH, Jahn A (1999) Extraction of EPS. In: Wingender J, Neu TR, Flemming H-C (eds) Microbial extracellular polymeric substances. Springer, Berlin, pp 49–72
71. Gallaher TK, Wu S, Webster P, Aguilera R (2006) Identification of biofilm proteins in non-typeable *Haemophilus influenzae*. BMC Microbiol 6:1–9
72. Curtis PD, Atwood J III, Orlando R, Shimkets LJ (2007) Proteins associated with the *Myxococcus xanthus* extracellular matrix. J Bacteriol 189:7634–7642
73. Dumas E, Meunier B, Berdague J-L et al (2008) Comparative analysis of extracellular and intracellular proteomes of *Listeria monocytogenes* strains reveals a correlation between protein expression and serovar. Appl Environ Microbiol 74:7399–7409
74. Paes Leme AF, Bellato CM, Bedi G et al (2008) Effects of Sucrose on the extracellular matrix of plaque-like biofilm formed in vivo, studied by proteomic analysis. Caries Res 42: 435–443
75. Cao B, Shi L, Brown RN et al (2011) Extracellular polymeric substances from *Shewanella sp.* HRCR-1 biofilms: characterization by infrared spectroscopy and proteomics. Environ Microbiol 13:1018–1031
76. Ivleva N, Wagner M, Horn H, Niessner R, Haisch C (2009) Towards a nondestructive chemical characterization of biofilm matrix by Raman microscopy. Anal Bioanal Chem 393: 197–206
77. Wagner M, Ivleva NP, Haisch C et al (2009) Combined use of confocal laser scanning microscopy (CLSM) and Raman microscopy (RM): investigations on EPS-matrix. Water Res 43:63–76
78. Böckelmann U, Janke A, Kuhn R et al (2006) Bacterial extracellular DNA forming a defined network like structure. FEMS Microbiol Lett 262:31–38
79. Guiton PS, Hung CS, Kline KA et al (2009) Contribution of autolysin and sortase A during *Enterococcus faecalis* DNA-dependent biofilm development. Infect Immun 77:3626–3638
80. Jurcisek JA, Bakaletz LO (2007) Biofilms formed by nontypeable *Haemophilus influenzae* in vivo contain both double-stranded DNA and type IV pilin protein. J Bacteriol 189:3868–3875
81. Vilain S, Pretorius JM, Theron J, Brözel VS (2009) DNA as an adhesin: *Bacillus cereus* requires extracellular DNA to form biofilms. Appl Environ Microbiol 75:2861–2868
82. Mann EE, Rice KC, Boles BR et al (2009) Modulation of eDNA release and degradation affects *Staphylococcus aureus* biofilm maturation. PLoS One 4:e5822
83. Lappmann M, Claus H, van Alen T et al (2010) A dual role of extracellular DNA during biofilm formation of *Neisseria meningitis.* Mol Microbiol 75:1355–1371
84. Seper A, Fengler VHI, Roier S et al (2011) Extracellular nucleases and extracellular DNA play important roles in *Vibrio cholerae* biofilm formation. Mol Microbiol 82:1015–1037
85. Goedeke J, Paul K, Lassak J, Thormann KM (2011) Phage-induced lysis enhances biofilm formation in *Shewanella oneidensis* MR-1. ISME J 5:613–626
86. Goedeke J, Heun M, Bubendorfer S et al (2011) Roles of two *Shewanella oneidensis* MR-1 extracellular endonucleases. Appl Environ Microbiol 77:5342–5351
87. Pinchuk GE, Ammons C, Culley DE et al (2008) Utilization of DNA as a sole source of phosphorus, carbon, and energy by *Shewanella* spp: ecological and physiological implications for dissimilatory metal reduction. Appl Environ Microbiol 74:1198–1208
88. Caldwell DE, Korber DR, Lawrence JR (1992) Confocal laser scanning microscopy and digital image analysis in microbial ecology. Adv Microb Ecol 12:1–67
89. Gorman SP, Mawhinney WM, Adair CG (1993) Confocal laser scanning microscopy of

adherent microorganisms, biofilms and surfaces. In: Denyer SP, Gorman SP, Sussman M (eds) Microbial biofilms: formation and control. Blackwell, London, pp 95–107

90. Manning PA (1995) Use of confocal microscopy in studying bacterial adhesion and invasion. Methods Enzymol 253:159–167
91. Lawrence JR, Korber DR, Wolfaardt GM, Caldwell DE (1996) Analytical imaging and microscopy techniques. In: Hurst CJ, Knudsen GR, McInerney MJ, Stetzenbach LD, Walter MV (eds) Manual of environmental microbiology. ASM, Washington, pp 29–51
92. Lawrence JR, Wolfaardt G, Neu TR (1998) The study of microbial biofilms by confocal laser scanning microscopy. In: Wilkinson MHF, Shut F (eds) Digital image analysis of microbes. Wiley, Chichester, pp 431–465
93. Lawrence JR, Neu TR (1999) Confocal laser scanning microscopy for analysis of microbial biofilms. Methods Enzymol 310:131–144
94. Palmer RJ Jr, Sternberg C (1999) Modern microcopy in biofilm research: confocal microscopy and other approaches. Curr Opin Biotechnol 10:263–268
95. Adair CG, Gorman SP, Byers LB et al (2000) Confocal laser scanning microscopy for examination of microbial biofilms. In: An YH, Friedman RJ (eds) Handbook of bacterial adhesion. Humana, Totowa, pp 249–256
96. Neu TR, Kuhlicke U, Lawrence JR (2002) Assessment of fluorochromes for two-photon laser scanning microscopy biofilms. Appl Environ Microbiol 68:901–909
97. Lawrence JR, Korber DR, Wolfaardt GM et al (2002) Analytical imaging and microscopy techniques. In: Hurst CJ, Crawford RL, Knudsen GR, McInerney MJ, Stetzenbach LD (eds) Manual of environmental microbiology. ASM, Washington, pp 39–61
98. Li Y, Dick WA, Tuovinen OH (2004) Fluorescence microscopy for visualisation of soil microorganisms: a review. Biol Fertil Soils 39:301–311
99. Schmid M, Rothballer M, Assmus B et al (2004) Detection of microbes by scanning confocal laser microscopy (SCLM). In: Kowalchuk GA, Bruijn FJ, Head IM, Akkermans ADL, Elsas JD (eds) Molecular microbial ecology manual. Kluwer Academic Publishers, Dordrecht, pp 875–910
100. Neu TR, Lawrence JR (2005) One-photon versus two-photon laser scanning microscopy and digital image analysis of microbial biofilms. Methods Microbiol 34:87–134
101. Palmer RJ Jr, Haagensen J, Neu TR, Sternberg C (2006) Confocal microscopy of biofilms: spatiotemporal approaches. In: Pawley JB (ed) Handbook of biological confocal microscopy. Springer, New York, pp 882–900
102. Lawrence JR, Korber D, Neu TR (2007) Analytical imaging and microscopy techniques. In: Hurst CJ, Crawford RL, Garland JL, Lipson DA, Mills AL, Stetzenbach LD (eds) Manual of environmental microbiology. ASM, Washington, DC, pp 40–68
103. Lawrence JR, Neu TR (2007) Laser scanning microscopy. In: Reddy CA, Beveridge TJ, Breznak JA, Marzluf GA, Schmidt TM, Snyder LR (eds) Methods for general and molecular microbiology. ASM, Washington DC, pp 34–53
104. Lawrence JR, Neu TR (2007) Laser scanning microscopy for microbial flocs and particles. In: Wilkinson KJ, Lead JR (eds) Environmental colloids: behavior, structure and characterisation. John Wiley, Chichester, pp 469–505
105. Neu TR, Lawrence JR (2010) Examination of microbial communities on hydrocarbons by means of laser scanning microscopy. In: Timmis KN (ed) Microbiology of hydrocarbons, oils, lipids and derived compounds. Springer, Heidelberg, pp 4073–4084
106. Neu TR, Manz B, Volke F et al (2010) Advanced imaging techniques for assessment of structure, composition and function in biofilm systems. FEMS Microbiol Ecol 72:1–21
107. Brown MJ, Lester JN (1980) Comparison of bacterial extracellular polymer extraction methods. Appl Environ Microbiol 40:179–185
108. Jahn A, Nielsen PH (1995) Extraction of extracellular polymeric substances (EPS) from biofilms using a cation exchange resin. Water Sci Technol 32:157–164
109. Zhang X, Bishop PL, Kinkle BK (1999) Comparison of extraction methods for quantifying extracellular polymers in biofilms. Water Sci Technol 39:211–218
110. Liu H, Fang HHP (2002) Extraction of extracellular polymeric substances (EPS) of sludges. J Biotechnol 95:249–256
111. Comte S, Guibaud G, Baudu M (2006) Relations between extraction protocols for activated sludge extracellular polymeric substances (EPS) and EPS complexation properties: part I. Comparison of the efficiency of eight EPS extraction methods. Enzyme Microb Technol 38:237–245
112. Klock J-H, Wieland A, Seifert R, Michaelis W (2007) Extracellular polymeric substances (EPS) from cyanobacterial mats: characterisation and isolation method optimisation. Marine Biol 152:1077–1085

113. Ras M, Girbal-Neuhauser E, Paul E et al (2008) Protein extraction from activated sludge: an analytical approach. Water Res 42:1867–1878

114. Aguilera A, Souza-Egipsy V, San Martin-Uriz P, Amils R (2008) Extraction of extracellular polymeric substances from extreme acidic microbial biofilms. Appl Microbiol Biotechnol 78:1079–1088

115. Wu J, Xi C (2009) Evaluation of different methods for extracting extracellular DNA from the biofilm matrix. Appl Environ Microbiol 75:5390–5395

116. D'Abzac P, Bordas F, van Hullebusch E et al (2010) Extraction of extracellular polymeric substances (EPS) from anaerobic granular sludge: comparison of chemical and physical extraction protocols. Appl Microbiol Biotechnol 85:1589–1599

117. Wei LL, Wang K, Zhao QL et al (2012) Fractional, biodegradable and spectral characteristics of extracted and fractionated sludge extracellular polymeric substances. Water Res 46:4387–4396

118. Sun M, Li WW, Yu HQ, Harada H (2012) A novel integrated approach to quantitatively evaluate the efficiency of extracellular polymeric substances (EPS) extraction process. Appl Microbiol Biotechnol 96:1577–1585

119. Zhang L, Ren H, Ding L (2012) Comparison of extracellular polymeric substances (EPS) extraction from two different activated sludges. Water Sci Technol 66:1558–1564

Chapter 5

Multiplex Fluorescence In Situ Hybridization (M-FISH) and Confocal Laser Scanning Microscopy (CLSM) to Analyze Multispecies Oral Biofilms

Lamprini Karygianni, Elmar Hellwig, and Ali Al-Ahmad

Abstract

Multiplex fluorescence in situ hybridization (M-FISH) constitutes a favorable microbiological method for the analysis of spatial distribution of highly variable phenotypes found in multispecies oral biofilms. The combined use of confocal laser scanning microscopy (CLSM) produces high-resolution three-dimensional (3D) images of individual bacteria in their natural environment. Here, we describe the application of M-FISH on early (*Streptococcus* spp., *Actinomyces naeslundii*) and late colonizers (*Fusobacterium nucleatum*, *Veillonella* spp.) of in situ-formed oral biofilms, the acquisition of CLSM images, as well as the qualitative and quantitative analysis of these digitally obtained and processed images.

Key words Multiplex fluorescence in situ hybridization (M-FISH), Confocal laser scanning microscopy (CLSM), Three-dimensional images, Multispecies oral biofilms, *Streptococcus* spp., *Fusobacterium nucleatum*

1 Introduction

The nonspecific nature of traditional bacterial identification techniques, such as 4′,6-diamidino-2-phenylindole (DAPI) staining, crystal violet (CV), and SYTO 9/propidium iodide fluorochrome uptake, has limited their use in oral biofilm research [1]. Among the more widely accepted approaches, the multiplex fluorescence in situ hybridization (M-FISH) method allows for the in situ analysis of the spatial and temporal dynamics of different bacterial populations within oral biofilms [2, 3]. By means of fluorescently labeled oligonucleotide probes specific 16S rRNA sequences are targeted that are barely influenced by the growth condition [4, 5]. The advantages of using M-FISH to spatially discriminate between various members of the microbial community involve the ability for identification of uncultured bacteria and the rapid manufacturing of new oligonucleotide probes [6].

Gianfranco Donelli (ed.), *Microbial Biofilms: Methods and Protocols*, Methods in Molecular Biology, vol. 1147,
DOI 10.1007/978-1-4939-0467-9_5, © Springer Science+Business Media New York 2014

The invention of novel techniques for the visualization of in situ- or in vivo-established oral biofilms has introduced a new era in the field of dental research [7–10]. The combined use of M-FISH with confocal laser scanning microscopy (CLSM) monitors the three-dimensional spatial distribution of different bacteria in multispecies oral biofilms and can quantify semiplanktonic or desorbed biofilms in their natural habitat [11–13]. The acquisition of high-resolution images is followed by image processing and data analysis, offering the chance to achieve a full segmentation of the image into subareas with evident biofilm formation. The biomass of the different targets within oral biofilms can be then quantified by the analysis of total fluorescent staining of the confocal micrographs using an appropriate image analysis program.

In this chapter, a five-color M-FISH assay was utilized with the aid of CLSM for the spatial organization analysis of important bacterial members representing early and late colonizers within oral biofilms formed in situ after different time periods (1, 2, 3, 5, and 7 days). In addition to a nucleotide probe which allowed detection of all eubacteria, specific nucleotide probes for *Streptococcus* spp., *Fusobacterium nucleatum*, *Actinomyces naeslundii*, and *Veillonella* spp. were simultaneously utilized to identify the aforementioned oral biofilm bacteria. The analysis of the digitally acquired and processed images succeeded in quantifying the total amount of the bacterial targets within the tested oral biofilms.

2 Materials

Unless indicated otherwise prepare and store all reagents at room temperature. Prepare all solutions using double-distilled water (H_2O). Carefully follow all waste disposal regulations when disposing waste materials.

2.1 Equipment for In Situ Oral Biofilm Acquisition

1. At least four healthy volunteers (*see* **Note 1**).
2. Upper-jaw individual acrylic appliances (*see* Fig. 1).
3. Six sterilized and BSE-free bovine enamel discs (diameter, 5 mm; 19.63-mm^2 surface area; height, 1.5 mm) (*see* **Notes 2** and **3** and Fig. 1).
4. Red wax (*see* **Note 4**).
5. 0.9 % saline solution (NaCl) (*see* **Note 5**).
6. 3 % sodium hypochlorite (NaOCl) (*see* **Note 3**).
7. 70 % ethanol (*see* **Note 3**).
8. Double-distilled water (H_2O) (*see* **Note 3**).

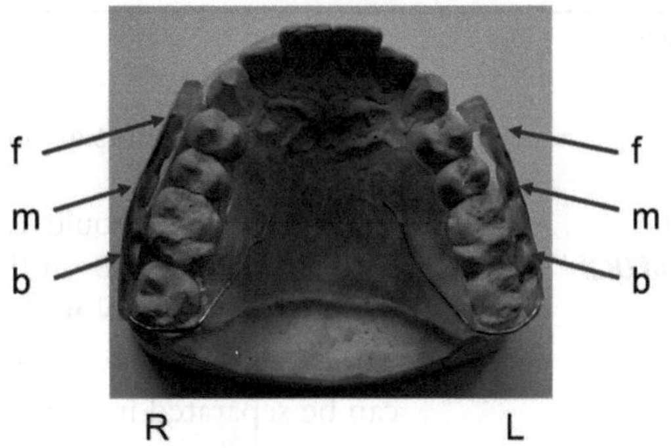

Fig. 1 Individual upper-jaw acrylic appliance with the enamel slabs in place at different locations. The specimens were positioned at the front (*f*), in the middle (*m*), and at the back (*b*), on both sides, right (*R*) and left (*L*), of the appliance. The exposed surfaces were fixed towards the tooth enamel by red wax. Reproduced from [9] with permission from ASM Journals

2.2 Multiplex Fluorescence In Situ Hybridization Components

1. Dulbecco's phosphate-buffered saline (PBS; 1.7 mM KH_2PO_4, 5 mM Na_2HPO_4, and 0.15 M sodium chloride, pH 7.2). Store at 4 °C.
2. Fix buffer: 4 % paraformaldehyde in PBS.
3. Fixative solution: Ethanol (50 %, v/v, in PBS). Store at 4 °C.
4. Permeabilization buffer: 7 mg lysozyme per ml 0.1 M Tris–HCl, 5 mM EDTA (pH 7.2). Store at 37 °C.
5. Dehydrating solution: Ethanol washes containing 50, 80, and 100 % ethanol.
6. Oligonucleotide probes: EUB 338 [14], E 79[15], FUS 664 [3], IF 201[16], and STR 405 [15] (*see* **Note 6**).
7. Hybridization buffer: 0.9 M NaCl, 20 mM Tris–HCl (pH 7.2), 25 % (v/v) formamide, and 0.01 % (w/v) sodium dodecyl sulfate.
8. Wash buffer: 20 mM Tris–HCl (pH 7.5), 5 mM EDTA, 159 mM NaCl, and 0.01 % (w/v) sodium dodecyl sulfate.

2.3 Confocal Laser Scanning Microscopy Equipment

1. Chambered coverglass (Lab-Tek II, Nalge Nunc International).
2. Confocal laser scanning microscope (μ-Slide, 8 Well, Ibidi, Martinsried, Germany) using a ×63 water immersion objective (HCX PL APO/bd.BL 63.0 × 1.2 W; Leica).
3. Image analysis program MetaMorph 6.3 × 7 (Molecular Devices Corporation).

3 Methods

Carry out all procedures at the temperatures specified each time.

3.1 In Situ Acquisition of Oral Biofilm

1. The volunteers should carry an individual acrylic appliance with six enamel slabs in the upper jaw over periods of 1, 2, 3, 5, and 7 days (*see* **Note 7**).
2. After each of the five tested time periods, all six enamel chips can be separated from the splint and further examined.

3.2 Multiplex Fluorescence In Situ Hybridization

1. Fix biofilms grown on enamel chips in 4 % paraformaldehyde in PBS for 12 h at 4 °C [4].
2. After fixation wash all specimens with PBS and fix again in solution containing ethanol (50 %, v/v, in PBS) for 12 h at 4 °C.
3. Wash specimens twice with PBS at room temperature.
4. Incubate in permeabilization solution for 10 min at 37 °C.
5. Dehydrate biofilms with a series of ethanol washes containing 50, 80, and 100 % ethanol for 3 min each at room temperature.
6. Incubate specimens with the oligonucleotide probes at a concentration of 50 ng each per 20 ml hybridization buffer in 96-well plates (Greiner Bio-One) at 46 °C for 2 h (*see* **Notes 8–10**).
7. Incubate specimens in wash buffer for 15 min at 48 °C.

3.3 Confocal Laser Scanning Microscopy

1. The analysis of the labeled biofilms can be conducted in a chambered coverglass by CLSM (*see* **Notes 11** and **12**).
2. The biofilm is scanned at several locations [17, 18]. As a result, the measured areas originate from three separate and representative locations on the oral biofilm-coated bovine enamel slabs (*see* **Note 13**).
3. With the aid of a zoom setting of 1.7 corresponding to physical dimensions of 140 × 140 μm for each image standard images are made. The area of each section is transformed into a digital image containing 1,024 × 1,024 pixels [19–21] (Fig. 2).
4. The biomass of the different targets within the oral biofilm and total fluorescent staining of the confocal micrographs are quantified and analyzed by the image analysis program MetaMorph 6.3 × 7 (Molecular Devices Corporation) (*see* **Note 14**).

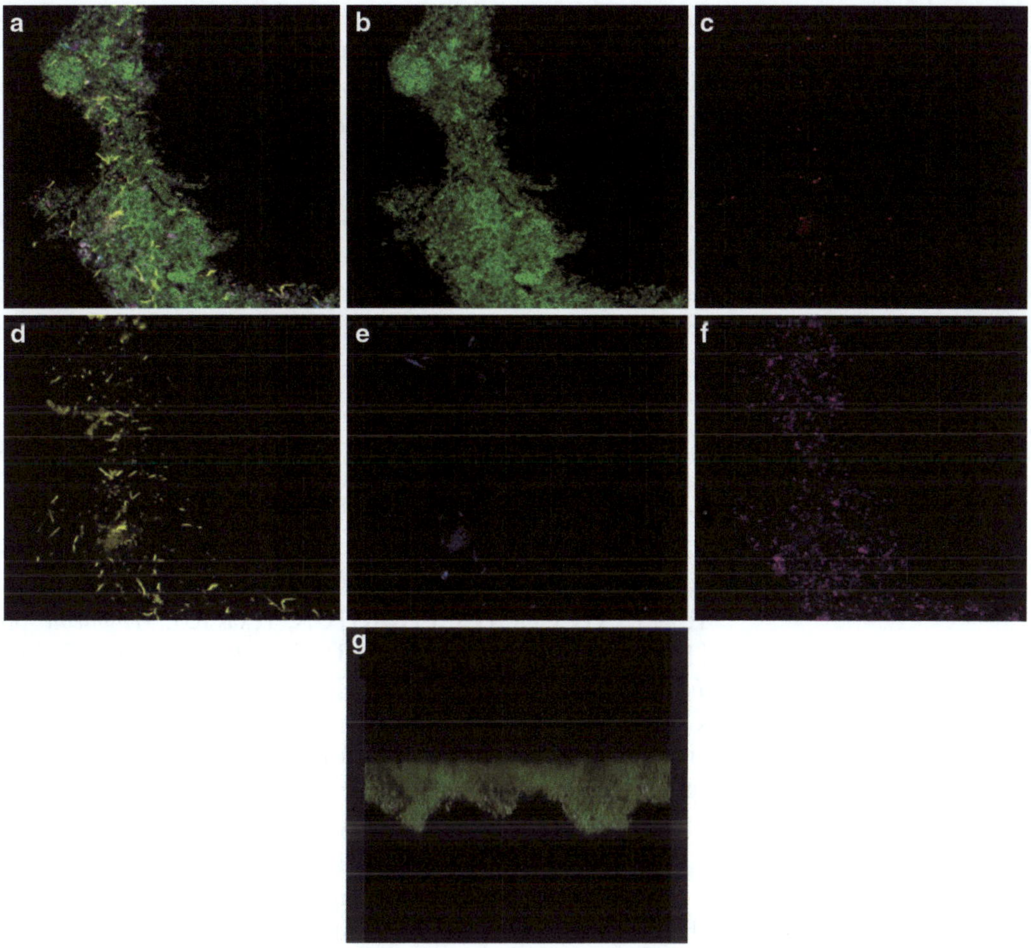

Fig. 2 Confocal micrographs of 3-day-old dental plaque biofilm hybridized with five different specific probes (EUB 338, E 79, FUS 664, IF 201, and STR 405). (**a**) Single optical section from five-channel image stack. (**b**–**f**) All channels of the stack. (**b**) *Green*, eubacteria-specific probe; (**c**) *red*, *Veillonella* spp.-specific probe; (**d**) *yellow*, *Fusobacterium nucleatum*-specific probe; (**e**) *blue*, *Actinomyces naeslundii*-specific probe; (**f**) *magenta*, *Streptococcus* spp.-specific probe. (**g**) Orthogonal slice of the plaque biofilm. Reproduced from [2] with permission from JMM Journal

4 Notes

1. Patients with conditions that meet the following criteria are excluded: (1) severe systemic disease, (2) pregnancy or lacta tion, (3) use of antibiotics or antibacterial mouth rinses within the last 3 months, and (4) DMFT index less than 3.
2. The buccal surfaces of bovine incisors are separated and modified into cylindrical enamel specimens. The enamel surfaces of all samples are then disinfected and polished by wet grinding with abrasive paper (400–4,000 grit). The absence of

bovine spongiform encephalopathy (BSE) is confirmed by the use of the IDEXX Laboratories BSE diagnostic kit (Ludwigsburg, Germany).

3. The protocol for disinfection of the enamel plates includes ultrasonication in NaOCl (3 %) for 3 min to remove the superficial smear layer, air-drying, and ultrasonication in 70 % ethanol for another 3 min. The disinfected samples are then ultrasonicated twice in double-distilled water for 10 min and, finally, stored in distilled water for 24 h to hydrate prior to exposure in the oral cavity [2].
4. The enamel specimens can be attached to the approximal sides with the aid of red wax.
5. After their exposure in the oral cavity the enamel slabs can be rinsed off with 0.9 % saline solution for 10 s.
6. The sequences, 5′-modifications, and target species of the oligonucleotide probes used are described as follows: EUB 338 (GCTGCCTCCCGTAGGAGT; fluorescein; eubacteria [14]), E 79 (AATCCCCTCCTTCAGTGA; Texas red; *Veillonella* spp. [15]), FUS 664 (CTTGTAGTTCCGC(C/T)TACCTC; Cy5; *Fusobacterium nucleatum* [3]), IF 201 (GCTACCGT CAACCCACCC; Pacific Blue; *Actinomyces naeslundii* [16]), and STR 405 (TAGCCGTCCCTTTCTGGT; Cy3; *Streptococcus* spp. [15]).
7. The subject maintains his regular diet and should carry the appliance intraorally throughout all tested time periods, except during meals as well as brushing and flossing his teeth. After their removal from the oral cavity the splints are kept in 0.9 % NaCl solution.
8. All commercially synthesized oligonucleotide probes are purified with the aid of high-performance liquid chromatography (HPLC) and 5′-end labeled with different fluorochromes. In a multiplex FISH assay the different fluorochromes are tested utilizing the bacterial strains used to verify the specificity of the different probes, and the 5′-modification is subsequently determined. EUB 338 is used for the visualization of the entire bacterial population within the oral biofilm.
9. In order to enhance the permeability of bacteria within the oral biofilm, cell wall-degrading enzymes such as mutanolysin can be used.
10. The sequence specificities of the oligonucleotide probes used were examined for the following strains: *A. naeslundii* DSM 17233T, *A. naeslundii* clinical isolate, *Actinomyces viscosus* clinical isolate, *Candida albicans* ATCC 90028, *Enterococcus faecalis* clinical isolate, *F. nucleatum* subsp. *nucleatum* ATCC 25586, *F. nucleatum* clinical isolate, *Fusobacterium necrophorum* ATCC 27852, *Lactobacillus brevis* DSM 20054, *Lactobacillus salivarius* DSM 20555, *Peptostreptococcus micros* ATCC 23195, *Prevotella*

nigrescens NCTC 9336, *Staphylococcus aureus* ATCC 25923, *Staphylococcus epidermidis* DSM 1798, *Streptococcus mutans* ATCC 25175, *Streptococcus oralis* ATCC 35037, *Streptococcus sanguis* DSM 20068, *Streptococcus salivarius* DSM 20067, *Streptococcus sobrinus* DSM 20381, *Veillonella parvula* DSM 2008T, and *V. parvula* clinical isolate [2].

11. Excitation of the FISH probes is carried out at the following wavelengths: 405 nm (Pacific Blue), 488 nm (fluorescein), 543 nm (Cy3), 594 nm (Texas red), and 633 nm (Cy5). Fluorescence emission of the probes can be measured at the following wavelengths: 406–473 nm (pacific blue), 495–565 nm (fluorescein), 552–592 nm (Cy3), 605–670 nm (Texas red), and 644–703 nm (Cy5) [12].
12. Confocal scanning is conducted sequentially for each image intending to minimize spectral overlap between the probes. Fluorescence response is principally demonstrated by a sigmoid curve. However, this depends on the concentration of fluorescent probes adhered to the ribosomes. Since the number of ribosomes within cells identified by FISH varies, the overall detected fluorescence response is sigmoid, but it can also be linear for individual cells.
13. After the determination of the upper and lower boundaries of the biofilm the thickest point is identified within each measured area. This procedure is repeated twice enabling the estimation of a mean thickness of the biofilm resulting from the three measurements. Sections of a thickness of approximately 0.5 μm each are generated at 2 μm intervals throughout the biofilm layers as biofilms are scanned from these three starting points (to avoid overlaps).
14. The EUB 338 corresponding fluorescent volume is set as 100 % of bacterial biomass in the biofilm. All other targets are calculated as percentage of the biomass calculated by EUB 338. The measurement of voxel intensities enables the image analysis program to calculate the biofilm composition from stacks of five-channel images. For each of the used fluorescent colors fluorescence intensity thresholds are manually set. Eighteen biofilm points, three on each of the six bovine enamel slabs, are analyzed for each time period and thus for each biofilm age. The analysis of the resulting biofilm contents reveals their statistical significance.

References

1. Almeida C, Azevedo NF, Santos S et al (2011) Discriminating multi-species populations in biofilms with peptide nucleic acid fluorescence in situ hybridization (PNA FISH). PLoS One 6:e14786
2. Al-Ahmad A, Wunder A, Auschill TM et al (2007) The in vivo dynamics of *Streptococcus* spp., *Actinomyces naeslundii*, *Fusobacterium nucleatum* and *Veillonella* spp. in dental plaque biofilm as analysed by five-colour multiplex

fluorescence in situ hybridization. J Med Microbiol 56:681–687

3. Thurnheer T, Gmür R, Guggenheim B (2004) Multiplex FISH analysis of a six-species bacterial biofilm. J Microbiol Methods 56:37–47
4. Amann RI, Ludwig W, Schleifer KH (1995) Phylogenetic identification and in situ detection of individual microbial cells without cultivation. Microbiol Rev 59:143–169
5. Amann R, Fuchs BM (2008) Single-cell identification in microbial communities by improved fluorescence in situ hybridization techniques. Nat Rev Microbiol 6:339–348
6. Foster JS, Palmer RJ Jr, Kolenbrander PE (2003) Human oral cavity as a model for the study of genome–genome interactions. Biol Bull 204:200–204
7. Bergmans L, Moisiadis P, Van Meerbeek B et al (2005) Microscopic observation of bacteria: review highlighting the use of environmental SEM. Int Endod J 38:775–788
8. Hannig C, Hannig M, Rehmer O et al (2007) Fluorescence microscopic visualization and quantification of initial bacterial colonization on enamel in situ. Arch Oral Biol 52:1048–1056
9. Karygianni L, Follo M, Hellwig E et al (2012) Microscope-based imaging platform for large-scale analysis of oral biofilms. Appl Environ Microbiol 78:8703–8711
10. Conrad C, Wünsche A, Tan TH et al (2011) Micropilot: automation of fluorescence microscopy-based imaging for systems biology. Nat Methods 8:246–249
11. Li J, Helmerhorst EJ, Leone CW et al (2004) Identification of early microbial colonizers in human dental biofilm. J Appl Microbiol 97:1311–1318
12. Al-Ahmad A, Follo M, Selzer AC et al (2009) Bacterial colonization of enamel in situ investigated using fluorescence in situ hybridization. J Med Microbiol 58:1359–1366
13. Hannig C, Follo M, Hellwig E et al (2010) Visualization of adherent micro-organisms using different techniques. J Med Microbiol 59:1–7
14. Amann RI, Binder BJ, Olson RJ et al (1990) Combination of 16S rRNA-targeted oligonucleotide probes with flow cytometry for analyzing mixed microbial populations. Appl Environ Microbiol 56:1919–1925
15. Paster BJ, Bartoszyk IM, Dewhirst FE (1998) Identification of oral streptococci using PCR-based, reverse-capture, checkerboard hybridization. Methods Cell Sci 20:223–231
16. Foster JS, Kolenbrander PE (2004) Development of a multispecies oral bacterial community in a saliva-conditioned flow cell. Appl Environ Microbiol 70:4340–4348
17. McLean JS, Ona ON, Majors PD (2008) Correlated biofilm imaging, transport and metabolism measurements via combined nuclear magnetic resonance and confocal microscopy. ISME J 2:121–131
18. Neu TR, Manz B, Volke F et al (2010) Advanced imaging techniques for assessment of structure, composition and function in biofilm systems. FEMS Microbiol Ecol 72:1–21
19. Pfeiffer S, Beese M, Boettcher M et al (2003) Combined use of confocal laser scanning microscopy and transmission electron microscopy for visualisation of identical cells processed by cryo-techniques. Protoplasma 222: 129–137
20. Wood SR, Kirkham J, Marsh PD et al (2000) Architecture of intact natural human plaque bio-films studied by confocal laser scanning microscopy. J Dent Res 79:21–27
21. Dige I, Nilsson H, Kilian M et al (2007) In situ identification of streptococci and other bacteria in initial dental biofilm by confocal laser scanning microscopy and fluorescence in situ hybridization. Eur J Oral Sci 115:459–467

Chapter 6

Field Emission Scanning Electron Microscopy of Biofilm-Growing Bacteria Involved in Nosocomial Infections

Claudia Vuotto and Gianfranco Donelli

Abstract

Scanning electron microscopy (SEM) provides useful information on the shape, size, and localization within the biofilm of single bacteria as well as on the steps of biofilm formation process, on bacterial interactions, and on production of extracellular polymeric substances.

When biofilms are constituted by microbial species involved in health care-associated infections, information provided by SEM can be fruitfully used not only for basic researches but also for diagnostic purposes.

The protocols currently used in our laboratory for biofilm investigation by SEM are reported here. Particularly, the procedures to fix, dehydrate, and metalize in vitro-developed biofilms or ex vivo clinical specimens colonized by biofilm-growing microorganisms are described as well as the advantages of the observation of these samples by field emission scanning electron microscopy.

Key words Scanning electron microscopy, FESEM, Biofilm ultrastructure, Medical devices

1 Introduction

Scanning electron microscopy (SEM) is a powerful technique to investigate bacterial colonization of biotic and abiotic surfaces and to reveal ultrastructural details on the bacteria–bacteria and bacteria–surface interactions. In fact, SEM is able to offer, at a higher resolution than a light microscope and at the needed magnification, a more detailed insight of the microbial sample under study. Among the currently available electron microscopy techniques employed to investigate bacterial and fungal biofilms, SEM is the most used one from about two decades [1–8]. By SEM investigations, researchers have understood that microbial biofilms are not just clusters of cells with more or less accumulated slime but well-structured communities of cells adherent to surfaces and embedded in a self-produced polymeric matrix.

In the case of biofilm-growing bacteria identified as causative agents of nosocomial infections, all the information provided by

Gianfranco Donelli (ed.), *Microbial Biofilms: Methods and Protocols*, Methods in Molecular Biology, vol. 1147,
DOI 10.1007/978-1-4939-0467-9_6, © Springer Science+Business Media New York 2014

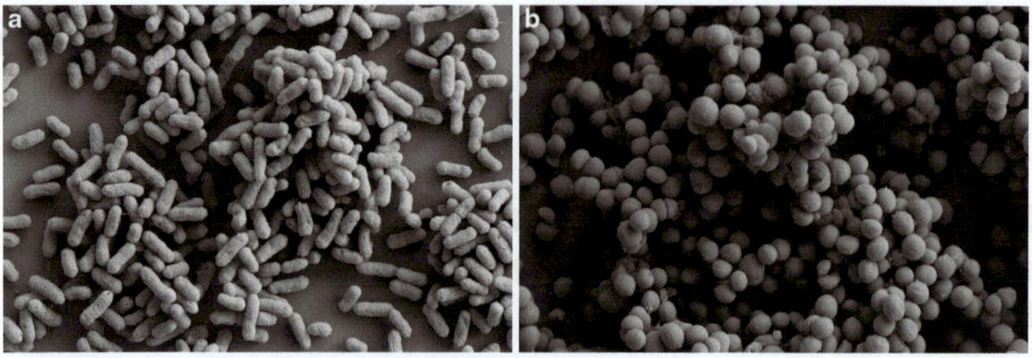

Fig. 1 FESEM micrographs of rod-shaped (**a**) *Escherichia coli* cells (10,000×) and coccoid (**b**) *Staphylococcus epidermidis* cells (15,000×) growing as single-species biofilms

Fig. 2 *Enterococcus faecalis* biofilm observed by FESEM at low (**a**, 2,000×) and high magnification (**b**, 20,000×) on the luminal surface of a Foley urinary catheter explanted from a patient suffering from neurogenic bladder

SEM can be exploited both for diagnostic purposes and translational researches in the fields of tissue infections [9–11] and device-associated infections, including central venous catheters [12–15], urinary catheters [16], biliary stents [17, 18], and orthopedic prostheses [19]. In fact, SEM observations can provide information on (1) the types of bacteria involved in the observed biofilm as presumed on the basis of their size and shape, e.g., rod-shaped bacilli versus cocci (Fig. 1a, b); (2) the occurrence of a single-, dual-, or multi-species biofilm (Figs. 2, 3, and 4); and (3) the detection of biofilm-forming bacteria growing in a viable but nonculturable (VBNC) state, not detectable by standard laboratory procedures.

Apart from its contribution in defining the biofilm-forming ability of clinical strains isolated from patients, SEM offers also the possibility to carry on translational researches, such as those focused on the investigation of bacterial killing and induction of cell modifications (shape loss, cellular debris, and pore formation) induced by disinfectants [20], antibiotics [21, 22] (Fig. 5), quorum sensing inhibitors [23], anti-biofilm compounds [24], and coatings [25].

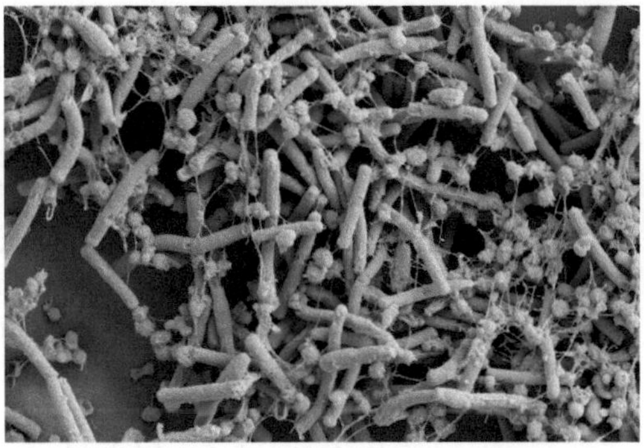

Fig. 3 FESEM micrograph of a dual-species biofilm formed in vitro by clinical strains of *Veillonella* spp. and *Clostridium difficile* (10,000×), isolated from explanted biliary stents

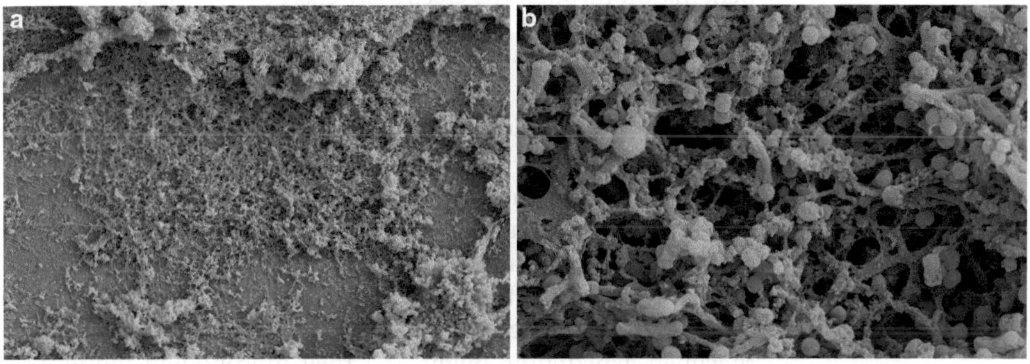

Fig. 4 Multi-species biofilms observed by FESEM at low (**a**, 2,000×) and high magnification (**b**, 15,000×) on the luminal surfaces of central venous catheters explanted from severely brain-injured patients

Fig. 5 FESEM observation of shape modifications and cell wall alterations induced by the antibiotic tigecycline on *Staphylococcus epidermidis* cells (10,000×) growing as biofilm

As for other techniques, SEM also has some limitations, mainly related to the need to fix, dehydrate, and metalize the biological sample before its observation. As a consequence, these preparation steps can slightly modify cell shape and alter the biofilm morphology.

As the fixation process is concerned, the exposure of the sample to low concentrations of fixative implies a prolonged duration of treatment, so often causing extraction of cell materials and shrinkage of cells. Thus, to minimize the damage to the cell structures, a fixation using the minimal effective concentration for a time as short as possible is recommended.

Anyway, a primary fixation with glutaraldehyde, followed by a secondary fixation with osmium tetroxide (OsO_4), is today considered the optimal procedure for preventing structural damages of sessile-growing bacteria.

In fact, glutaraldehyde is the most effective fixative in maintaining the fine structure of bacteria because of its ability to cross-link proteins in a rapid and irreversible way, while the post-fixation with OsO_4 preserves lipids and the nonpolar nature promotes its penetration through the electrically charged cell surface.

The addition of electrolytes (especially $CaCl_2$) to the fixative solutions has been demonstrated to supply beneficial effects including an improvement in the maintenance of the cell shape, a reduction in the extraction of cellular materials, and a membrane stabilization.

To gradually dehydrate the specimen before its examination in a scanning electron microscope, graded solvents (alcohol, acetone, and xylene) must be utilized, since the water content of the sample is not compatible with the vacuum needed for the electron beam functioning. However, as biofilms are particularly rich of water, the specimen dehydration might alter their morphology and give rise to some artifacts and structural alterations. For example, the extracellular polymeric substances, when investigated by SEM [26], often appear as fibers (Fig. 3) connecting bacteria [27] or as a thick slimy matrix that surrounds the cells [28]. As an alternative approach to overcome the structural changes in microbial cells due to the dehydration process, environmental scanning electron microscopy (ESEM) has also been applied to the investigation of bacterial biofilms [29].

To obtain an optimal resolution, dehydrated biofilm samples have to be first critical point dried [30] or treated with hexamethyldisilazane (HMDS) solvent [31] and then coated with heavy metals, typically a gold–palladium alloy or platinum, to increase their electrical conductivity.

The recent introduction of the field emission scanning electron microscope (FESEM), able to work at reduced accelerating voltage (range from 0.2 to 30 kV), has allowed to obtain cleaner, high-quality, less electrostatically distorted images of samples [32]. FESEM, having a spatial resolution down to 1 nm, that is 3–6 times better than the traditional SEM, has been recently used to investigate biofilm-growing bacteria [27, 33–35].

2 Materials

1. Stock 10× phosphate-buffered saline (PBS) solution: Dissolve 80 g NaCl, 2 g KCl, 14.4 g $Na_2HPO_4 \cdot 2H_2O$, and 2.4 g KH_2PO_4 in 800 ml of distilled water, and then bring the solution volume up to 1 l. Autoclave to sterilize.
2. Working PBS solution: Dilute the stock solution tenfold and filter.
3. Buffer solution sodium cacodylate 0.2 M: Dissolve 42.8 g of cacodylic acid sodium salt trihydrate in 800 ml of demineralized water, and then add demineralized water to a volume of 1,000 ml (*see* **Note 1**). Finally, adjust pH of buffer to 7.2 by using concentrated HCl (36–38 %) (*see* **Note 2**).
4. Glutaraldehyde 25 % solution: 25 wt % in water (*see* **Note 3**).
5. Osmium tetroxide solution: Dilute 4 % OsO_4 solution in ddH_2O (*see* **Note 4**) to concentration at least twice that of desired final concentration [final concentration usually 1 % (*see* **Note 5**)]. Since OsO_4 dissolves slowly in water, the process can be accelerated by shaking the solution.
6. Working fixative solutions: Dilute to a tenth the glutaraldehyde solution (25 %) in 0.1 M sodium cacodylate buffer, obtained by mixing 1 vol of sodium cacodylate stock solution and 1 vol of demineralized water (*see* **Note 6**). Mix 1 vol of stock OsO_4 solution with 1 vol of sodium cacodylate 0.2 M.
7. Gradients of ethanol/water: 30 % vol/vol, 50 % vol/vol, 70 % vol/vol, 80 % vol/vol, 95 % vol/vol, and absolute ethanol solutions (100 % vol/vol).
8. HMDS: 98 % solution (*see* **Note 7**).
9. Aluminum specimen stub: 13 mm diameter.
10. Conductive paint: Colloidal silver paint.
11. Spectrophotometer.
12. Incubator at 37 °C.
13. Refrigerator at 4 °C.
14. 24-well polystyrene microtiter plates.
15. 12-well polystyrene microtiter plates.
16. Critical point drier (CPD) (e.g., Quorum Technologies Ltd).
17. Rotary-pumped sputter coater (e.g., Quorum Technologies Ltd) suitable for non-oxidizing metals, such as gold (Au) and platinum (Pt).
18. FESEM (e.g., SIGMA, Zeiss).

3 Methods

3.1 In Vitro Biofilm Formation

This protocol can be easily adapted to different species by using specific media by adjusting the OD of the bacterial inoculum and by selecting the appropriate time of incubation.

1. Plate bacteria on selected agar medium and grow at 37 °C overnight.
2. Use single colony to inoculate 3 ml of broth, and grow this culture overnight at 37 °C without shaking.
3. Adjust the bacterial culture to the desired optical density (most often $OD_{600\ nm} = 0.1$) by using a spectrophotometer.
4. Put sterile round glass cover slips of 1 cm diameter on the bottom of one or more wells, according to the number of culture samples, of a 24-well polystyrene microtiter plate.
5. Fill each well with 1.8 ml of broth supplemented with glucose (1 % w/v) and 200 μl of bacterial inoculum.

3.2 Preparation of Glass Cover Slips Covered by Biofilm for FESEM Investigation

1. To enable biofilm development, a 24–48-h incubation (this time is adequate for most of the bacterial species) at 37 °C in static conditions is recommended.
2. To eliminate non-adherent bacteria, rinse the wells three times with 2 ml of PBS.
3. Remove the glass cover slip from the bottom of the well, and put it in an empty well of the microplate (*see* **Note 8**).

3.3 Preparation of Biliary Stents, Central Venous Catheters, and Urinary Catheters Removed from Patients for FESEM Investigation

1. Immediately after device removal, cut under sterile conditions approx. 2 cm segments from the distal, central, and proximal portion, respectively (*see* **Note 9**).
2. Bisect each segment along its long axis, and place each half segment into wells of a 12-well polystyrene microtiter plate filled with sterile PBS (pH 7.4).

From this step onwards, the procedure can be considered in principle the same regardless of the sample type. When needed, the changes in the procedure will be specified.

3.4 Sample Fixation

1. Add 2.5 % glutaraldehyde solution (*see* **Note 10**) at room temperature (*see* **Note 11**) for 30 min in the wells (*see* **Note 12**) containing biofilm samples.
2. Remove the glutaraldehyde solution, and rinse the sample three times with sodium cacodylate 0.1 M buffer (*see* **Note 13**).
3. Postfix by adding OsO_4 solution (1 %) in the wells (*see* **Note 14**) containing biofilm samples for 20 min at room temperature in the dark (*see* **Note 15**).
4. Remove the OSO_4 solution, and rinse the sample three times for 2–3 min with sodium cacodylate 0.1 M buffer.

3.5 Dehydration with Ethanol Gradients

Wash the sample with:

1. 30 % ethanol for 10 min (not to be used for device segments) (*see* **Note 16**).
2. 50 % ethanol for 10 min (15 min for device segments).
3. 70 % ethanol for 10 min (15 min for device segments) (*see* **Note 17**).
4. 85 % ethanol for 10 min (15 min for device segments).
5. 95 % ethanol for 10 min (15 min for device segments).
6. 100 % ethanol three times for 10 min (15 min for device segments).

3.6 Final Dehydration by Critical Point Drying or HMDS Solution

To reduce damages to biological structures related to the formation of forces of surface tension, alcohol dehydration must be followed by the application of one of these methods allowing a direct shift from the liquid phase to the gas phase: critical point drying or dehydration with HMDS.

3.6.1 Critical Point Drying

1. Insert the sample between two sheets of tissue paper held together with staples, and indicate with a pencil the biofilm side on the paper.
2. Dip the sample in 100 % ethanol.
3. Precool the chamber.
4. Introduce the sample within the chamber.
5. Transfer in and unload from the chamber the liquid CO_2 (*see* **Note 18**).
6. Fill partially the chamber with liquid CO_2 (1/2 the volume total).
7. Heat the system (raising temperature and pressure) up to the achievement of the CO_2 critical point values (35 °C and 1,200 psi).
8. Reduce temperature and pressure values (*see* **Note 19**).
9. Remove the sample from the chamber.

3.6.2 Dehydration with HMDS

This procedure is particularly recommended when the sample is constituted by a segment of medical device:

1. Remove 100 % ethanol, and add HMDS 98 % for 1 h 30 min at room temperature.

3.7 Mounting on Stubs

This step consists in preparing the sample on an aluminum stub (*see* **Note 20**) by running the following procedure:

1. Put colloidal silver paint on the upper base of the stub (*see* **Note 21**).
2. Paste the biofilm sample on the stub (*see* **Notes 22** and **23**).

3. Achieve points of contact between the sample and the stub by depositing drops of silver paints on 3–4 different areas at the edge of the sample (*see* **Note 24**).

3.8 Sputtering

High-quality images for non-conductive samples are obtained by coating biofilms mounted on stubs with a metal layer (gold, platinum, etc.) able to provide a better interaction between the sample and the electron beam.

1. Place the samples on the table of the chamber (anode).
2. To determine the thickness of the gold layer, set the instrument parameters as follows: 0.05 mbar for the vacuum chamber, 2.5 kV as working voltage, 5 cm distance between anode and cathode, 20 mA current intensity, and 120 s of sputtering time (*see* **Note 25**).
3. Start the sputter coater, and let in a flow of inert gas (argon).
4. When the sputtering of the sample is completed, leave the instrument off for 3–4 min while it will slowly vent and then open the chamber. At this point the samples are ready for FESEM observation.

3.9 Sample Investigation by FESEM

1. Place stubs containing biofilm samples in the stub holder.
2. Fill the FESEM chamber with argon.
3. Open the door of the chamber, and slide the holder into the platform.
4. Close the chamber, and start the vacuum pump by the software.
5. When a green check mark appears in the vacuum information slot click "EHT on" to establish the beam.
6. Set the EHT voltage (*see* **Note 26**).
7. Select the kind of electrons (secondary or backscattered) that have to be used to acquire images (*see* **Note 27**).
8. Move the sample stage from its default loading position to the inspection area by using the "stage navigation system" in TV modality (*see* **Note 28**).
9. Switch to "Normal modality," and start to observe the biofilm sample by using the controls for magnification, focus, contrast, and brightness (*see* **Note 29**).
10. When a sample image is to be recorded, reduce scan speed of approx. 70 % with respect to that used to observe the sample and acquire micrograph by "line integration" (about ten lines).
11. Save the micrograph.
12. Turn off EHT and vent by using the software. This can take a few minutes.

13. Open the chamber door, and remove the stub holder.
14. Close the door, and start again the vacuum pump.
15. When the required vacuum in the chamber is obtained, turn of the software and the server and switch to "stand by."

4 Notes

1. Sodium cacodylate is the best buffer to dilute glutaraldehyde, since it is able to avoid glutaraldehyde polymerization for a longer time. The arsenic content of sodium cacodylate is enough to represent a health hazard. Thus, for the weighing of reagent and the preparation of the buffer solution, fume hood and gloves to protect hands should be used.
2. This buffer is effective in the pH range 7.2–7.4.
3. To reduce deterioration, purified 25 % glutaraldehyde solutions have to be stocked at −20 °C.
4. OsO_4 is a toxic and hazardous substance. Due to the extreme volatility (11 mm vapor pressure at 25 °C), OsO_4 solutions must be prepared in small aliquots and maintained at 4 °C in brown bottles with glass stopper to avoid its reduction under light exposure.
5. 1 % is the optimal concentration to preserve highly hydrated specimens such as microbial biofilms.
6. To avoid polymerization, the freshly prepared final 2.5 % glutaraldehyde solution must be stored at 4 °C and should be used whenever possible.
7. HMDS is a flammable liquid that may be harmful if swallowed, inhaled, or absorbed through skin. It may cause severe irritation or burns to skin, eyes, and respiratory tract.
8. Use sharply pointed tweezers and be careful not to break the cover slip or scratch the surface to avoid the removing of the formed biofilm. It is necessary to always maintain upward the same side of the cover slip.
9. During this step it is important to avoid the detachment of the biofilm grown within the device lumen. The use of a very sharp scalpel is recommended in order to make as little mechanical pressure as possible on the device.
10. All the following steps must be performed in a fume hood.
11. The penetration of glutaraldehyde within the sample is definitely faster at room temperature than in the cold.
12. Glutaraldehyde solution has to be added to cover the sample, and the same procedure must be followed for all the solutions used henceforth.

13. If necessary, the procedure can be discontinued at this step and the sample can be stored at +4 °C till the day after.
14. OsO_4 is used as post-fixative and not as primary fixative since it is not able to preserve proteins and carbohydrates but only lipids.
15. OsO_4 must be maintained in the dark as it can be reduced by light.
16. To avoid crushing of bacteria, never leave the sample dry after each dehydration step.
17. If necessary, the procedure can be discontinued at this step and the sample can be stored at +4 °C till the day after.
18. The number of CO_2 washing steps depends on the nature of sample. Usually, three cycles of 2–3 min each are enough for microbial biofilm samples.
19. This operation must be carried out taking great care not to cross, going down, the critical point, which could lead to the recondensation of CO_2. The samples, so dehydrated, are very hydrophilic and damp environments have to be avoided.
20. Stubs must be cleaned by dipping in 100 % alcohol or acetone and handled with specific stub tweezers.
21. Silver paint is used both to paste sample to the stub and to facilitate the conduction of electrons.
22. According to the sample size, stubs of different diameters are available. Stubs should be a little bit larger than the sample.
23. Device segments must be fixed very well on the stubs to avoid astigmatic images.
24. Points of contact are needed to assure the formation of a continuous conductive system, necessary to fulfill the requirements of electrical conductivity and chemical and physical stability.
25. In case of device segments, an increase in time of sputtering up to 150 s is recommended.
26. FESEM is able to minimize sample damage due to the time of beam exposure by decreasing the EHT voltage level up to 0.2 kV. However, to obtain high-quality micrographs of our biofilm samples, an EHT of 2–5 kV is recommendable.
27. The beam electrons explore through lines and sequential points the surface of the sample giving rise basically to secondary electrons and backscattered electrons. Both these types of electrons, properly collected by specific detectors, are usable for the formation of images. The signal most frequently used in biofilm investigation is the one generated by the secondary electrons, which better reveals the morphology of the sample surface.

28. The TV modality is based on the use of a CCD camera located on the back of the chamber in order to monitor the mechanical movements of the stage in real time. The use of the navigation system in this modality helps to prevent collision sample stage and detectors and improves the speed at which you can find your sample on the stub.
29. To initially find the focus, select a sample area of scarce interest to avoid beam-related contamination of the sample area of major interest. In fact, the sample exposure to the beam for a long time causes the deposition of material (e.g., carbon) in the long-term scanned region, so giving rise to poor-quality images.

Acknowledgements

Authors are indebted with Dr. Maria Pia Balice, Director of the Clinical Chemistry and Microbiology Laboratory for kindly providing bacterial strains and explanted medical devices, and gratefully acknowledge Dr. Antonino Salvia, Director, and Dr. Angelo Rossini, Vice Director of the Medical Services of the Fondazione Santa Lucia in Rome for their advices on the clinical issues.

References

1. Giangrande A, Allaria P, Torpia R et al (1993) Ultrastructure analysis of Tenckhoff chronic peritoneal catheters used in continuous ambulatory peritoneal dialysis patients. Perit Dial Int 13:S133–S135
2. McLaughlin-Borlace L, Stapleton F, Matheson M, Dart JK (1998) Bacterial biofilm on contact lenses and lens storage cases in wearers with microbial keratitis. J Appl Microbiol 84:827–838
3. Fux CA, Quigley M, Worel AM et al (2006) Biofilm-related infections of cerebrospinal fluid shunts. Clin Microbiol Infect 12(4):331–337
4. Sangetha S, Zuraini Z, Suryani S, Sasidharan S (2009) In situ TEM and SEM studies on the antimicrobial activity and prevention of *Candida albicans* biofilm by Cassia spectabilis extract. Micron 40:439–443
5. Hannig C, Follo M, Hellwig E, Al-Ahmad A (2010) Visualization of adherent micro-organisms using different techniques. J Med Microbiol 59:1–7
6. Lattif AA, Mukherjee PK, Chandra J et al (2010) Characterization of biofilms formed by *Candida parapsilosis*, *C. metapsilosis*, and *C. orthopsilosis*. Int J Med Microbiol 300:265–270
7. Grin I, Schwarz H, Linke D (2011) Electron microscopy techniques to study bacterial adhesion. Adv Exp Med Biol 715:257–269
8. Hingston PA, Stea EC, Knøchel S, Hansen T (2013) Role of initial contamination levels, biofilm maturity and presence of salt and fat on desiccation survival of *Listeria monocytogenes* on stainless steel surfaces. Food Microbiol 36:46–56
9. Perloff JR, Palmer JN (2005) Evidence of bacterial biofilms in a rabbit model of sinusitis. Am J Rhinol 19:1–6
10. Vitkov L, Hermann A, Krautgartner WD et al (2005) Chlorhexidine-induced ultrastructural alterations in oral biofilm. Microsc Res Tech 68:85–89
11. Carron MA, Tran VR, Sugawa C, Coticchia JM (2006) Identification of *Helicobacter pylori* biofilms in human gastric mucosa. J Gastrointest Surg 10:712–717
12. Passerini L, Lam K, Costerton JW, King EG (1992) Biofilms on indwelling vascular catheters. Crit Care Med 20:665–673
13. Donelli G, De Paoli P, Fadda G, CVC Study Group et al (2001) A multicenter study on central venous catheter-associated infections in Italy. J Chemother 13:251–262

14. Nishikawa K, Takasu A, Morita K et al (2010) Deposits on the intraluminal surface and bacterial growth in central venous catheters. J Hosp Infect 75:19–22
15. Aybar Y, Ozaras R, Besirli K et al (2012) Efficacy of tigecycline and vancomycin in experimental catheter-related *Staphylococcus epidermidis* infection: microbiological and electron microscopic analysis of biofilm. Int J Antimicrob Agents 39:338–342
16. Stickler DJ (2008) Bacterial biofilms in patients with indwelling urinary catheters. Nat Clin Pract Urol 5:598–608
17. Donelli G, Guaglianone E, Di Rosa R et al (2007) Plastic biliary stent occlusion: factors involved and possible preventive approaches. Clin Med Res 5:53–60
18. Guaglianone E, Cardines R, Vuotto C et al (2010) Microbial biofilms associated with biliary stent clogging. FEMS Immunol Med Microbiol 59:410–420
19. Bernthal NM, Stavrakis AI, Billi F et al (2010) A mouse model of post-arthroplasty *Staphylococcus aureus* joint infection to evaluate in vivo the efficacy of antimicrobial implant coatings. PLoS One 5:e12580
20. Takeuchi Y, Guggenheim B, Filieri A, Baehni P (2007) Effect of chlorhexidine/thymol and fluoride varnishes on dental biofilm formation in vitro. Eur J Oral Sci 115:468–472
21. Donelli G, Francolini I, Romoli D et al (2007) Synergistic activity of dispersin B and cefamandole nafate in inhibition of staphylococcal biofilm growth on polyurethanes. Antimicrob Agents Chemother 51:2733–2740
22. Ferreira JA, Carr JH, Starling CE et al (2009) Biofilm formation and effect of caspofungin on biofilm structure of Candida species bloodstream isolates. Antimicrob Agents Chemother 53:4377–4384
23. Siddiqui MF, Sakinah M, Singh L, Zularisam AW (2012) Targeting N-acyl-homoserine-lactones to mitigate membrane biofouling based on quorum sensing using a biofouling reducer. J Biotechnol 161:190–197
24. Wang W, Tao R, Tong Z et al (2012) Effect of a novel antimicrobial peptide chrysophsin-1 on oral pathogens and *Streptococcus mutans* biofilms. Peptides 33:212–219
25. Donelli G, Francolini I, Ruggeri V et al (2006) Pore formers promoted release of an antifungal drug from functionalized polyurethanes to inhibit Candida colonization. J Appl Microbiol 100:615–622
26. Barnes AM, Ballering KS, Leibman RS et al (2012) *Enterococcus faecalis* produces abundant extracellular structures containing DNA in the absence of cell lysis during early biofilm formation. MBio 3:e00193-12
27. Novais A, Vuotto C, Pires J et al (2013) Diversity and biofilm-production ability among isolates of *Escherichia coli* phylogroup D belonging to ST69, ST393 and ST405 clonal groups. BMC Microbiol 21(13):144
28. Combrouse T, Sadovskaya I, Faille C et al (2013) Quantification of the extracellular matrix of the *Listeria monocytogenes* biofilms of different phylogenic lineages with optimization of culture conditions. J Appl Microbiol 114:1120–1131
29. Priester JH, Horst AM, Van de Werfhorst LC et al (2007) Enhanced visualization of microbial biofilms by staining and environmental scanning electron microscopy. J Microbiol Methods 68:577–587
30. Carr JH, Anderson RL, Favero MS (1996) Comparison of chemical dehydration and critical point drying for the stabilization and visualization of aging biofilm present on interior surfaces of PVC distribution pipe. J Appl Bacteriol 80:225–232
31. Araujo JC, Téran FC, Oliveira RA et al (2003) Comparison of hexamethyldisilazane and critical point drying treatments for SEM analysis of anaerobic biofilms and granular sludge. J Electron Microsc (Tokyo) 52:429–433
32. Cazaux J (2005) Recent developments and new strategies in scanning electron microscopy. J Microsc 217:16–35
33. Raulio M, Wilhelmson A, Salkinoja-Salonen M, Laitila A (2009) Ultrastructure of biofilms formed on barley kernels during malting with and without starter culture. Food Microbiol 26:437–443
34. Donelli G, Vuotto C, Cardines R, Mastrantonio P (2012) Biofilm-growing intestinal anaerobic bacteria. FEMS Immunol Med Microbiol 65:318–325
35. Bridier A, Meylheuc T, Briandet R (2013) Realistic representation of *Bacillus subtilis* biofilms architecture using combined microscopy (CLSM, ESEM and FESEM). Micron 48:65–69

Chapter 7

Experimental Approaches to Investigating the Vaginal Biofilm Microbiome

Marc M. Baum, Manjula Gunawardana, and Paul Webster

Abstract

Unraveling the complex ecology of the vaginal biofilm microbiome relies on a number of complementary techniques. Here, we describe the experimental approaches for studying vaginal microbial biofilm samples with a focus on specimen preparation for subsequent analysis. The techniques include fluorescence microscopy, fluorescence in situ hybridization, and scanning and transmission electron microscopy. Isolation of microbial DNA and RNA from these samples is covered along with a brief discussion of chemical analysis methods.

Key words Vaginal microbiome, Microbial biofilms, Fluorescence microscopy, Fluorescence in situ hybridization, Scanning electron microscopy, Transmission electron microscopy, Genomic microbial DNA, RNA, Correlative microscopy

1 Introduction

Bacteria colonize surfaces and establish a sessile mode of growth in biofilms, chemically and morphologically heterogeneous matrices of extracellular polymeric substances (EPS). Microbial biofilms are implicated in chronic infections [1] and are more resistant to antimicrobial agents and the immune system than their planktonic counterparts [2, 3]. The normal vaginal microbiota makes up a complex, diverse community that is believed to play an important protective role in maintaining the health of a woman, sexual partner, or newborn [4]. Microorganisms in the vaginal tract are believed to grow predominantly as sessile polymicrobial communities encapsulated in biofilms [5–7]. The Human Microbiome Project [8, 9] is improving our understanding of how vaginal bacteria in healthy individuals maintain a balanced community [10, 11] and the pathogenic capabilities of key species that mediate poor health outcomes. Vaginal bacterial biofilms have been associated with a bacterial vaginosis and mortality resulting from tampon-related toxic shock syndrome in menstruating women. In previous studies, we have demonstrated that polymicrobial biofilms grow

Gianfranco Donelli (ed.), *Microbial Biofilms: Methods and Protocols*, Methods in Molecular Biology, vol. 1147,
DOI 10.1007/978-1-4939-0467-9_7, © Springer Science+Business Media New York 2014

in vivo on the surface of intravaginal rings (IVRs) implanted in pig-tailed macaques [12] and women [13].

Gaps in our understanding of medical microbial ecology largely stem from our reliance on culture-dependent microbiological methods, which typically can identify less than 1 % of the bacterial cells in a given ecosystem [14]. This seminal realization, known as the "great plate count anomaly," has led to tremendous advances over the past 20 years through the development and application of environmental metagenomics, microanalytical methodologies, novel cultivation methods, and the coupling of stable and radiogenic isotopes with molecular analysis of biosignatures [15]. The strict requirement for pure cultures has been alleviated by the refinement of experimental approaches aimed at improving our understanding of the composition, structure, and function of vaginal microbial biofilm communities. The approaches covered here are based on a combination of established imaging techniques using instrumentation accessible to most research laboratories, including:

- *Fluorescence microscopy*: Allows community structure to be studied by selectively labeling different components of the biofilm EPS with a range of fluorescent probes [16–19].
- *Fluorescence in situ hybridization (FISH)*: FISH is an established molecular, cultivation-independent technique that detects nucleic acid sequences by a fluorescently labeled probe, which hybridizes specifically to its complementary target sequence within the intact cell [14].
- *Scanning and transmission electron microscopy (SEM and TEM)*: Electron microscopy allows the biofilm microbial community structure to be examined at high resolution. We have developed a method to combine FISH and SEM data to label bacteria (FISH) and provide high-resolution information on the reference space (SEM) [20].
- *Molecular analysis*: Genomic DNA and RNA isolation, amplification, and sequencing.
- *Chemical analysis*: Colorimetry, nondestructive spectroscopy, and chromatography with tandem mass spectrometric detection techniques.

The corresponding protocols described below were validated in our laboratories for vaginal microbial biofilm samples.

2 Materials

All solutions should be prepared with deionized water, unless otherwise specified, using analytical grade reagents. Prepare and store all reagents at room temperature unless otherwise noted. Diligently follow all appropriate regulations and precautions when handling reagents and disposing of waste materials.

2.1 Fluorescence Microscopy

1. Method optimization and imaging were carried out using an EVOS fl digital inverted fluorescence microscope (AMG).
2. Samples are prepared and viewed in a Lab-Tek II Chamber #1.5 German Coverglass System. We typically use the 8-chamber system (Model 155409, Nalge Nunc International Corp.).
3. 20 μM SYTO® 63 in deionized H_2O, 0.5 % w/v fluorescein isothiocyanate isomer I (FITC) in PBS, a 5,000× stock solution SYPRO® Orange in DMSO, 10 μg/mL Calcofluor White in sodium phosphate buffer (10 mM, pH 7.5), 500 μg/mL Nile red in acetone, and 10 mM 4′,6-diamidino-2-phenylindole dihydrochloride (DAPI) in deionized H_2O.
4. 1× PBS (pH 7.2), phosphate buffer (0.05 M, pH 7.4).
5. Aqueous glycerol solution (75 %, v/v).
6. Aqueous acetic acid solution (7.5 %, v/v).
7. Aqueous potassium hydroxide solution (10 %, w/v).
8. Deionized H_2O.
9. Timer.
10. Kimwipes.
11. Fluorescence microscope.

2.2 Fluorescence In Situ Hybridization (FISH)

1. Method optimization and imaging were carried out using an EVOS fl digital inverted fluorescence microscope (AMG).
2. Samples are prepared and viewed in a Lab-Tek II Chamber #1.5 German Coverglass System. We typically use the 8-chamber system (Model 155409, Nalge Nunc International Corp.).
3. Published oligonucleotide probe sequences [21] are employed when possible. We typically order our probes from Integrated DNA Technologies, Inc., and request purification by HPLC.
4. Hybridization buffer: 0.9 M NaCl, 0.02 M Tris-HCl buffer (pH 7.4), and 0.001 % SDS in deionized H_2O (*see* **Note 1**).
5. Washing buffer: 0.175 M NaCl, 0.02 M Tris-HCl buffer (pH 7.4), and 0.001 % SDS in deionized H_2O.
6. 1× PBS (pH 7.2).
7. Aqueous ethanol solution (50 %, v/v).
8. Deionized H_2O.
9. Hybridization oven.
10. Timer.
11. Kimwipes.
12. Fluorescence microscope.

2.3 Electron Microscopy

1. Transmission electron microscope.
2. Scanning electron microscope.
3. Critical-point dryer.
4. Ultramicrotome.
5. 45° diamond knife for routine sectioning.
6. Formvar-coated metal specimen grids.
7. Specimen stubs.
8. 2.5 % v/v glutaraldehyde in 100 mM phosphate buffer (pH 7.2).
9. 1 % w/v aqueous osmium tetroxide.
10. 1 % w/v aqueous tannic acid.
11. Saturated aqueous thiocarbohydrazide (TCH).
12. Maleate buffer (50 mM, pH 5.2).
13. 3 % potassium ferricyanide.
14. Dilution series of ethanol or acetone.
15. 100 % ethanol or acetone.
16. Epoxy resin for specimen embedding.
17. 60 °C oven.
18. Embedding molds.
19. Fine jewelers forceps (e.g., Dumont tweezers styles 5 and 7).
20. Parafilm.
21. 3 % w/v aqueous uranyl acetate.
22. Aqueous lead citrate.
23. Deionized H_2O.

3 Methods

3.1 Sample Preservation and Storage

The vaginal biofilm sample collection, preservation, and storage strategy are fundamentally important in determining the range and quality of the subsequent measurements. Typically samples need to be divided upon collection and stored in a variety of preservative media, as described below:

1. *Electron microscopy*: 2.5 % v/v glutaraldehyde in phosphate buffer (pH 7.2) (*see* **Note 2**).
2. *Fluorescence microscopy/FISH*: 50 % v/v ethanol in deionized H_2O.
3. *Microbial DNA/chemical analysis*: Flash freeze in liquid nitrogen.
4. *Microbial RNA*: Aurum lysis buffer (Bio-Rad) [22].

5. Samples preserved in **steps 1** and **2** should be stored and transported at 4 °C.
6. Samples preserved in **steps 3** and **4** should be stored and transported at −80 °C.

3.2 Fluorescence Microscopy

1. The combination of multiple, selective fluorescent probes allows the principal chemical components to be imaged (*see* **Note 3**).
2. Numerous fluorescent probes have been reported for microbial biofilm imaging [16–19]. The most useful probes for vaginal biofilm analysis are described below.
3. We have used confocal laser scanning and inverted fluorescence microscopes to image vaginal microbial biofilms with similar results [12, 13].
4. Biofilms are stained after hybridization for FISH, directly in the chamber slide, unless otherwise noted.
5. Vortex agitation is not used to mix the samples. Sample washing is carried out gently using a 1 mL pipet and by blotting with a clean Kimwipes. (*See* **Note 4**).
6. Samples containing the fluorescent stains are incubated in the dark.
7. Prior to staining, gently remove the residual liquid from the chamber slide sample.
8. *SYTO® 63* (*total cells*): Add stock solution to just cover the sample.
9. Incubate for 30 min at room temperature.
10. Gently remove stain solution and wash sample with 1× PBS (pH 7.2) until no background stain is visible (2–3 washes are typical).
11. *FITC* (*proteins, amino sugars*): Wash sample three times with phosphate buffer (0.05 M, pH 7.4).
12. Immerse the sample in 0.5 mL phosphate buffer (0.05 M, pH 7.4) and add FITC stock solution (5 μL).
13. Incubate for 10 min at room temperature.
14. *SYPRO® Orange* (*proteins*) (Fig. 1): Dilute the stock SYPRO Orange solution 1:5,000 with aqueous acetic acid (7.5 %, v/v) with vigorous mixing.
15. Add diluted stock solution to just cover the sample.
16. Incubate for 5 min at room temperature.
17. Gently remove stain solution and wash sample with 1× PBS (pH 7.2) until no background stain is visible (2–3 washes are typical).
18. *Calcofluor White* (*cellulose and chitin*): Place sample onto a clean glass slide (*see* **Note 5**).

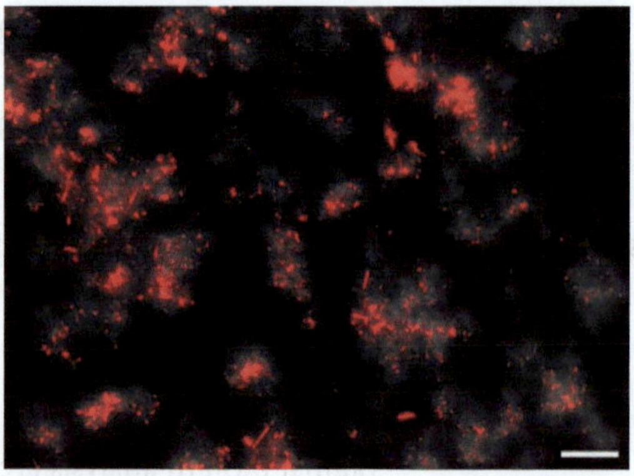

Fig. 1 FISH micrograph of *Lactobacillus gasseri* (ATCC 33323) biofilms grown in vitro. The sample was fixed overnight at 4 °C in 50 % ethanol. The cells were hybridized with universal bacterial probe EUB-338-Cy5 (*red*), and the biofilm proteins were labeled with SYPRO Orange (*blue*). The scale bars is 20 μm

19. Add on drop of Calcofluor White stock solution and one drop of potassium hydroxide solution (10 % w/v).
20. Place coverslip over the specimen and leave for 1 min prior to visualization.
21. *Nile red* (*lipids, hydrophobic sites*): Dilute the stock solution with aqueous glycerol (75 % v/v) to a final concentration of 2.5 μg/mL.
22. Add just enough of the diluted stock solution to just cover the sample.
23. Incubate for 1–2 min at room temperature.
24. Gently remove stain solution and wash sample with 1× PBS (pH 7.2) until no background stain is visible (2–3 washes is typical).
25. *DAPI* (*nucleic acids*): Dilute the DAPI stock solution to 300 nM in PBS.
26. Equilibrate the sample briefly in PBS.
27. Add the diluted DAPI solution (300 μL) to the sample.
28. Incubate 1–5 min at room temperature.
29. Gently remove stain solution and wash sample with 1× PBS (pH 7.2) until no background stain is visible (2–3 washes are typical).

3.3 Fluorescence In Situ Hybridization (FISH)

1. Our validated FISH method is based on a combination of literature procedures [23–25] (Fig. 2).
2. Fix specimen in 50 % ethanol for at least 18 h at 4 °C (once the sample is fixed, it can be stored at 4 °C for several months without appreciable deterioration).

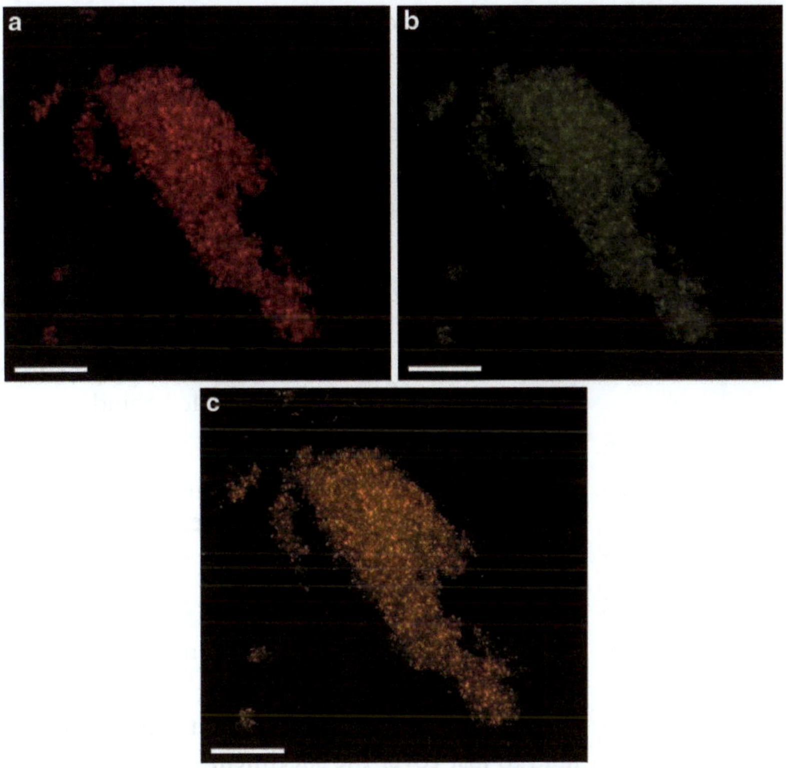

Fig. 2 FISH micrographs of *Lactobacillus gasseri* (ATCC 33323) biofilms grown in vitro. The cells were hybridized with (**a**) universal bacterial probe EUB-338-Cy5 (*red*) and (**b**) *Lactobacillus*-specific probe Lac-158-Cy3 (*green*) [21]. (**c**) Cells that hybridized with both probes appear *yellow/orange*. The images clearly show that universal coverage was obtained with both probes. The scale bars are 50 μm

3. On the day when hybridization is planned, prepare the hybridization and washing buffers (*see* **Note 1**) according to the probe specification and incubate at hybridization temperature (we typically use 47 °C) until needed (*see* **Note 6**).
4. Remove sample from cooler and thaw to room temperature.
5. Gently remove the liquid with a pipet, taking care not to disturb the fragile biofilm. Remove the residual 50 % ethanol by blotting with a clean Kimwipes (*see* **Note 4**).
6. Gently add a predetermined volume of PBS solution (pH 7.2) along the inside walls of the tube to rinse the biofilm taking care not to disturb the fragile structure. Never add solutions directly to the specimen. With our specimens (microfuge tubes and chamber slides), 500 μL is typically sufficient.
7. Remove the PBS solution as in **step 5**.
8. Remove hybridization buffer from the incubator and rapidly (to avoid cooling of the buffer) add 400–500 μL to the sample (*see* **Note 7**).

9. Immediately add the FISH probe(s) (50–200 pmol) to the sample.
10. Mix probe and hybridization buffer by gently aspirating the supernatant fluid up and down in the pipet, 3–5 times.
11. In all subsequent steps, protect the sample from light to avoid probe photobleaching.
12. Incubate at 47 °C for 90 min.
13. Quickly add deionized H_2O (1 mL, room temperature) to the sample.
14. Using a 1 mL pipet, remove all the liquid from the tube, leaving only the hybridized sample behind.
15. Repeat **steps 13** and **14**.
16. Add washing buffer directly from incubator using the same volume as used in **step 8** with the hybridization buffer.
17. Incubate for 20 min at 47 °C.
18. Remove the washing buffer with a 1 mL pipet.
19. Add deionized H_2O (1 mL, room temperature) to the sample.
20. Remove most of the H_2O with a 1 mL pipet. Lease enough fluid to just cover the specimen.
21. Store sample at 4 °C until it is to be imaged (*see* **Notes 8–11**).

3.4 Scanning Electron Microscopy (SEM)

1. Fix specimen in 2.5 % glutaraldehyde in 100 mM phosphate buffer (2.5 % v/v) for 2 h (*see* **Notes 2** and **12–16**).
2. Wash thoroughly with deionized H_2O to remove aldehyde and phosphate from the specimens.
3. Soak in 1 % w/v aqueous tannic acid solution for 1 h.
4. Wash thoroughly with deionized H_2O.
5. Postfix in 1 % w/v aqueous osmium tetroxide for 1 h.
6. Wash thoroughly with deionized H_2O.
7. Treat specimens with aqueous saturated TCH for 1 h.
8. Wash thoroughly with deionized H_2O.
9. Treat with aqueous 1 % w/v osmium tetroxide for 1 h.
10. Wash thoroughly with deionized H_2O.
11. Dehydrate using the following ethanol series: 30 % ethanol (2 × 10 min), 50 % ethanol (2 × 10 min), 70 % ethanol (2 × 10 min), 95 % ethanol (2 × 10 min), and 100 % ethanol (2 × 10 min).
12. Finally, change in 100 % dry ethanol (*see* **Note 17**).
13. Dry the specimen using a critical-point dryer or hexamethyldisilazane (HMDS) (*see* **Note 18**).

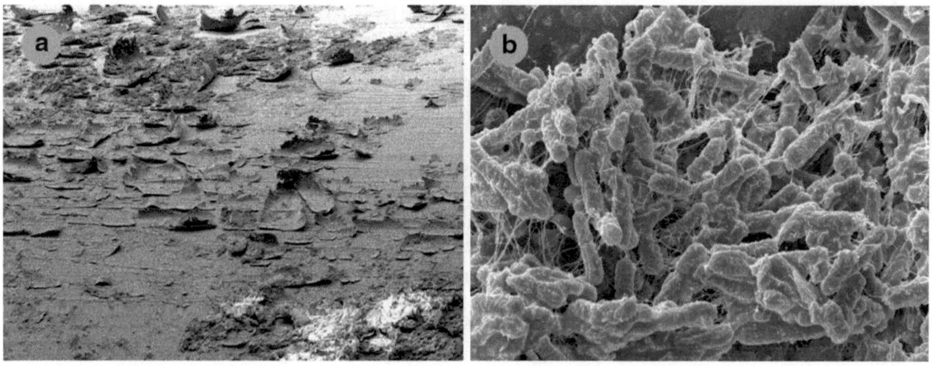

Fig. 3 Day 14 microbial biofilms forming on intravaginal rings, delivering the antiherpetic drug acyclovir, worn by women with recurrent genital herpes [13, 26]. (**a**) SEM image of epithelial cell monolayer on the ring surface; scale bar = 200 μm. (**b**) SEM image of nanowire-linked microbial biofilm cluster; scale bar = 20 μm

14. Mount dried specimen on aluminum specimen stub and examine in a scanning electron microscope (*see* **Note 19**) (Fig. 3).
15. Specimens should be stored in the presence of desiccant (*see* **Notes 20–22**).

3.5 Transmission Electron Microscopy (TEM)

1. For examination of morphology, fix specimen in glutaraldehyde buffered in 100 mM phosphate buffer (2.5 % v/v/, pH 7.2) for up to 3 h (*see* **Note 23**).
2. Take care not to cause handling trauma during any relevant dissection (e.g., no crushing, pulling, squeezing, etc.).
3. Wash with phosphate buffer (100 mM, pH 7.2), at least 3 × 10 min.
4. Postfix with 1 % w/v aqueous osmium tetroxide on wet ice for 2 h.
5. Wash thoroughly with deionized H_2O.
6. Postfix again with 1 % aqueous osmium tetroxide. It is possible to reuse the first solution. For the second fixation, add aqueous potassium ferrocyanide solution (3 % w/v) for a final concentration of 0.3 % reduced osmium. Leave for 2 h.
7. Wash with deionized H_2O 3 × 10 min (*see* **Notes 24–27**).
8. Wash with sodium maleate buffer (50 mM, pH 5.2) (*see* **Note 28**).
9. En bloc stain with 1 % w/v uranyl acetate in maleate buffer (50 mM, pH 5.2) for 1 h (*see* **Note 29**).
10. Wash with sodium maleate buffer (50 mM, pH 5.2), two changes, 10 min each.
11. Wash with deionized H_2O 3 × 10 min (*see* **Notes 30** and **31**).
12. Dehydrate with an ethanol series: 30 % ethanol (2 × 10 min), 50 % ethanol (2 × 10 min), 70 % ethanol (2 × 10 min), 95 % ethanol (2 × 10 min), and 100 % ethanol (2 × 10 min).

13. Ensure that the final ethanol steps are carried out using dried 100 % ethanol.
14. Replace ethanol with propylene oxide, and leave for two changes, 10 min each.
15. Replace with fresh propylene oxide for 10 min.
16. Replace with epoxy resin/propylene oxide (1:1) and leave overnight. At this stage, the epoxy resin should not contain catalyst (*see* **Note 29**).
17. Replace 1:1 mixture with 1:3 propylene oxide/epoxy resin mix and leave for 3–6 h.
18. Open caps for 1 h to let propylene oxide evaporate.
19. Replace with fresh resin for 4 h with rotation on mixing wheel.
20. Prepare fresh resin containing catalyst and place in embedding molds.
21. Place specimen in molds with unique identifier labels, orientate for best sectioning position, and transfer to 60 °C oven overnight.
22. Section the embedded specimen in an ultramicrotome equipped with a 45° angle diamond knife.
23. Collect the section onto Formvar-coated specimen grid and contrast by floating section-side down on drops of 3 % w/v aqueous uranyl acetate placed on Parafilm.
24. Wash with deionized H_2O.
25. Repeat the staining procedure using aqueous lead citrate (*see* **Notes 32–34**).
26. Dry the grids and examine in the transmission electron microscope (*see* **Note 35**).

3.6 Genomic DNA Extraction

1. The PowerBiofilm™ DNA Isolation Kit is recommended for genomic DNA extraction (*see* **Notes 36–38**).
2. Follow the manufacturer's instructions without modifications.
3. The RNeasy Mini Kit is recommended for total RNA isolation.
4. Follow the manufacturer's instructions without modifications.
5. Include an on-column DNase digestion step with the RNase-free DNase kit (*see* **Note 39**).

3.7 Chemical Analysis

A detailed discussion of the chemical analysis of vaginal microbial biofilm samples is outside the scope of this book chapter as these endeavors usually are hampered by the small sample sizes available for study. A number of destructive (colorimetric and fluorometric) methods have been described [27] and can provide semiquantitative biofilm chemical composition. SEM coupled with energy dispersive spectroscopy (SEM-EDS) is a powerful tool to obtain elemental microanalysis overlaid with high-resolution SEM images.

Nondestructive methods such as Fourier transform infrared (FTIR) spectroscopy and 2-D nuclear magnetic resonance (NMR) spectroscopy can be useful techniques for determining and comparing the principal components of vaginal microbial biofilms [27–35].

4 Notes

1. Formamide can be added to the hybridization buffer if required.
2. Rapid freezing, ideally using a high-pressure freezer, is the preferred method of sample preservation for electron microscopy examination but is rarely feasible in practice due to logistical constraints at the collection point and the need for specialized, expensive equipment.
3. Mucins, highly glycosylated large proteins (10–40 MDa) secreted by epithelial cells, make up an entangled viscoelastic gel that forms part of the cervicovaginal secretions [36]. To differentiate these proteins from biofilm EPS, mucins are labeled specifically with commercial monoclonal antibodies in biofilm cryostat sections using standard protocols [37].
4. Care should be taken to ensure that sample remains hydrated throughout.
5. Calcofluor White staining can reduce the signal intensity of FISH probes.
6. The fixation and hybridization times were optimized for our specimens. Due to variability in the vaginal microbial biofilm composition across subjects, it is recommended to vary these times as part of protocol optimization.
7. Minimize the amount of time the hybridization buffer is out of the incubator to avoid cooling of the solution to room temperature.
8. When imaging the samples, move from low to high magnification. Use the low magnification to align the sample as much as possible and thereby minimize photobleaching.
9. Minimize the time spent under high magnification to minimize photobleaching.
10. Magnification at 60× is usually sufficient to obtain images of single microbial cells. Oil immersion optics and higher magnification provide little benefit due to high photobleaching rates.
11. While a number of commercial photobleaching suppressants exist, they are not recommended as they can add a haze to the sample image and reduce the probe intensity. It is recommended to practice obtaining high-quality images in a short time period instead.

12. High-pressure freezing produces immediate immobilization of small specimens in a frozen state. These specimens can be sectioned and examined by electron microscopy while still frozen [38–42]. This approach produces optimal ultrastructural preservation but requires specialized specimen preparation and imaging equipment, as well as considerable technical skills.
13. High-pressure frozen specimens can be processed for embedding in resin by dehydration at low temperature using a method called freeze substitution [43]. Freeze-substituted specimens can be embedded subsequently in low-temperature resin such as Lowicryl HM20 [43, 44] or warmed for embedding in epoxy resin [45]. Such an approach makes it possible for high-pressure frozen specimens to be examined using routine resin sectioning.
14. Although chemical fixation is a convenient collection method and dehydration at ambient temperature is a routinely applied protocol in almost all electron microscopy laboratories, it is not an ideal approach for preserving bacterial cells or bacterial biofilms [46, 47].
15. A feasible approach for preparing bacterial biofilms that does not require high-pressure freezing and avoids chemical fixation is freezing by immersion in a cryogen such as liquid propane or ethane [47]. The cryogen can be prepared in advance and taken to the collection site while frozen in liquid nitrogen. Immediately prior to sample freezing, the cryogen can be warmed by immersion of a warm metal block, and, when liquid, the specimen can be immersed. The frozen specimen is then transferred to 100 % dry ethanol or acetone on dry ice for subsequent processing or transport. Further processing consists of transferring to fresh solvent (optional contents include 2 % glutaraldehyde, 1 % osmium tetroxide, 1 % uranyl acetate, or suitable combinations). Immersion fixation will not produce optimal ultrastructure of bacterial cells or bacterial biofilms, but the preservation will be improved when compared with chemical fixation [47].
16. Formaldehyde or glutaraldehyde both work well, but glutaraldehyde will make the sample harder [48]. Best results are obtained if specimens are fixed as soon as possible. Handling of the specimen should be kept to a minimum, both before and after fixation. Specimen drying should be avoided at all stages during collecting and subsequent processing.
17. The second approach for dehydrating chemically fixed material at low temperature requires the specimens to be frozen in the presence of cryoprotectant. Chemically fixed specimens are soaked in dimethylformamide, 30 % v/v glycerol, or 2.3 M sucrose and then immersed in liquid cryogen (propane, ethane, or liquid nitrogen). The frozen specimens are freeze

substituted in solvent. If sucrose is used as the cryoprotectant, then methanol must be used as the freeze substitution medium [49]. Methanol is able to solubilize the sucrose in the specimens, whereas if ethanol or acetone is used, the sucrose crystallizes and interferes with subsequent processing.

18. If immersion fixation is not an option, perhaps due to the hazardous properties of gaseous propane and ethane, an aldehyde fixation is the only option for preserving bacteria and biofilms. The ultrastructure can be improved by dehydration at low temperature. Two approaches are possible. The first, called the progressive lowering of temperature (PLT) method, gradually takes chemically fixed specimens through washing and dehydration steps while lowering the temperature. Fully dehydrated specimens can be held at low temperature, infiltrated with low-temperature embedding resin, and polymerized in resin by ultraviolet light [50]. Alternatively, cold, dehydrated specimens can be gradually warmed to ambient temperature and either embedded in epoxy resin for examination by transmission electron microscopy or further dried using a critical-point drier for examination by scanning electron microscopy. If a critical-point drier is not available, then specimens can be dried in the presence of HMDS, a reasonable substitute for critical-point drying [51].
19. Metal coating of specimens using a sputter coater may be required to reduce charging in the microscope. Coating with platinum produces a finer layer, but any metal coating over the specimen surface can potentially cover essential fine structural details.
20. For electron microscopy, specimen preparation is a compromise between applying ideal preparation protocols and practical considerations. For example, the best approach for preserving subcellular morphology usually is high-pressure freezing [47]. However, such an approach is challenging to apply to specimens removed from human subjects due to logistical considerations. Specimens have to be transferred immediately to a high-pressure freezer for immediate freezing by a trained operator.
21. In extreme circumstances, specimens collected for SEM can be air-dried immediately after collection. The dry specimens can be attached to specimen stubs and examined in the SEM, either with or without a metal coating. The specimen ultrastructure will not be optimal, and damage caused by air-drying will be present. However, the specimen can be observed at low magnification to document structures that might not be obvious by light microscopy. Specimens that have been processed for FISH can also be simply air-dried for CLEM, and if fine, ultrastructural detail is not important.
22. Specimens processed for FISH (Subheading 3.3) subsequently can be imaged by SEM [20]. Once light microscopy or confocal

images have been obtained, the specimens are carefully removed from the glass slide or coverslip. The fully hydrated specimens then are processed by fixation in buffered glutaraldehyde (2.5 %), dehydrated, and critical-point dried. When mounted onto the specimen stub, the orientation is noted so that the same fields examined by light microscopy are imaged in the SEM. This approach to correlative light and electron microscopy (CLEM) is not common, but offers an opportunity for imaging large specimens where the fluorescent signal is overlaid on an SEM image of the intact specimens (or reference space).

23. Fix specimens immediately after collection and avoid drying. If specimens are to be used for immunolabeling, then fixation in phosphate-buffered formaldehyde (4 %, v/v, pH 7.2) alone may be more useful. However, specimens should then be embedded in acrylic resins (e.g., Lowicryl HM20 or LR White). Dehydration and resin infiltration times can be reduced with the assistance of a microwave processor [52]. Microwave processing is useful for embedding in epoxy resins and LR White resin. Semi-thin sections of embedded biofilms mounted on glass substrate can be immunolabeled for light microscopy examination, and thin sections of embedded biofilms labeled for TEM examination.

24. The protocol has been written to give regular contrast to the specimens and can be modified in many ways to affect contrast. Some suggested changes are described below.

25. Different fixatives will affect the final appearance of the specimens. Substituting sodium cacodylate (100 mM, pH 7.2) for the phosphate buffer will produce less extraction of cell cytoplasm and thus result in decreased contrast in the TEM. The use of Good's buffers (TRIS, HEPES, PIPES, etc.) [53] will result in decreased extraction and thus reduced contrast.

26. Shorter fixation times, or the use of selective detergents, also can be used to manipulate retention of cellular contents and thus affect contrast.

27. Shorten the postfixation step in osmium tetroxide. In **step 4** (Subheading 3.5), incubate for 3–4 h in 1 % osmium tetroxide. Jump to **step 7**.

28. **Steps 9–11** (Subheading 3.5): Omit the sodium maleate buffer and use aqueous uranyl acetate for slightly less contrast. Alternatively, leave the specimens overnight at 4 °C in saturated uranyl acetate in 70 % methanol for more contrast.

29. **Step 12** (Subheading 3.5): Instead of using ethanol, dehydrate in graded acetone series. The propylene oxide steps then can be omitted, substituting acetone for the propylene oxide. If acetone is used, ensure it is completely removed from the resin before polymerization starts.

30. The incubation times given here are only suggested guidelines. They can be changed to fit different protocols and specimen types. Small pellets of cells will require less incubation at each step than will tissues with tightly packed cells such as nerve or muscle.
31. If at any time the protocol needs to be halted and continued the next day, it is possible to store the samples at any of the washing steps. Place the samples in the appropriate washing solution and store it at 4 °C.
32. *Reynold's lead citrate* [54]: Dissolve lead nitrate (1.33 g) in 30 mL of deionized H_2O in a 50 mL volumetric flask. Add sodium citrate (1.76 g) and mix. Shake the suspension for 1 min and then leave to stand for 30 min with intermittent shaking. The lead nitrate is being converted to lead citrate during this time. Add NaOH solution (4 g in 100 mL, 1 N, 8 mL) and mix well. Filter before use.
33. *Venable and Coggeshall's lead citrate* [55]: Weigh out portions of lead citrate in 10 mL tubes. The amounts can vary between 0.1 and 0.4 g, with the larger amounts producing stronger staining. When needed for staining, add 1 mL of 1 N NaOH to a tube containing the aliquot of lead citrate and dissolve the solid. Then add 9 mL of deionized H_2O, filter, and use. Use carbonate-free NaOH and fresh deionized H_2O or H_2O that has been boiled to remove dissolved CO_2.
34. *Staining protocol for lead citrate*: Prepare CO_2-free, deionized H_2O by boiling and cooling H_2O or by sparging nitrogen gas through the H_2O. Place drops of the lead citrate stain on clean Parafilm. Float grids, section-side down, on the drops of stain (one grid per drop). Incubate for recommended, or optimal, time. Fill four 10 mL beakers with CO_2-free deionized H_2O; in the first, add one drop of 1 N NaOH solution. Remove each grid from the drop of stain and wash by rapidly immersing, sequentially, in the four beakers of deionized H_2O. Start with the beaker containing H_2O with 1 drop of NaOH. Dry and examine in the TEM.
35. If the vaginal biofilms under study are on the surface of a medical device (e.g., intravaginal ring, intrauterine device) and can be embedded in epoxy resin to produce thin sections for examination in the TEM, they should produce high-quality images. However, the epoxy resin will most likely not infiltrate the medical device, causing difficulties in sectioning the polymer at ambient temperature. A convenient alternative approach is to remove the embedded biofilm from the medical device substrate, mark the interface on the device (e.g., using nail polish), and replace the device with fresh resin that is subsequently polymerized. The embedded biofilm can be sectioned

easily using routine methods. If the medical device is required to be in the section, remove as much of the material as possible, leaving the biofilm-device interface intact, and either re-embed in fresh resin or attempt to section the interface. Soft polymers can be successfully sectioned at low temperatures using a cryoultramicrotome.

36. The PowerBiofilm™ DNA Isolation Kit (MO BIO Laboratories, Inc.) can accommodate the low sample masses (50–200 mg) typically available from vaginal specimens.
37. The above DNA isolation kit embodies three important features relevant to vaginal microbial biofilms: Mechanical shearing breaks up the cross-linked biofilm, heat-activated enzymatic digestion dissolves most of the polysaccharides, and the protein/inhibitor removal step usually affords high-quality genomic microbial DNA.
38. The removal of PCR inhibitors sometimes may require a different DNA isolation strategy. Under these circumstances, the InstaGene Matrix (Bio-Rad Laboratories, USA) kit is recommended.
39. A discussion of the methods for analyzing DNA and RNA isolated from the vaginal specimens is beyond the scope of this chapter. The first report of culture-independent characterization of vaginal microbial communities dates to 2004 [56]. A number of subsequent studies have surveyed the vaginal microbiome using culture-independent approaches to census bacterial community composition, usually to improve our understanding of the etiology underlying bacterial vaginosis (BV) [11, 57–62]. Analyzing the metatranscriptomes of vaginal microbial communities is more challenging and not as well developed as methods for determining community composition and structure. McNulty and colleagues have described methods for microbial RNA-Seq analysis of the fecal metatranscriptomes from gnotobiotic mice and humans [63]. Total RNA was purified and converted into the corresponding barcode-ligated double-stranded cDNA (dscDNA) [63, 64]. Multiplex sequencing was performed using the Illumina platform according to published protocols [63].

Acknowledgment

Research reported in this publication was supported, in part, by the National Institute of Allergy and Infectious Diseases of the National Institutes of Health under Award Number R01AI100744. The content is solely the responsibility of the authors and does not necessarily represent the official views of the National Institutes of Health.

References

1. Costerton JW, Stewart PS, Greenberg EP (1999) Bacterial biofilms: a common cause of persistent infections. Science 284:1318–1322
2. Stewart PS (2002) Mechanisms of antibiotic resistance in bacterial biofilms. Int J Med Microbiol 292:107–113
3. Davies D (2003) Understanding biofilm resistance to antibacterial agents. Nat Rev Drug Discov 2:114–122
4. Hillier SL (2005) The complexity of microbial diversity in bacterial vaginosis. N Engl J Med 353:1886–1887
5. Donlan RM, Costerton JW (2002) Biofilms: survival mechanisms of clinically relevant microorganisms. Clin Microbiol Rev 15:167–193
6. Romero R, Schaudinn C, Kusanovic JP et al (2008) Detection of a microbial biofilm in intraamniotic infection. Am J Obstet Gynecol 198:135.e1–135.e5
7. Morales DK, Hogan DA (2010) *Candida albicans* interactions with bacteria in the context of human health and disease. PLoS Pathog 6:e1000886
8. Turnbaugh PJ, Ley RE, Hamady M et al (2007) The Human Microbiome Project. Nature 449:804–810
9. Peterson J, Garges S, Giovanni M et al (2009) The NIH Human Microbiome Project. Genome Res 19:2317–2323
10. Gajer P, Brotman RM, Bai GY et al (2012) Temporal dynamics of the human vaginal microbiota. Sci Transl Med 4:132ra52
11. Ravel J, Gajer P, Abdo Z et al (2011) Vaginal microbiome of reproductive-age women. Proc Natl Acad Sci U S A 108:4680–4687
12. Gunawardana M, Moss JA, Smith TJ et al (2011) Microbial biofilms on the surface of intravaginal rings worn in non-human primates. J Med Microbiol 60:828–837
13. Keller MJ, Malone AM, Carpenter CA et al (2012) Safety and pharmacokinetics of acyclovir in women following release from a silicone elastomer vaginal ring. J Antimicrob Chemother 67:2005–2012
14. Amann RI, Ludwig W, Schleifer KH (1995) Phylogenetic identification and *in situ* detection of individual microbial cells without cultivation. Microbiol Rev 59:143–169
15. Orphan VJ (2009) Methods for unveiling cryptic microbial partnerships in nature. Curr Opin Microbiol 12:231–237
16. Lawrence JR, Nie TR, Swerhone GDW (1998) Application of multiple parameter imaging for the quantification of algal, bacterial and exopolymer components of microbial biofilms. J Microbiol Methods 32:253–261
17. Wigglesworth-Cooksey B, Cooksey KE (2005) Use of fluorophore-conjugated lectins to study cell-cell interactions in model marine biofilms. Appl Environ Microbiol 71:428–435
18. Bahulikar RA, Kroth PG (2007) Localization of EPS components secreted by freshwater diatoms using differential staining with fluorophore-conjugated lectins and other fluorochromes. Eur J Phycol 42:199–208
19. Lawrence JR, Swerhone GDW, Kuhlicke U, Neu TR (2007) *In situ* evidence for microdomains in the polymer matrix of bacterial microcolonies. Can J Microbiol 53:450–458
20. Schaudinn C, Carr G, Gorur A et al (2009) Imaging of endodontic biofilms by combined microscopy (FISH/cLSM—SEM). J Microsc 235:124–127
21. Boskey ER, Moench TR, Hees PS, Cone RA (2003) A self-sampling method to obtain large volumes of undiluted cervicovaginal secretions. Sex Transm Dis 30:107–109
22. Dezzutti CS, Hendrix CW, Marrazzo JM et al (2011) Performance of swabs, lavage, and diluents to quantify biomarkers of female genital tract soluble mucosal mediators. PLoS One 6:e23136
23. Jespers V, Francis SC, van de Wijgert J, Crucitti T (2011) Methodological issues in sampling the local immune system of the female genital tract in the context of HIV prevention trials. Am J Reprod Immunol 65:368–376
24. Loy A, Maixner F, Wagner M, Horn M (2007) ProbeBase—an online resource for rRNA-targeted oligonucleotide probes: new features 2007. Nucleic Acids Res 35:D800–D804
25. Miller A, Spagnuolo RA, Baras V, Pyles RB (2010) High-throughput automated extraction of RNA using the Aurum™ total RNA 96 kit. Bio-Rad Laboratories Tech Notes, Bulletin 0618
26. Amann RI, Krumholz L, Stahl DA (1990) Fluorescent oligonucleotide probing of whole cells for determinative, phylogenetic, and environmental studies in microbiology. J Bacteriol 172:762–770
27. Roller C, Wagner M, Amann R et al (1994) In situ probing of gram positive bacteria with high DNA G + C content using 23S-ribosomal-RNA-targeted oligonucleotides. Microbiology 140:2849–2858
28. Glockner FO, Fuchs BM, Amann R (1999) Bacterioplankton compositions of lakes and oceans: a first comparison based on fluorescence *in situ* hybridization. Appl Environ Microbiol 65:3721–3726
29. Baum MM, Kainović A, O'Keeffe T et al (2009) Characterization of structures in biofilms formed by a *Pseudomonas fluorescens* isolated from soil. BMC Microbiol 9:103

30. Filippov MP, Kohn R (1974) Determination of composition of alginates by infrared spectroscopic method. Chem Zvesti 28: 817–819
31. Sartori C, Finch DS, Ralph B, Gilding K (1997) Determination of the cation content of alginate thin films by FTIR spectroscopy. Polymer 38:43–51
32. Beech I, Hanjagsit L, Kalaji M et al (1999) Chemical and structural characterization of exopolymers produced by *Pseudomonas sp.* NCIMB 2021 in continuous culture. Microbiology 145:1491–1497
33. Karadenizli A, Kolayli F, Ergen K (2007) A novel application of Fourier-transformed infrared spectroscopy: classification of slime from Staphylococci. Biofouling 23:63–71
34. Marcotte L, Kegelaer G, Sandt C et al (2007) An alternative infrared spectroscopy assay for the quantification of polysaccharides in bacterial samples. Anal Biochem 361:7–14
35. Serra D, Bosch A, Russo DM et al (2007) Continuous nondestructive monitoring of *Bordetella pertussis* biofilms by Fourier transform infrared spectroscopy and other corroborative techniques. Anal Bioanal Chem 387: 1759–1767
36. Lai SK, Wang YY, Hida K et al (2010) Nanoparticles reveal that human cervicovaginal mucus is riddled with pores larger than viruses. Proc Natl Acad Sci U S A 107: 598–603
37. Mills SE (2012) Histology for pathologists, 4th ed. Lippincott Williams & Wilkins, Baltimore
38. Dubochet J (1995) High-pressure freezing for cryoelectron microscopy. Trends Cell Biol 5:366–368
39. Al-Amoudi A, Chang JJ, Leforestier A et al (2004) Cryo-electron microscopy of vitreous sections. EMBO J 23:3583–3588
40. Al-Amoudi A, Norlen LPO, Dubochet J (2004) Cryo-electron microscopy of vitreous sections of native biological cells and tissues. J Struct Biol 148:131–135
41. Al-Amoudi A, Diez DC, Betts MJ, Frangakis AS (2007) The molecular architecture of cadherins in native epidermal desmosomes. Nature 450:832–838
42. Al-Amoudi A, Frangakis AS (2013) Three-dimensional visualization of the molecular architecture of cell-cell junctions *in situ* by cryo-electron tomography of vitreous sections. Methods Mol Biol 961:97–117
43. Giddings TH (2003) Freeze-substitution protocols for improved visualization of membranes in high-pressure frozen samples. J Microsc 212:53–61
44. Carlemalm E, Villiger W, Hobot JA et al (1985) Low-temperature embedding with lowicryl resins—2 new formulations and some applications. J Microsc 140:55–63
45. McDonald KL, Webb RI (2011) Freeze substitution in 3 hours or less. J Microsc 243: 227–233
46. Dubochet J, McDowall AW, Menge B et al (1983) Electron microscopy of frozen hydrated bacteria. J Bacteriol 155:381–390
47. Webster P, Wu S, Webster S et al (2004) Ultrastructural preservation of biofilms formed by non-typeable *Hemophilus influenzae*. Biofilms 1:165–182
48. Levanon D, Stein H (1999) Tannic acid and thiocarbohydrazide as structural reinforcement agents in the preparation of rabbit knee articular cartilage for the scanning electron microscope. Histochem J 31:71–73
49. Van Genderen IL, Vanmeer G, Slot JW et al (1991) Subcellular-localization of Forssman glycolipid in epithelial MDCK cells by immuno-electronmicroscopy after freeze-substitution. J Cell Biol 115:1009–1019
50. Robertson D, Monaghan P, Clarke C, Atherton AJ (1992) An appraisal of low-temperature embedding by progressive lowering of temperature into Lowicryl-HM20 for immunocytochemical studies. J Microsc 168: 85–100
51. Bray DF, Bagu J, Koegler P (1993) Comparison of hexamethyldisilazane (HMDS), peldri II, and critical-point drying methods for scanning electron microscopy of biological specimens. Microsc Res Tech 26:489–495
52. Webster P (2007) Microwave-assisted processing and embedding for transmission electron microscopy. In: Kuo J (ed) Electron microscopy: methods and protocols, 2nd edn. Humana Press, Totowa, NJ, pp 47–65
53. Good NE, Winget GD, Winter W et al (1966) Hydrogen ion buffers for biological research. Biochemistry 5:467–477
54. Reynolds ES (1963) Use of lead citrate at high ph as an electron-opaque stain in electron microscopy. J Cell Biol 17:208–212
55. Venable JH, Coggesha R (1965) A simplified lead citrate stain for use in electron microscopy. J Cell Biol 25:407–408
56. Zhou X, Bent SJ, Schneider MG et al (2004) Characterization of vaginal microbial communities in adult healthy women using cultivation-independent methods. Microbiology 150:2565–2573
57. Fredricks DN, Fiedler TL, Marrazzo JM (2005) Molecular identification of bacteria associated with bacterial vaginosis. N Engl J Med 353:1899–1911

58. Hyman RW, Fukushima M, Diamond L et al (2005) Microbes on the human vaginal epithelium. Proc Natl Acad Sci U S A 102:7952–7957
59. Oakley BB, Fiedler TL, Marrazzo JM, Fredricks DN (2008) Diversity of human vaginal bacterial communities and associations with clinically defined bacterial vaginosis. Appl Environ Microbiol 74:4898–4909
60. Spear GT, Sikaroodi M, Zariffard MR (2008) Comparison of the diversity of the vaginal microbiota in HIV-infected and HIV-uninfected women with or without bacterial vaginosis. J Infect Dis 198:1131–1140
61. Hummelen R, Fernandes AD, Macklaim JM et al (2010) Deep sequencing of the vaginal microbiota of women with HIV. PLoS One 5:e12078
62. Srinivasan S, Liu CZ, Mitchell CM et al (2010) Temporal variability of human vaginal bacteria and relationship with bacterial vaginosis. PLoS One 5:e10197
63. McNulty NP, Yatsunenko T, Hsiao A et al (2011) The impact of a consortium of fermented milk strains on the gut microbiome of gnotobiotic mice and monozygotic twins. Sci Transl Med 3:106ra106
64. Rey FE, Faith JJ, Bain J et al (2010) Dissecting the *in vivo* metabolic potential of two human gut acetogens. J Biol Chem 285: 22082–22090

Chapter 8

Imaging Bacteria and Biofilms on Hardware and Periprosthetic Tissue in Orthopedic Infections

Laura Nistico, Luanne Hall-Stoodley, and Paul Stoodley

Abstract

Infection is a major complication of total joint arthroplasty (TJA) surgery, and even though it is now as low as 1 % in some hospitals, the increasing number of primary surgeries translates to tens of thousands of revisions due to prosthetic joint infection (PJI). In many cases the only solution is revision surgery in which the hardware is removed. This process is extremely long and painful for patients and is a considerable financial burden for the health-care system. A significant proportion of the difficulties in diagnosis and treatment of PJI are associated with biofilm formation where bacteria attach to the surface of the prosthesis and periprosthetic tissue and build a 3-D biofilm community encased in an extracellular polymeric slime (EPS) matrix. Bacteria in biofilms have a low metabolic rate which is thought to be a major contributor to their recalcitrance to antibiotic treatment. The diagnosis of biofilm infections is difficult due to the fact that bacteria in biofilms are not readily cultured with standard clinical microbiology techniques. To identify and visualize in situ biofilm bacteria in orthopedic samples, we have developed protocols for the collection of samples in the operating room, for molecular fluorescent staining with 16S rRNA fluorescence in situ hybridization (FISH), and for imaging of samples using confocal laser scanning microscopy (CLSM). Direct imaging is the only method which can definitively identify biofilms on implants and complements both culture and culture-independent diagnostic methods.

Key words Orthopedic samples, Hardware, Prosthesis, Tissues, Membranes, Biofilm, Molecular fluorescent imaging, FISH, Confocal laser scanning microscopy (CLSM)

1 Introduction

Periprosthetic joint infection (PJI) is a devastating complication that is of major concern to patients, surgeons, and payers because of the resulting high rate of disability, the difficult and lengthy treatment, and high costs involved. The success rate for total hip arthroplasties (THA) for periods exceeding 10 years is greater than 95 % [1], with a similar percentage for total knee arthroplasties (TKAs). However, the ever increasing number of replacements, which was approximately 300,000 THAs and 600,000 TKAs in the United States in 2009 [2], translates to around 45,000 revisions

Gianfranco Donelli (ed.), *Microbial Biofilms: Methods and Protocols*, Methods in Molecular Biology, vol. 1147,
DOI 10.1007/978-1-4939-0467-9_8, © Springer Science+Business Media New York 2014

per year, assuming continuing rates of implantation and failure. For hips, aseptic loosening accounts for approximately half of all revisions, followed by instability (10–17 %) and infection or septic loosening (6–8 %) [1, 3]. In TKA the rate of septic loosening is almost double (15 %) [3]. Infection is of greater concern in the early lifetime of the implant and causes 24 % of failures within 5 years after primary THA [1]. Thus, by conventional clinical diagnostic methods, infection could account for over 110,000 hip and knee revisions per year in the United States alone. However, recent reports suggest that difficulty in detecting PJI through conventional culturing methods resulting from culture-negative results may result in the misdiagnosis of a significant number of PJIs as "aseptic loosening." Pathogen-specific and broad-spectrum PCR have been used to directly detect pathogen DNA in orthopedic samples as an alternative to culture [4–7]. Rasouli et al. [5] found that 17 of 65 (26 %) revised TKAs originally diagnosed as aseptic loosening were actually infected, by using a broad-spectrum PCR technique to detect DNA of bacterial pathogens extracted from intraoperative periprosthetic tissue and fluid. It is suspected that one of the difficulties with culturing is that the pathogens are present as biofilms residing on the hardware and periprosthetic tissue [6–8] and that biofilm bacteria are not readily cultured. It is not clear, however, whether culture-negative results are due to the presence of residual antibiotics and incorrect sampling (the bacteria are on the surfaces and not necessarily the fluids) or whether biofilm bacteria represent a dormant-like slow-growing phenotype [9]. While sonication or vortexing of tissue and hardware certainly increases the sensitivity of culture suggesting that false-negative culture might indeed be a sampling issue with sonication more effectively dislodging biofilm bacteria than a swab [9], the sonication procedure might also have a dilution effect on residual antibiotics as well as stimulating microbial activity.

Moreover, not only are biofilm infections difficult to detect; they are also genuinely challenging to treat because they may be up to 1,000 times more resistant to antibiotics than their planktonic (free-living) counterparts. It is estimated that about 13 million biofilm-related infections occur every year in the United States [7]. In vitro and direct examination of clinical specimens indicates that many species of pathogen can colonize virtually any medical material and type of implant as well as the surrounding host tissues [10].

Currently, molecular methods appear to be better suited to assess the presence of biofilm bacteria than culturing. Post et al. [11] demonstrated that PCR effectively detected the presence of bacterial DNA in cultural-negative middle-ear effusions. However, there are potential difficulties with the diagnosis of orthopedic infections. For example, one concern is that PCR-based methods are too sensitive and can provide false-positive results by amplifying the DNA of contaminant or colonizing (nonpathogenic) bacteria.

Clinical microbiology labs must make a distinction between contaminating and potential opportunistic organisms in the case of orthopedic implants. For example, *Staphylococcus epidermidis*, a common skin-colonizing bacterium, although a common contaminant in clinical culture, is capable of biofilm development which represents its only virulence factor. Thus, in situ demonstration of *S. epidermidis* in a biofilm facilitates this distinction. Similarly, while increased culture-positive results following sonication imply the presence of attached biofilm bacteria, this technique also fails to directly demonstrate biofilm. Currently, the only way to clearly diagnose bacteria being present in a biofilm is by direct observation. To this end, we have applied a variation of the "full circle" approach used by Amann et al. [12] to identify non-culturable bacteria in the natural environment to clinical specimens. In this approach, pathogens are detected by PCR, and their presence is confirmed by direct microscopic examination. By using this method, not only do we confirm the presence of bacteria, but we can also determine whether they are present as a biofilm aggregates as well as confirm the localization of infection (implant, cement, perioperative tissue, fluid). Our working hypothesis is that microscopic evidence of adhered biofilms, even after various rinse steps, is strong evidence of a "growth in place" process and that these bacteria are unlikely to be transient contaminants. Using this technique, Hall-Stoodley et al. [13] showed that pediatric otitis media with effusion (OME) is a biofilm infection, and Nistico et al. [14] found the most common middle-ear pathogens also formed biofilms in pediatric adenoidal tissues. Biofilms have additionally been detected on surgical sutures [15] and orthopedic prosthesis [6] using fluorescent viability stains and fluorescent in situ hybridization (FISH) in conjunction with confocal laser scanning microscopy (CLSM).

In order to identify bacterial pathogens that are directly associated with recovered orthopedic hardware, periprosthetic tissue, "membrane" (the "slimy" tissue often found at the surgical site), and surrounding fluid, we have developed protocols for the collection of samples in the operating room (OR) specifically designed for visualizing bacteria and biofilms using CLSM following molecular fluorescent DNA staining and 16S rRNA FISH. In addition, we also use polysaccharide-specific probes in order to visualize the extracellular polymeric slime (EPS) matrix, a hallmark characteristic of biofilms. Our method allows us to scan fully hydrated surfaces from a top-down view, rather than relying on thin sections, in which biofilm might be easily missed, in order to detect if any aggregated bacteria are present in the sample. In the following described protocols, we concentrate on imaging bacteria and determining if they are present in a biofilm context.

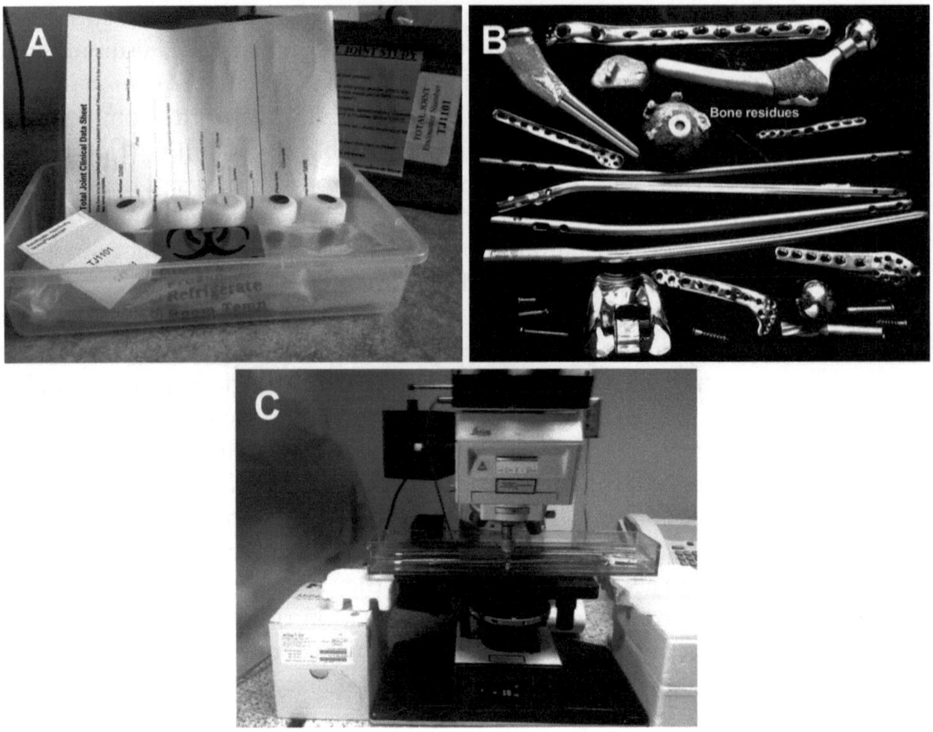

Fig. 1 Sample collection. (**a**) Sample collection kit consisting of (1) a plastic box with a lid pre-labeled with a coded patient identification number, (2) written instructions for sample collection, (3) several labeled 15 ml sterile plastic containers with sterile buffer solutions or RNAlater for the various types of samples (tissue, membrane, cement, aspirates, and smaller hardware such as screws), (4) a labeled large biohazard bag for the implant itself, and (5) a clinical data sheet. (**b**) Various kinds of metal implants. (**c**) Positioning and stabilization of a metal rod sample on the microscope stage for imaging with a 63× water immersion objective

1.1 Overview of the Experimental Protocol

1.1.1 Collection, Preservation, and Cataloging

The first part of the process is the collection, preservation, and cataloging of the specimens and samples. Keeping track of samples from an orthopedic revision surgery is very important since there are usually several types of samples of widely varying sizes, some requiring different preservation methods. Thus, the surgical team must first be provided with a sample collection kit (Fig. 1a). The importance of coordinating the specimen collection protocols and their rationale with the OR team cannot be underestimated. The kit that we have developed is comprised of a plastic box with a lid pre-labeled with a coded patient identification number and instructions specific for the sample collection. The sample kit box contains (1) several labeled 15 ml sterile specimen containers with sterile buffer solution for different samples (tissue, membrane, cement, aspirates, and smaller hardware such as screws), (2) a labeled large biohazard bag for the implant itself (Fig. 1b), and (3) a clinical data sheet. In addition, we routinely include containers of RNAlater® (Invitrogen) for collection of periprosthetic tissue to complement the direct microscopic observation with nucleic acid

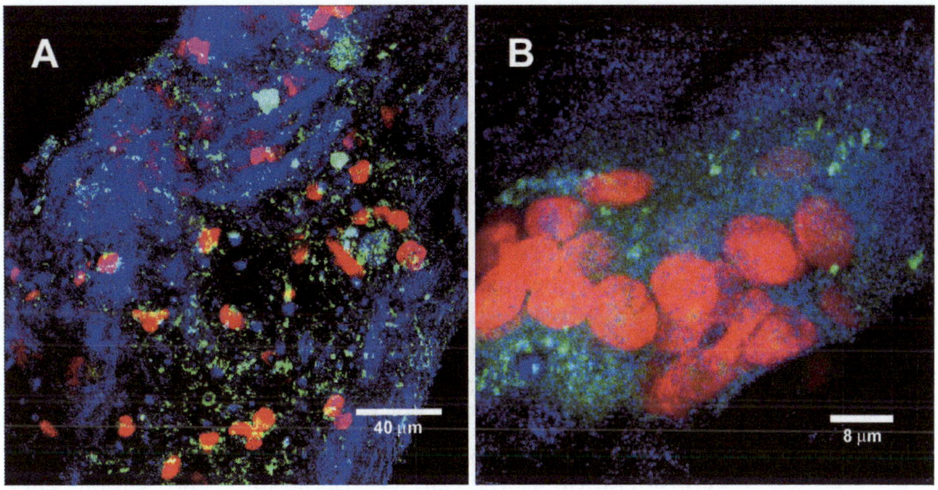

Fig. 2 (**a**) and (**b**) Periprosthetic tissue stained with the LIVE/DEAD kit. Live bacteria (*green*) were present as single cells and as biofilm clusters. In *panel a*, a few patches of dead (*red*) bacteria can be seen. In both images, the nuclei of the tissue cells can be seen stained red with propidium iodide

detection methods such as PCR-based methods. The samples must be immediately refrigerated at 4 °C or on ice upon collection to arrest bacterial growth and should be processed within 1–2 h.

1.1.2 Staining of Bacteria and Biofilm EPS

To assess biofilm bacterial viability using differential staining, the samples should be stained with fluorescent markers as soon as possible after collection. For those samples allocated for FISH, immediate fixation of the sample is required. The fixed samples can be stored at −20 °C for long periods of time (months), and FISH can be performed when convenient. However, once FISH has been performed, samples must be examined as quickly as possible to prevent deterioration of the permeabilized bacteria. If both viability staining and FISH techniques are to be used, the samples must be subdivided.

To assess the presence of bacterial biofilms nonspecifically, generic fluorescent bacterial DNA stains can be used such as DAPI and the SYTO probes. The most commonly used staining method to assess bacterial viability is the LIVE/DEAD® BacLight™ viability kit which contains two different nucleic acid stains, SYTO 9 and propidium iodide, to distinguish live bacteria with intact plasma membranes (green) from dead bacteria with compromised membranes (red). These DNA dyes also stain human DNA, which is useful for identifying human cells, which are readily distinguished from bacteria due to their size and morphology (Fig. 2). We have found that host cells initially take up the SYTO 9 and appear green but after a few minutes may take up the propidium iodide and appear red, regardless of original viability. Thus, the BacLight LIVE/DEAD kit is not a reliable viability indicator for host cells. In some cases, particularly if there is a dense concentration of host cells, the brightness of the fluorescent staining of these cells can "mask" the signal from adjacent bacteria. To address this pitfall, the LIVE/DEAD probes can be diluted by 1:5 or 1:10; however, the staining will obviously not be as bright. To stain the EPS of the biofilm, we have developed a lectin cocktail that binds to the carbohydrate residues present in the EPS [16]. A cocktail is used since bacteria produce a wide variety of carbohydrates and there are no generic carbohydrate stains.

One of the main challenges in staining the implants is their size. While it is possible to cover the surface of a screw with fluorescent stain, complete coverage is often not possible with larger specimens such as stems and necks. For these larger specimens, we recommended applying the stain and associated rinses to specific locations on the specimen where biofilm is suspected. Another consideration is the surface topography and sample curvature since the CLSM objectives typically have a relatively short focal depth (μm to mm) and therefore may be unable to image into screw holes. Additionally, imaging may be blocked by protrusions on the hardware. In our experience, we have found that the locations likely to harbor biofilm are the parts of metal or cement with visible residual tissue, membrane, or bone attached. Anecdotally, we have also noticed that the threads of the screws are more likely to harbor biofilm.

1.1.3 Staining Host Cells and Matrix

Cytological fluorescent stains, such as phalloidin, can be used to stain F-actin for visualization of the cell cytoskeleton in tissue (*see* **Note 1**).

1.2 Limitations of the Methods

FISH is a powerful imaging technique for the identification and mapping of bacteria on clinical specimens ex vivo. FISH also provides phylogenetic information, morphology, and distribution. However, FISH does have limitations, including difficulties in the detection of slow-growing bacteria (with a low number of ribosomes and subsequently low-signal fluorescence), and importantly differences in the permeability of bacterial cell to FISH probes. Permeability is of concern particularly for specimens containing both Gram-negative and Gram-positive bacteria since an aggressive lysozyme treatment for Gram-positive cells such as *S. aureus* may lead to complete lysis of Gram-negative bacteria. Conversely, a less aggressive permeabilization technique may leave Gram-positive bacteria unstained and, therefore, undetected. To overcome this problem, specimens are frequently fixed, embedded, and sectioned before being immobilized on gelatin-coated slides so that bacterial ribosomes remain relatively localized. However, we prefer to view the specimens in plan view rather than thin sections to provide a more accurate representation of biofilm distribution and ultrastructure. In adenoid tissues, we found that thin H&E

sections could easily overlook biofilm [9]. However, it may be advisable to use conventional histological techniques in addition to FISH since we suspect that the fixation, permeabilization, and washing steps used in the FISH protocol without cryopreservation or embedding may cause some biofilm detachment.

Finally, in its current format, FISH as an approach for detecting pathogenic biofilms in samples and indeed confocal microscopy generally are highly specialized and time-consuming techniques that are not likely to be adopted in the near future for standard clinical microbiology diagnostics. Currently, direct imaging is the only definitive method of identifying bacterial biofilms, and the power of FISH and confocal imaging is that they can be used to independently corroborate more rapid and standard diagnostic methods such as culture and PCR.

2 Materials

2.1 Sample Collection

1. Clear plastic rectangular box with lid 8″ × 15″.
2. Sterile Nalgene sample storage pots, 15.0 ml.
3. Biohazard Lab Guard polyethylene specimen transport bags, 12″ × 15″ bag.
4. Precoded labels.
5. Clinical data sheet.
6. Sterile phosphate buffer solution (PBS).

2.2 Bacterial DNA Staining to Assess Bacterial Presence and Viability

1. Sterile Hank's buffer saline solution (HBSS) with Ca^{2+} and Mg^{2+}.
2. DNA stains such as 0.5–1.0 μg DAPI/ml of HBSS with Ca^{2+} and Mg^{2+}, 3 μl of SYTO 59 or LIVE/DEAD® BacLight™ viability kit (propidium iodide, SYTO 9) in 1 ml of HBSS with Ca^{2+} and Mg^{2+} (*see* **Notes 1** and **4**).

2.3 Lectin Cocktail Stain for the Visualization of the Carbohydrate Component of the EPS of the Biofilm

1. Sterile Hank's buffer saline solution (HBSS) with Ca^{2+} and Mg^{2+}.
2. Concanavalin A, conjugated with Alexa Fluor fluorescent stain (*see* **Notes 1** and **4**). 50 μl of the stock solution1mg/ml.
3. Lectin SBA from Glycine max (soybean), conjugated with Alexa Fluor fluorescent stain (*see* **Notes 1** and **4**), 125 μl of the stock solution 1 mg/ml.
4. Wheat germ agglutinin, conjugated with Alexa Fluor fluorescent stain (*see* **Notes 1** and **4**), 125 μl of the stock solution 1 mg/ml.
5. Lectin PNA from *Arachis hypogaea* (peanut), conjugated with Alexa Fluor fluorescent stain (*see* **Notes 1** and **4**), 50 μl of the stock solution 1 mg/ml.

6. Lectin GS-II from *Griffonia simplicifolia*, conjugated with Alexa Fluor fluorescent stain (*see* **Notes 1** and **4**), 75 μl of the stock solution 1 mg/ml.

   Mix the 5 lectins (total volume 425 μl) in a 2 ml tube. Add 575 μl of HBSS with Ca^{2+} and Mg^{2+} to reach a total volume of 1 ml. Microcentrifuge briefly to pelletize out any protein aggregates which may be present according to manufacturer instruction. Decant the supernatant to a new 2 ml tube and discard the pellet. The lectin cocktail is ready to use.

2.4 Cytology Fluorescent Stain: F-Actin Stain for Visualization of the Cell Cytoskeleton in the Tissue

1. Sterile Hank's buffer saline solution (HBSS) with Ca^{2+} and Mg^{2+}.
2. Triton 0.1 % in PBS: Add 0.1 ml of Triton X 100 to 99.9 ml of PBS.
3. Phalloidin conjugated with Alexa Fluor fluorescent stain (*see* **Notes 1** and **4**). Add 25 μl of the stain to 1 ml of PBS.

2.5 Sample Fixation

1. Phosphate buffered saline (PBS): 7.6 g NaCl, 0.4 g NaH_2PO_4, 1.3 g Na_2HPO_4, 800 ml MQ water (MQ); adjust to pH 7.2, fill to 1 L, and autoclave.
2. Paraformaldehyde (4 % PFA): Dilute electron microscopy grade 16 % paraformaldehyde stock solution with PBS; filter sterilize. Use fresh PFA (stored at 4 °C) for 1 week maximum. A concentrated stock solution of 16 % can be kept long term in the dark at room temperature and diluted with PBS prior to use. PFA is a hazardous material which gives off irritating fumes, and precautions must be taken in handling. Disposal should be in accordance with all local, state, and federal regulations (*see* **Note 2**).
3. Ethanol–PBS solutions 50:50 % and 80:20 %; filter sterilize.
4. Containers of different sizes for fixed sample storage.
5. Bench centrifuge.

2.6 Liquid Sample (Aspirate) Immobilization

1. Gelatin-coated slides: Soak Shandon multi-spot microscope slides in a solution of 0.075 % gelatin and 0.01 % $CrK(SO_4)_2$ heated to 70 °C for 1 min and then air-dry in vertical position.

2.7 Additional Permeabilization for Gram+ Bacteria (If Required)

1. Tris–HCl 1 M (2-amino-2-(hydroxymethyl)propane-1,3-diol–HCl): 121.1 g Tris, 800 ml MQ; add concentrated HCl to pH 8.0 (about 42 ml). Autoclave.
2. Ethylenediaminetetraacetic acid (EDTA) 0.5 M: 186.1 g EDTA, 800 ml MQ; to dissolve, stir and adjust pH to 8.0; fill to 1 L. Autoclave.
3. Solution of 0.1 mg/ml lysozyme in a buffer of 0.1 M Tris–HCl and 0.05 M EDTA. Filter sterilize. Store in aliquots in freezer at −20 °C.

4. Ethanol (molecular biology grade); filter sterilize.
5. Ethanol–PBS solutions: 50 % and 80 %; filter sterilize.
6. Sterile disposable scalpel.
7. Petri dishes 60 × 15 mm.

2.8 FISH

2.8.1 In Situ Hybridization Buffer

1. NaCl, 5 M: 292.2 g NaCl, 800 ml MQ, fill to 1 L.
2. Tris–HCl 1 M, pH 8.0.
3. Formamide deionized. Formamide is a hazardous material releasing irritating fumes, and precautions must be taken when handling. Disposal should be in accordance with all local, state, and federal regulations (*see* **Note 2**).
4. Sodium dodecyl sulfate (SDS) 10 %: 100 g SDS (electrophoresis grade), 900 ml ultrapure MQ water, heated to 68 °C; adjust pH to 7.2, with concentrated HCl, and make up to 1 L. SDS is a hazardous material, and precautions must be taken in handling. Dispose in accordance with all local, state, and federal regulations (*see* **Note 2**).

   Mix 360 μl of 5 M NaCl, 40 μl of 1 M Tris–HCl, pH 8.0, *x* μl of formamide deionized (the optimal formamide concentration is determined for each probe) in a sterile 2 ml Eppendorf snap cap tube. Fill to 2 ml with MQ. Add 2 μl of 10 % SDS (*see* **Note 3**).
5. Microscope slide.
6. Incubator.
7. Water bath.

Fluorescently Conjugated 16S rRNA Probes (Table 3, Note 4)

2.8.2 Washing Buffer

1. 1 M Tris–HCl, pH 8.0.
2. 5 M NaCl.
3. 0.5 M EDTA.

Mix 1 ml of 1 M Tris–HCl pH 8.0, *y* μl of 5 M NaCl (*y* depends on the formamide concentration in the hybridization buffer), *z* μl of 0.5 M EDTA (for *z see* Table 1) in a 50 ml sterile conical tube. Fill to 50 ml with MQ. Add 50 μl of 10 % SDS (*see* **Note 3**).

2.9 Microscopy

1. Shandon multi-spot microscope slides.
2. Petri dishes 60 × 15 mm.
3. Cover glass, No 1 ½, 24 × 50 mm^2.
4. Silicone Sealer (*see* **Note 5**).
5. Antibleaching mounting oil.

Table 1
NaCl and EDTA concentrations in the washing buffer to achieve required stringency conditions

% Form amide (hybridiz. buffer)	5 M NaCl (= *y*) μl	0.5 M EDTA (= *z*) μl
0	9,000	–
5	6,300	–
10	4,500	–
15	3,180	–
20	2,150	500
25	1,490	500
30	1,020	500
35	700	500
40	460	500
45	300	500
50	180	500
55	100	500
60	40	500
65	–[a]	500
70	–	350
75	–	250
80	–	175
85	–	125
90	–	88
95	–	62

[a]Enough NaCl in EDTA

6. Alkaline mounting oil: 50 % PBS pH 8.5–glycerol.
7. Immersion oil for fluorescence microscopy Type DF for use with gelatin-coated slides with liquid samples (either broth culture for probe validation or clinical aspirates).
8. Confocal laser scanning microscopic (CLSM) imaging system. In our case, we used a Leica DM RXE microscope attached to a TCS SP2 AOBS confocal system (Leica Microsystem, Exton, PA).
9. Microscope objectives, 10× dry, 20× dry, and 63× 0.90 n.a. WUVHCX APO LU-V-I water immersion objective with 2.2 mm working distance.

3 Methods

Before proceeding with specimen collection, it is essential to only do so under approval by the appropriate institutional review board(s) or ethics committee(s) and to have documentation of voluntary informed consent if required.

3.1 Sample Collection

1. Label the lid of the rectangular plastic box with instructions for the collection of the samples and the patient coded ID. Each patient has a unique code and "encounter number" so that specimens collected from the same patients but from multiple surgeries could be identified when the code was unlocked.
2. Label the sterile sample storage pot with the appropriate label (tissue, aspirate, membrane, etc.) and the patient coded ID.
3. Label the biohazard plastic bag with the patient ID.
4. Place 10 ml of PBS in each of the sterile sample storage pot. For samples requiring DNA or RNA extraction for PCR phylogenetic analysis instead of PBS, place 10 ml of RNA later in each of the sterile sample storage pot. We have demonstrated that FISH staining can be successfully conducted on samples stored in RNA later (data not shown).

3.2 Additional Tissue Staining

3.2.1 DNA Staining for the Visualization of Biofilm and the Tissue Cell Nucleus in the Tissue

1. Remove the sample from the PBS and place in a suitable sized container. This step also serves as a rinse step to remove loosely attached bacteria.
2. Cover the sample, if small (i.e., less than approximately 1 cm^2), with the working solution of DAPI, propidium iodide, or a SYTO stains for 20 min at room temperature (RT) in the dark (*see* **Note 1**). If the sample is larger, add the stain to the locations where biofilm is suspected. Usually one to three drops from a P20 pipette tip are sufficient to wet an area of 1 cm^2.
3. Rinse the sample twice with HBSS and then let the tissue stay moist with HBSS covering the surface.

3.2.2 F-Actin Stain for Visualization of the Cell Cytoskeleton in the Tissue

4. Incubate the sample with 0.1 % Triton for 3–5 min to increase the tissue and host cell permeability.
5. Rinse the sample 3 times with HBSS.
6. Cover the sample, if small (< approximately 1 cm^2), with the working solution of phalloidin fluorescently conjugated (*see* **Note 1**) and stain for 25 min at RT in the dark. If the sample is larger, add the stain in the places where most likely biofilm can be found.
7. Rinse the sample twice with HBSS and then let the tissue stay moist with HBSS covering the surface.

3.3 Fixation

3.3.1 Fixation of Gram-Negative or Unknown Bacteria in Suspension (Liquid Samples Such as Aspirate)

Ideally, FISH requires different fixation methods for Gram-positive and Gram-negative bacteria. However, because of limitations of time and available samples, it is not always possible to fix using two different protocols. To facilitate the detection of both Gram-positive and Gram-negative bacteria simultaneously using FISH, we developed a single fixation process:

1. Remove the PBS (or RNA later and in this case rinse with PBS) and add a freshly prepared solution of 4 % PFA to sample (2–4 % final concentration).
2. Incubate for 1–12 h at 4 °C depending on the size of sample.
3. Spin down biomass in the bench centrifuge 1 ml Eppendorf tube using a bench centrifuge and remove supernatant (discard in PFA/formamide waste). Resuspend the pellet in PBS.
4. Spin down biomass and remove supernatant a second time.
5. Repeat **steps 4** and **5**.
6. Finally, resuspend in 50 % ethanol–PBS solution.
7. Store at −20 °C.

3.3.2 Fixation of Gram-Negative or Unknown Bacteria in Tissue, Membrane, Screws, and Implants

1. Remove the PBS (or RNA later and in this case rinse with PBS) and add a freshly prepared solution of 4 % PFA to the sample to achieve a 2–4 % final concentration.
2. Incubate at 4 °C for 2–12 h (depending on sample size).
3. Remove PFA carefully using pipette aspiration (discard in the PFA/formamide waste).
4. Add PBS in at least 2× the volume of PFA used in **step 1**.
5. Incubate for 10 min at RT.
6. Remove the PFA by pipette aspiration (discard in the PFA/formamide waste).
7. Repeat **steps 4–6**.
8. Add 50 % ethanol–PBS.
9. Store at −20 °C.

3.3.3 Fixation of Gram-Positive Bacteria in Suspension (Liquid Bacterial Culture f or Probe Testing, Aspirate)

1. Spin down biomass and remove supernatant as described above.
2. Add 50 % ethanol–PBS in intended amount (cell concentration can be increased or decreased by adding the appropriate volume).
3. Store at −20 °C (incubate at least 24 h).

3.3.4 Fixation of Gram-Positive Bacteria in Tissue, Membrane, Screws, and Implants

1. Add 50 % ethanol–50 % PBS solution to the drained (but not dried) sample so that it is completely covered.
2. Store at −20 °C (incubate at least 24 h).

3.4 Immobilization of Liquid Samples

1. Add 10–20 μl of the fixed stored sample per well on gelatin-coated Shandon multi-spot microscope slide.
2. Allow to dry at RT in a closed container for a few hours.
3. Wash for a few seconds with MQ water.
4. Air-dry.

3.5 Additional Permeabilization for Gram-Positive or Unknown Bacteria

If Gram-positive bacteria are suspected or have been cultured, an additional treatment for permeabilization of bacterial cells is necessary:

1. Add 20 μl/ml of lysozyme solution on the microscope slide immobilized samples.
2. Place the slide in a 50 ml conical tube with a tissue paper gently wadded in the bottom and soaked, but not dripping, to maintain a humid environment such that the sample does not dry out.
3. Incubate at 37 °C for up to 3 h (*see* **Note 6**).
4. Remove lysozyme solution by gentle pipetting.
5. Rinse twice with a few drops of PBS by gentle pipetting.
6. Add 20 μl 50 % ethanol–50 % PBS solution for 3 min. Remove.
7. Add 20 μl 80 % ethanol–20 % PBS for 3 min. Remove.
8. Add 20 μl 100 % ethanol–PBS for 3 min. Remove.
9. Allow to air-dry at RT.

In case of tissue and membrane (containing unknown bacteria), place the fixed and stored sample in a Petri dish, cut a 0.5–1 mm thick section using a sterile scalpel, and store the residual sample at −20 °C for further analysis. Place the tissue section on a slide and treat it as described in Subheading 3.5, using adequate volumes.

In the case of screws or implants (containing unknown bacteria), place the fixed and stored sample in an adequate size container and treat as described above in this Subheading 3.5, using adequate volumes.

3.6 Fluorescent in Situ Hybridization (FISH)

The fluorescently conjugated 16S rRNA probes are applied to fixed and permeabilized bacterial cells under stringent hybridization conditions. The probes specifically hybridize with their complementary target sequences in the ribosomes within the bacterial cells, making them fluorescent. Stringency conditions were modified from Manz et al. [17] and achieved by changing the formamide from 0 to 50 % in incremental steps of 5 % together with the salt concentrations but keeping the hybridization and wash temperatures constant. The optimal stringency is usually considered the highest formamide concentration that doesn't result in loss of fluorescent intensity on the target bacterial cells using a pure culture as a representative species. Hybridization to nontarget bacteria should not occur. Controls for probe evaluation should always include the target probe, the domain-level probe (EUB338) [18], the nonsense probe (NONEUB338) [19], and a no-probe control to check for sample autofluorescence [20].

3.6.1 Probe Selection

FISH is a targeted approach, meaning that probes are selected for a genus or species of interest or informed from culture or PCR results. Generally, a good starting point is to search for probes which have already been described and published. If species-specific probes are not available, generic domain or genus probes may be used; however, it must be kept in mind that these results are less compelling for pathogen identification. For common pathogens, it is likely that 16S FISH probes have already been designed and optimized and reported in the literature. probeBase (http://www.microbial-ecology.net/probebase/) is an excellent resource to search for available probes. Discussion regarding the design of probe sequences for genera or species not reported in the literature is outside the scope of the present protocol. 16S rRNA probes can be selected to hybridize at various phylogenetic levels down to species level with the most commonly used for surgical site infections being domain bacteria for all bacteria, genus, and species. Once a probe sequence has been identified, several companies provide synthesis and conjugation with various fluorescent dyes including DAPI (4′,6-diamidino-2-phenylindole); the sulfoindocarbocyanines Cy3, Cy5, and Cy7; and FAM (6-carboxy-fluorescine). Multiple fluorescent probes can be used simultaneously or sequentially if they are conjugated with different fluorescent dyes (*see* **Note 4**) [21].

3.6.2 Probe Reconstitution and Staining

This section is modified from a protocol described for FISH on adenoid tissues [22].

1. Usually, the probe stock is delivered in a lyophilized state. First, suspend in 100 μl of MQ water according to manufacturer instructions. The probe concentration can be verified by absorbance of a 1:100 diluted stock solution at 260 nm (1 A_{260} = 20 μg/ml DNA). The Cy3 dye has maximum absorbance around 550 nm, and a A_{260}/A_{550} ratio of approximately 1 is expected for a monolabeled 18-mer probe. Prepare aliquots of working solution probes at 50 ng/μl and store at −20 °C (*see* **Note 7**).
2. Use 8 μl of hybridization buffer and 1 μl of fluorescently labeled probe as a working solution. For large samples, use a larger volume of the solution with the same ratio of hybridization buffer and fluorescently labeled probe. Multiple probes can be used simultaneously if they use the same formamide concentration; in this case, add 1 μl of each to 8 μl of hybridization buffer. In the case where three probes are to be used at

the same time, the total volume of the probes should be reduced to 2 μl. Probes that require different formamide concentrations have to be hybridized sequentially. Start with the probe with the higher formamide concentration and move to those with lower formamide concentrations.

3. Place a folded paper towel approximately the size of a microscope slide in a 50 ml Falcon conical tube and pour the remaining hybridization buffer onto a piece of wadded paper towel or Kimwipe which is positioned underneath the glass slide. This creates a moist environment reducing evaporation of buffer from the sample during hybridization. Apply 10 μl of the hybridization buffer–probe mixture onto each well in the slide for a liquid sample immobilized on multi-spot microscope gelatin-coated slide. Apply a larger volume of the mixture (maintain the same ratio) to cover the sample (tissue, membranes, screws, and implants). Place the slide with the sample in the tube, screw on cap, place horizontally in the hybridization oven, and hybridize for 90–120 min at 46 °C. For large samples, place the sample in an adequate size container, screw on cap, place in the hybridization oven, and hybridize for 90–120 min at 46 °C (*see* **Note 8**).
4. Quickly wash off the hybridization buffer with pre-warmed washing buffer (*see* **Note 9**).
5. Incubate the samples in the remaining pre-warmed washing buffer for 15 min in a water bath at 48 °C (*see* **Note 8**).
6. Rinse the samples with MQ water and allow to dry at RT in the dark.

3.7 Mounting the Specimens

In this section, we will provide protocols we use for examining liquid samples such as aspirate as well as tissue biopsies which can vary in size from a few millimeters to very large implants.

3.7.1 Mounting Microscopy Slides

Mount slides in antibleaching oil and cover with a 24 × 50 mm^2 cover slip. If using fluorescein stains, the PBS pH should be adjusted to 8.5 with 0.1 mM NaOH to maximize the fluorophore brightness.

3.7.2 Mounting Tissue, Membrane, Screw, and Implant

Gently blot absorb the excess liquid from the tissue, membrane, and screw using a paper towel or a Kimwipe. Place a smear (approximately 0.5 ml) of silicone sealer on the bottom of a Petri plate 60 × 15 mm and attach the tissue section to it by gravity or if necessary some gentle pressure applied at the edges of the specimen. Allow to dry for 10 min. For large metal pieces, use adequate size containers. Since FISH staining of the entire component will not be economically viable, localized areas can be stained by applying stains and rinses in drops to an area of interest.

3.8 Confocal Laser Scanning Microscopy

Once the samples are stained, they are ready for imaging with CLSM. The large size and sometimes irregular shape of the implants can make imaging challenging. For example, intramedullary rods (nails) can commonly be up to 45 cm long. The main issues are overloading the stage and limited space to position the specimen under the microscope objective. We used custom-made polycarbonate containers and side supports to accommodate the overhang| (Fig. 1c). It is important to consider the weight-bearing capacity of the stage so as not to cause damage. Some microscopes, such as the Leica system, have a removable piezoelectric stage, which should be removed when imaging heavy samples. For inverted microscopes, weight is less of an issue although imaging can be difficult for non-flat surfaces which might be held too far away from the objective to allow imaging.

3.8.1 Tissue, Membrane, and Screw Mounted in Petri Plate or Larger Containers

The sample attached to the Petri plate and submerged under PBS can be observed with a 63× water long working distance immersion objective. The advantage of using this objective is that the biofilm can be observed on relatively rough surfaces since the long working distance minimizes hitting any "high spots" on the specimen with the objective. For larger hardware, we recommend to lower the microscope stage as much as possible, stabilize the container by possibly adding supports to hold the overhang and take some of the weight bearing off of the stage (Fig. 1c), and if necessary remove all the microscope objectives not in use so that they do not hit the implant or sides of the containers. Generally, we recommend first performing a surveillance scan with a long working distance 10 or 20× air objective. While individual cells are not readily resolved at these magnifications, small clusters of biofilm can appear as pinpoints of bright light. After identifying promising areas, move to the 63× water immersion objective for higher resolution. Even though the resolution is not as great as higher numerical aperture objectives, such as oil immersion objectives, single cells and biofilm clusters are readily resolved. We often use the zoom function for the final image collection.

The selection of compatible stains and the appropriate laser lines in accordance with the spectral properties of the fluorochromes is extremely important for successful imaging when using multiple staining. When performing multiple fluorescent generic DNA stains and cytology stain or FISH labels simultaneously, it is important to choose fluorophores whose excitation/emission wavelengths have minimal overlap (*see* **Notes 1** and **4**) to avoid cross talk, which occurs when several fluorescence signals are recorded in a single detection channel so that they cannot be separated into individual images. To avoid cross talk, sequential scanning is an option which allows greater flexibility in choice of fluorophores. Sequential sequencing allows dyes with overlapping emission spectra but different excitation wavelengths to be

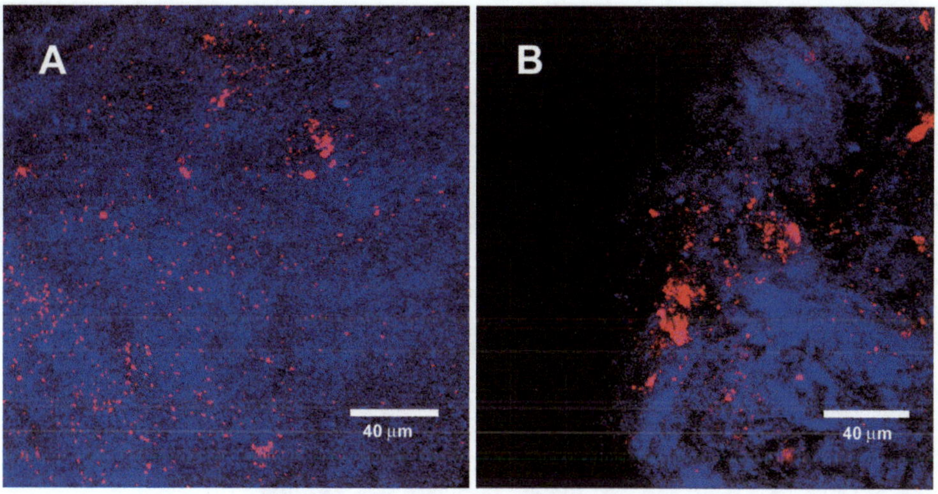

Fig. 3 Examples of FISH staining. (**a**) Periprosthetic membrane stained with the *Propionibacterium acnes* FISH probe conjugated with the Cy3 fluorophore. (**b**) Periprosthetic tissue stained with the universal bacterial FISH probe EUB338 conjugated with the CY3 fluorophore. In both images, the *blue* is the reflected light from the sample surface and in *red* are the bacteria as single cells and as biofilm clusters

differentiated by capturing images excited with different laser lines in rapid sequence. In addition to the detection of fluorescence signals, it is also sometimes useful to collect transmitted light, providing the specimen is not opaque, or reflected light to show the surface of the specimen. For reflected imaging, we use the 488 nm laser line and set a "blue" detector to collect the reflected light. The reflected image was optimized using the acousto-optical beam splitter (AOBS) function, and the optimal contrast was achieved by using high-gain and high-background detector settings.

Figs. 3 and 4 are show CLSM images of tissue, membrane, screw, and implants. FISH was performed with several 16S rRNa probes (Table 2), and other dyes were used in different combinations (Table 3) to visualize the tissue and the localization of the bacteria on hardware and periprosthetic tissue.

4 Notes

1. Several fluorescent generic DNA dyes can be used to stain the tissue sample. Since the colors that can be observed simultaneously are limited, if the same color is used during FISH and in the tissue staining, it will be difficult to discriminate the bacteria in the tissue. If FISH is performed using different probes conjugated with different colors, it is preferable to observe the sample, annotate what bacteria are present, and then stain the tissue. It is recommended to choose the colors of the generic tissue stains such that it will be possible to see at least the most relevant (specific) probe used during FISH.

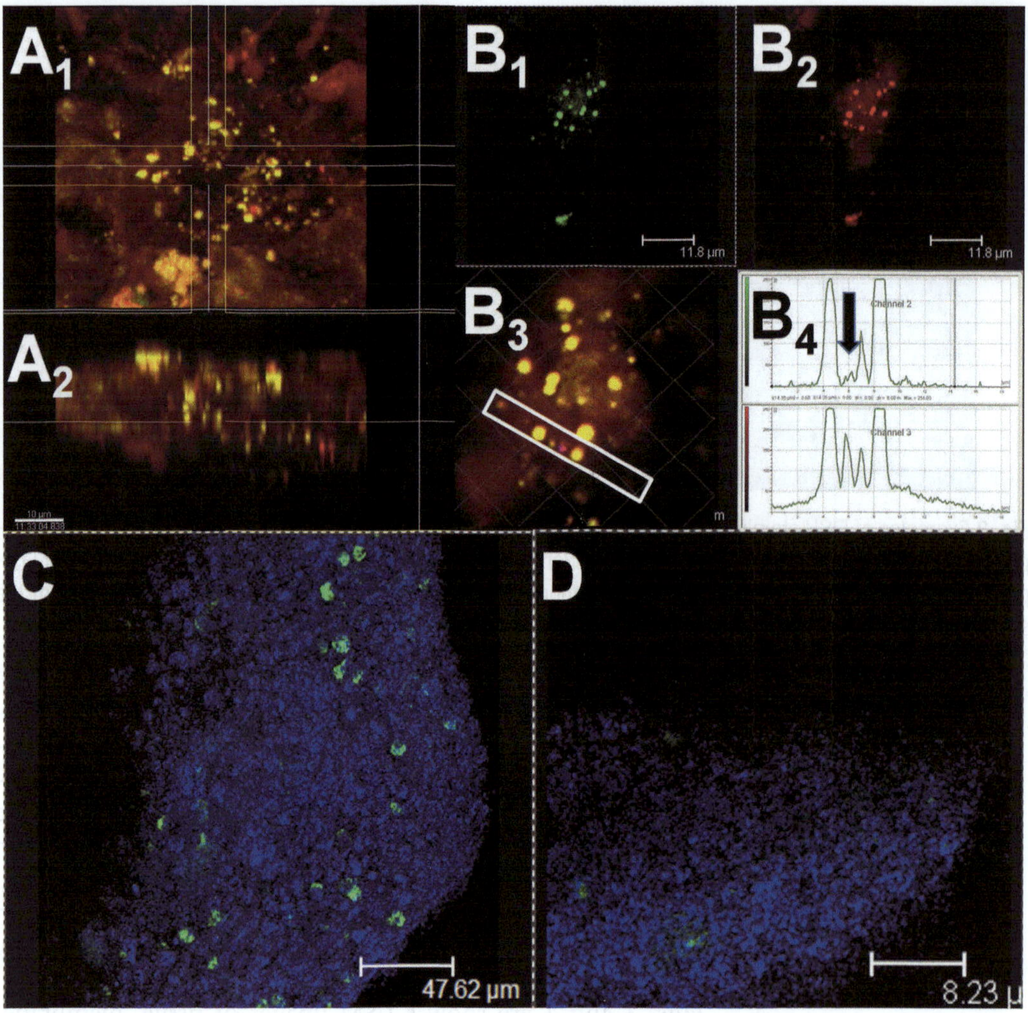

Fig. 4 Various combinations of FISH imaging of bacteria and biofilm on periprosthetic tissue form an ankle revision [7]. (**a**) *S. aureus* cells in biofilm clusters stained with FISH probe sauCy3 (*green*) and the nucleic acid stain SYTO 59 (*red*) to appear as *yellow*. SYTO 59 also stained host cells (*red*). A_1 and A_2 are plan and side views, respectively. (**b**) Dual staining of a biofilm cluster with sauCy3 and SYTO 59. B_1 *S. aureus* cells in stained *green* with the FISH probe and B_2 all cells stained *red*. Diffuse *red* staining between cells is possibly extracellular DNA (DNA) in the EPS. B_3 combined channels and B_4 show the fluorescent intensity profiles across the *box* shown in B_2 showing that 3 of the cells were stained with both sauCy3 (top profile) and 4 cells were stained with SYTO 9. One of the cells was stained with SYTO 9 only (*black arrow*) indicating it was not *S. aureus* and the biofilm was polymicrobial. (**c**) *S. aureus* stained with sauCy3 (*green*) on the tissue visualized by reflected imaging of 488 nm (*blue*) light. (**d**) Sample adjacent to "C" stained with the NONEUB-Cy5 probe (*red*). No bacteria were evident showing that the hybridization conditions were stringent for *S. aureus* specificity. There was some diffuse signal which was probably autofluorescence

2. It is recommended that chemical goggles, appropriate protective gloves, and masks are worn.
3. It is recommended to add the SDS last in the hybridization and washing buffer to avoid precipitation with the concentrated NaCl and to avoid foam formation.

Table 2
FISH probe and 16S target sequences and bacterial target

Probe	16S sequence	Target	References
Eub338	GCTGCCTCCCGTAGGAGT	All bacteria	Amann et al. [18]
NONEUB338	ACTCCTACGGGAGGCAGC	Nonsense sequence	Manz et al. [17]
Sta	TCCTCCATATCTCTGCGC	*Staphylococcus* spp.	Trebesius et al. [23]
Str	CACTCTCCCCTTCTGCAC	*Streptococcus* spp.	Trebesius et al. [23]
Pae997	TCT GGA AAG TTC TCA GCA	*Pseudomonas* spp.	Amann et al. [24]
Sau	GAAGCAAGCTTCTCGTCCG	*Staphylococcus aureus*	Kempf et al. [25]
Spn	GTG ATG CAA GTG CAC CTT	*Streptococcus pneumonia*	Kempf et al. [25]
PseaerA	GGT AAC CGT CCC CCT TGC	*Pseudomonas aeruginosa*	Hogardt et al. [26]
PAC 16S 598	GCC CCA AGA TTA CAC TTC CG	*Propionibacterium acnes*	Poppert et al. [27]
ENF 191	GAAAGCGCCTTTCACTCTTATGC	*Enterococcus faecalis*	Wellinghausen et al. [28]
TRE II	GCTCCTTTCCTCATTTACCTTTAT	*Treponema denticola*	Moter et al. [29]
pB 394	ATGCGGTCCAAAATGTTATCCGG	*Bacillus cereus*	Liu et al. [30]

4. Several fluorophores can be used during FISH, provided that the microscope is equipped with the appropriate set of excitation and emission filters. Fluorophores that have similar spectral characteristics cannot be used at the same time, but 2 morphologically distinct organisms can be marked with the same color.
5. Silicone sealer or a few drops of 10 % agarose dissolved in water can be used to attach the tissue section to the Petri dish before the attachment. It is always important to gently absorb the excess liquid from the tissue section using a paper towel. The tissue can be attached when the agarose is still soft and needs to dry very well before buffer is added for the microscope observation with a water immersion objective. The best product to use to obtain a firm attachment will depend on the kind and consistency of the tissue (moistness, texture, etc.).
6. If possible, prior testing should be performed to find the optimal time (in fact, a shorter time may better preserve the morphology of Gram-negative bacteria that might be present in unknown samples).

Table 3
Fluorescent probes commonly used showing the maximum excitation wavelength (Ex), the maximum fluorescence emission wavelength (Em), the confocal laser line used to excite the stain, the nominal false color that we commonly assign, and the target and/or conjugate. Our general convention was to image the surface of the implant/tissue/membrane with reflected light, normally depicted in *blue*, and the bacteria were normally shown in *red*, regardless of fluorophore

Fluorescent probe or imaging technique	Ex (nm)	Em (nm)	Laser (nm)	Color	Target and conjugate for FISH probes the name of the probe, i.e., "Sau"
General stains					
SYTO 9	483	500	488	Green	All bacteria and host nuclei (live bacteria when used in conjunction with propidium iodide).[a] In some cases, extracellular DNA (eDNA) in the EPS can be detected
Propidium Iodide	538	617	488 or 543 or 594	Red	All bacteria and host nuclei in fixed samples. Dead bacteria when used in conjunction with SYTO 9 on unfixed samples. In some cases, eDNA can be detected
SYTO 59	622	644	543	Red	All bacteria and host nuclei. In some cases, eDNA can be detected
Transmitted light	NA	NA	Any	Gray	Tissue components and bacteria
Reflected light	NA	NA	488	Blue	Surface of tissue, foreign body surfaces, and bacteria. In some cases, cytoskeleton is readily distinguishable
Cytology					
Phalloidin–Alexa488	493	519	488	Blue	Cytoskeleton—F-actin
FISH stains[a]					
Cy3	550	564	543	Green	*Streptococcus* genus "Str"
Cy3	550	564	543	Green	*S. pneumoniae* "Spn"
Cy3	550	564	543	Green	*Bacillus cereus*
Cy3	550	564	543	Green	*Staphylococcus* genus "Sta"
Cy5	649	666	633	Red	*S. aureus* "Sau"
Cy5	649	666	633	Red	*Pseudomonas* spp.
FAM-5 (pH 9)	492	518	488	Green	*Pseudomonas aeruginosa*
Cy3	550	564	543	Green	*Propionibacterium acnes*
Cy5	649	666	633	Red	*Enterococcus faecalis*
Cy3	550	564	543	Green	*Treponema denticola*
Cy3	550	564	543	Green	Nonsense probe (control) "NonEub"

[a]See table 2 for the probe sequence. We have probes conjugated to different colors allowing us to select an appropriate fluorescence with respect to other fluorophores being utilized to stain the same specimen

7. It is recommended that small aliquots (50 μl) of probe working solution be prepared. (Hybridization signals became dim and the background high when using a repeatedly thawed and frozen probe.)

8. The temperature is critical and must be accurate and steady. It is recommended that the temperature be checked periodically in both the hybridization oven and the water bath with a second mercury thermometer.
9. Transfer the samples rapidly from the hybridization oven to the washing buffer to prevent cooling that can lead to nonspecific probe binding.

Acknowledgments

We thank S. Conti, MD, G. Altman MD, D. Altman MD, and N. Sotereanos, MD, Orthopedic Department, Allegheny General Hospital, Pittsburgh, PA, for providing the samples; S. Kathju, MD, PhD, University of Pittsburgh School of Medicine, Pittsburgh, PA and from the Center for Genomic Sciences, Allegheny-Singer Research Institute, Pittsburgh, PA; G.D. Ehrlich, PhD, C.J. Post, MD, PhD, and J.W. Costerton, PhD, for protocol development and provision of resources; and Mary O'Toole for her help in the preparation of the manuscript.

References

1. Ulrich SD, Seyler TM, Bennett D et al (2008) Total hip arthroplasties: what are the reasons for revision? Int Orthop 32:597–604
2. Daigle ME, Weinstein AM, Katz JN et al (2012) The cost-effectiveness of total joint arthroplasty: a systematic review of published literature. Best Pract Res Clin Rheumatol 26: 649–658
3. Sadoghi P, Liebensteiner M, Agreiter M et al (2013) Revision surgery after total joint arthroplasty: a complication-based analysis using worldwide arthroplasty registers. J Arthroplasty 28:1329–1332
4. Gomez E, Cazanave C, Cunningham SA et al (2012) Prosthetic joint infection diagnosis using broad-range PCR of biofilms dislodged from knee and hip arthroplasty surfaces using sonication. J Clin Microbiol 50:3501–3508
5. Rasouli MR, Harandi AA, Adeli B et al (2012) Revision total knee arthroplasty: infection should be ruled out in all cases. J Arthroplasty 27:1239–1243
6. Stoodley P, Nistico L, Johnson S et al (2008) Direct demonstration of viable Staphylococcus aureus biofilms in an infected total joint arthroplasty. A case report. J Bone Joint Surg Am 90:1751–1758
7. Stoodley P, Conti SF, DeMeo PJ et al (2011) Characterization of a mixed MRSA/MRSE biofilm in an explanted total ankle arthroplasty. FEMS Immunol Med Microbiol 62:66–74
8. Palmer M, Costerton W, Sewecke J et al (2011) Molecular techniques to detect biofilm bacteria in long bone nonunion: a case report. Clin Orthop Relat Res 469:3037–3042
9. Hall Stoodley L, Stoodley P, Kathju S et al (2012) Towards diagnostic guidelines for biofilm-associated infections. FEMS Immunol Med Microbiol 65:127–145
10. Hall-Stoodley L, Costerton JW, Stoodley P (2004) Bacterial biofilms: from the environment to infectious disease. Nat Rev Microbiol 2:95–108
11. Post JC, Preston RA, Aul JJ et al (1995) Molecular analysis of bacterial pathogens in otitis media with effusion. JAMA 273: 1598–1604
12. Amann R, Ludwig V, Schleifer KH (1995) Phylogenetic identification and in situ detection of individual microbial cells without cultivation. Microbiol Rev 59:143–169
13. Hall-Stoodley L, Hu FZ, Gieseke A et al (2006) Direct detection of bacterial biofilms on the middle-ear mucosa of children with chronic otitis media. JAMA 296:202–211
14. Nistico L, Kreft R, Gieseke A et al (2011) Adenoid reservoir for pathogenic biofilm bacteria. J Clin Microbiol 49:1411–1420

15. Kathju S, Nistico L, Hall-Stoodley L et al (2009) Chronic surgical site infection due to suture-associated polymicrobial biofilm. Surg Infect (Larchmt) 10:457–461
16. Hall-Stoodley L, Nistico L, Sambanthamoorthy K et al (2008) Characterization of biofilm matrix, degradation by DNase treatment and evidence of capsule downregulation in Streptococcus pneumoniae clinical isolates. BMC Microbiol 8:173
17. Manz W, Amann R, Ludwig W et al (1992) Phylogenetic oligodeoxynucleotide probes for the major subclasses of proteobacteria: problems and solutions. Syst Appl Microbiol 15: 593–600
18. Amann RI, Krumholz L, Stahl DA (1990) Fluorescent-oligonucleotide probing of whole cells for determinative, phylogenetic, and environmental studies in microbiology. J Bacteriol 172:762–770
19. Pernthaler A, Pernthaler J, Amann R (2002) Fluorescence in situ hybridization and catalyzed reporter deposition for the identification of marine bacteria. Appl Environ Microbiol 68:3094–3101
20. Hugenholtz P, Gene WT, Blackall LL (2001) Design and evaluation of 16S rRNA-targeted oligonucleotide probes for fluorescence in situ hybridization. In: Lieberman BA (ed) Steroid receptors methods: protocols and assays. Humana Press, Totowa, NJ, pp 29–41
21. Thurnheer T, Gmür R, Guggenheim B (2004) Multiplex FISH analysis of a six-species bacterial biofilm. J Microbiol Methods 56:37–47
22. Nistico L, Gieseke A, Stoodley P et al (2009) Fluorescence in situ hybridization for the detection of biofilm in the middle ear and upper respiratory tract mucosa. Methods Mol Biol 493:191–215
23. Trebesius K, Leitritz L, Adler K et al (2000) Culture independent and rapid identification of bacterial pathogens in necrotising fasciitis and streptococcal toxic shock syndrome by fluorescence in situ hybridisation. Med Microbiol Immunol 188:169–175
24. Amann R, Snaidr J, Wagner M et al (1996) In situ visualization of high genetic diversity in a natural microbial community. J Bacteriol 178:3496–3500
25. Kempf VA, Tresbesius K, Autenrieth IB (2000) Fluorescent in situ hybridization allows rapid identification of microorganisms in blood cultures. J Clin Microbiol 38:830–838
26. Hodgart M, Trebesius K, Geiger AM et al (2000) Specific and rapid detection by fluorescent in situ hybridization of bacteria in clinical samples obtained from cystic fibrosis patients. J Clin Microbiol 38:818–825
27. Poppert S, Riecker M, Essig A (2010) Rapid identification of *Propionibacterium acnes* from blood cultures by fluorescence in situ hybridization. Diagn Microbiol Infect Dis 66: 214–216
28. Wellinghausen N, Bartel M, Essig A et al (2007) Rapid identification of clinically relevant *Enterococcus species* by fluorescence in situ hybridization. J Clin Microbiol 45: 3424–3426
29. Moter A, Leist G, Rudolph R et al (1998) Fluorescence in situ hybridization shows spatial distribution of as yet uncultured treponemes in biopsies from digital dermatitis lesions. Microbiology 144:2459–2467
30. Liu WT, Mirzabekov AD, Stalh DA (2001) Optimization of an oligonucleotide microchip for microbial identification studies: a non-equilibrium dissociation approach. Environ Microbiol 3:619

Chapter 9

Animal Models to Evaluate Bacterial Biofilm Development

Kim Thomsen, Hannah Trøstrup, and Claus Moser

Abstract

Medical biofilms have attracted substantial attention especially in the past decade. Animal models are contributing significantly to understand the pathogenesis of medical biofilms. In addition, animal models are an essential tool in testing the hypothesis generated from clinical observations in patients and preclinical testing of agents showing in vitro antibiofilm effect. Here, we describe three animal models — two non-foreign body *Pseudomonas aeruginosa* biofilm models and a foreign body *Staphylococcus aureus* model.

Key words Chronic lung infection, Chronic wound infection, *Pseudomonas aeruginosa*, *Staphylococcus aureus*, Animal models

1 Introduction

Biofilm infections have been shown or suggested to be present as chronic lung infections in cystic fibrosis, orthopedic alloplastic-related infections, vascular catheter-related infections, chronic infections in non-healing wounds, urinary tract catheter-related infections, endocarditis, etc. [1, 2]. Human biofilm infections are challenging due to difficulties in diagnosing (antibiotic treatments prior to sampling, adherent bacteria, dormant stages/persisters) and treating (antibiotic binding in biofilms, increased beta-lactamase production, upregulation of efflux pumps, inactive bacterial stages, different physiologic regions in the biofilms, increased MICs, increased mutation rates and hypermutators) making the biofilms recalcitrant to complete elimination by antibiotic treatments [1, 3]. In conjunction to the resilience to the host responses and the frequent consequent collateral tissue damages, biofilm infections demand further understanding and new additional treatment options [1]. Representative animal models are indispensible in this context. Establishing a representative model requires thoroughly clinical insight to avoid misinterpretations and to enable the model to answer the relevant questions. As an example cystic fibrosis is a complex multiorgan disease with pancreatic insufficiency,

Gianfranco Donelli (ed.), *Microbial Biofilms: Methods and Protocols*, Methods in Molecular Biology, vol. 1147,
DOI 10.1007/978-1-4939-0467-9_9, © Springer Science+Business Media New York 2014

diabetes mellitus, hepatic insufficiency, malabsorption, and male infertility besides the chronic lung infections. Obviously it is almost impossible to encompass all these parameters in a single model. Therefore, establishing an animal model should be designed to comprise the evaluation of all relevant parameters.

Endpoints can be chosen as clinical relevant in vivo measurements (weight loss, temperature, wound size, lung function testing, various illumination techniques (CT or MR scan or systems like the IVIS spectrum)). Mortality is not allowed as an intended endpoint in the majority of animal studies, and so-called human endpoints for euthanasia strategy to avoid unnecessary suffering of the animals are mandatory. The human endpoints can additionally serve as a clinical score system of the mice, thus providing alternative experimental endpoints. Typical endpoints of biofilm infection models after killing of an animal are quantitative bacteriology, in situ hybridization techniques to demonstrate the biofilms in relevant material, macroscopic and microscopic pathologies, inflammatory markers and measurements on adaptive immune responses.

Animal models increasingly involve larger animals, which possibly reflect the human physiology, and scientists are obliged to use the best suitable animal species. However, rodents especially mice are still the most used species in infection models especially due to abundant tools available for clinical relevant measurements and of course the relative low cost in purchasing and handling these animals. When establishing a mouse model, it is important to be aware that different inbred mouse strains may have significantly different courses of the same infection and immunologic responses to the same infectious challenge [4–6]. In the same line is the choice of infectious agent. Obviously it should be an agent relevant for the biofilm infection investigated (e.g., *P. aeruginosa* in chronic lung infection or wound infection models and *S. aureus* in orthopedic alloplastic-related infection models) and the model should be reproducible. Additionally, the imperative identification of biofilm-like structures in the animal tissue or on removed foreign body and recognition of the biofilm-like nature of the infection (chronic, relative resistant to the host response and antibiotic treatment) are specific mandatory characteristics for biofilm animal models. Finally, the choice of bacterial strain is a decisive point in establishing an animal infection model. Often the dilemma is the choice between a characterized often used strain for in vitro experiment and a selection of possibly more medical relevant clinical strains for the infection model being investigated—this is a crucial question with no clear answer. In many situations, however, a thorough characterization of a suitable clinical isolate is unachievable.

For the majority of biofilm infections, the patients have either a disease disposing for biofilm infections (e.g., genetic defect in cystic fibrosis or a physical condition resulting in the structural damage of the skin enabling for formation of the biofilm infection) or an implanted foreign body serving as adhesion material for the biofilm infection. Therefore, many animal models involve either a structural damage like creating a wound in the skin of the animal for chronic wound models or implantation of a foreign body like in orthopedic alloplastic models [7, 8]. Animal host response may challenge the establishment of infection when using planktonic bacteria as initial inoculum. The imitation of chronic lung infections in CF with inoculation of planktonic PA through the nasal or intratracheal route in rodents commonly results in a transient pulmonary colonization which is promptly cleared by host response. Thus, the manipulation may also be on the bacterial side by stressing the bacteria in minimal growth conditions to improve adhesion or embedding bacteria in a polysaccharide suspension to mimic bacterial growth in an extracellular matrix [9–11].

Although not comprehensive, the three animal models of biofilm infections in this chapter present different issues of biofilm infection models, including bacterial embedment, structural damage, or foreign body implantation, all disposing for biofilm formation. For further inspiration we would like to refer to a recent thorough and extensive review on biofilm models by Lebeaux and colleagues [2]. The three present animal models or modifications of what is presented here are meant as a guide on "how to do it" and have been useful in adding understanding to various topics within biofilm infections like impact of host responses, bacterial virulence, quorum sensing system, vaccinations, treatments (antibiotic and non-antibiotics), and immune modulations.

2 Materials

2.1 Chronic Lung Infection Model

2.1.1 Bead Preparation

1. Encapsulation Unit Nisco Var J30 (Nisco Engineering AG, Zurich, Switzerland) (*see* **Note 1**).
2. Syringe pump (Graseby 3100, Watford, UK).
3. *P. aeruginosa* strain PAO579 (*see* **Note 2**).
4. Protanal LF 10/60 dissolved in 0.9 % NaCl to an alginate concentration of 1 % and sterile filtered.
5. Gelling bath (0.1 M, pH 7.0 Tris–HCl buffer containing 0.1 M $CaCl_2$).
6. Magnetic stirrer (Fig. 1).

2.1.2 Infection Procedure

1. BALB/c female mice 11 weeks old.
2. Barrier facilities.
3. A 1:1 mixture of etomidate and midazolam.
4. Bead-tipped needle (20 G).

Fig. 1 Setup for making bacterial beads

2.2 Evaluation

1. Pentobarbiturate with lidocaine.
2. Homogenizer.
3. Sterile PBS.
4. 4 % W/V formaldehyde solution.
5. Agar plates.

2.3 Chronic Wound Infection Model

2.3.1 Burn Wound [12]

1. Female 12 week old C3H/HeN or BALB/c mice.
2. 1:1 mixture of etomidate and midazolam.
3. An electrical clipper to shave the hair on the back of each mouse.
4. A sledge.
5. Fire blanket with a window (1.7 × 2.6 cm) corresponding to 6 % of the total body area.
6. A metal plate with a window (1.7 × 2.6 cm).
7. Hot air blower.
8. Buprenorphine.
9. Electrical heating carpets.

2.3.2 Bead Preparation

Wild-type *P. aeruginosa* PAO1.

2.3.3 Evaluation

1. Pentobarbiturate with lidocaine.
2. Homogenizer.
3. Sterile PBS.

4. 4 % W/V formaldehyde solution.
5. Agar plates.

2.4 Chronic Osteomyelitis Model in Mice [13, 14]

2.4.1 Mice

1. Inbred C57BL/6 mice (6–8 weeks old) from Jackson Laboratories (Bar Harbor, ME).
2. Bacterial strains: different *S. aureus* strains including M2 [15] and USA300 LAC strain.

2.4.2 Infection Procedure

3. Scalpel.
4. Scalpel holder.
5. Forceps.
6. Needle driver.
7. Steel wire snips.
8. Scissors.
9. Stainless steel pins.
10. 0.25 mm insect pins.
11. Sterile sutures (4-0 coated vicryl).
12. Scale for measuring of weight.
13. Anesthetic—Ketaset and Xylazine.
14. Syringes (1 ml).
15. Needles for anesthesia.
16. 70 % alcohol.
17. Betadine iodine.
18. Alcohol swabs for superior cleansing action.
19. Sterile gauze sponges (4×4 cm).
20. 500 ml beaker.
21. Disposable gown without latex.
22. Respirators.
23. Latex gloves.
24. Isothermal pad (heat block).
25. Eye moisturizer.
26. Shaver.

2.4.3 Evaluation

1. Polytron PT 1200 handheld homogenizer.
2. 50 ml centrifuge tubes.
3. 70 % ethanol.
4. Sterile PBS.
5. Blood agar plates.
6. Sterile pipette tips.
7. Tubes for serial dilutions and plating.

3 Methods

3.1 Chronic Lung Infection Model

3.1.1 Bead Preparation

1. Bacterial isolate propagated from a freeze culture for 18 h and grown 18 h at 37 °C in Ox broth.
2. The overnight culture is centrifuged at 4 °C and 4,400 × *g* and the pellet re-suspended in 5 ml serum-bouillon.
3. The bacterial culture is diluted 1:20 in seaweed alginate solution.
4. The solution is transferred to a 10 ml syringe and placed in the syringe pump. The J30 uses a pressure chamber containing a needle that controls the flow of alginate. The pressure chamber is controlled by the pressure controlling unit. The pressure set point is fixed with a potentiometer. The J30 Unit is equipped with two connections one for the alginate and one for the airflow that drives the alginate from the needle through the exit orifice into the gelling bath (0.1 M, pH 7.0 Tris–HCl buffer containing 0.1 M $CaCl_2$). A magnetic stirrer is placed underneath the gelling bath to prevent the beads from sticking together during gelling. Distance between nozzle and gelling bath of 11 cm and 280 rpm magnetic stirrer is kept constant. Five mL of alginate beads is made. The beads are left for stabilization at magnetic stirring in the gelling bath for 1 h.
5. The beads are washed two times in 0.9 % NaCl containing 0.1 M $CaCl_2$. After wash 20 ml 0.9 % NaCl containing $CaCl_2$ is added. Serial dilutions of bead solution are made and cultured on a modified Conradi-Drigalski medium, selective for gram-negative rods. After overnight incubation at 37 °C, the number of colony-forming units (CFU) is determined and used for adjustment of challenge solution.
6. Mice are allowed to acclimatize for 1 week before use and should have free access to chow and water and under observation of trained personnel.
7. Mice are anesthetized s.c. with 1:1 mixture of etomidate and midazolam (10 ml/kg body weight) and tracheotomized with a 23 G needle after incision with a scissor below the pharynx region.
8. 0.04 ml of seaweed alginate beads embedded with *P. aeruginosa* PAO579 is installed in the left lung of BALB/c mice using a bead-tipped needle (20 G).
9. Mice are sacrificed using an overdose of pentobarbiturate.
10. Half the number of lungs are collected aseptically and transferred to 5 ml of sterile PBS and kept on ice until homogenization for quantitative bacteriology. Homoginates are serially diluted and 100 µl are plated on agar plates (Conradi-Drigalski medium). The number of CFU determined after overnight

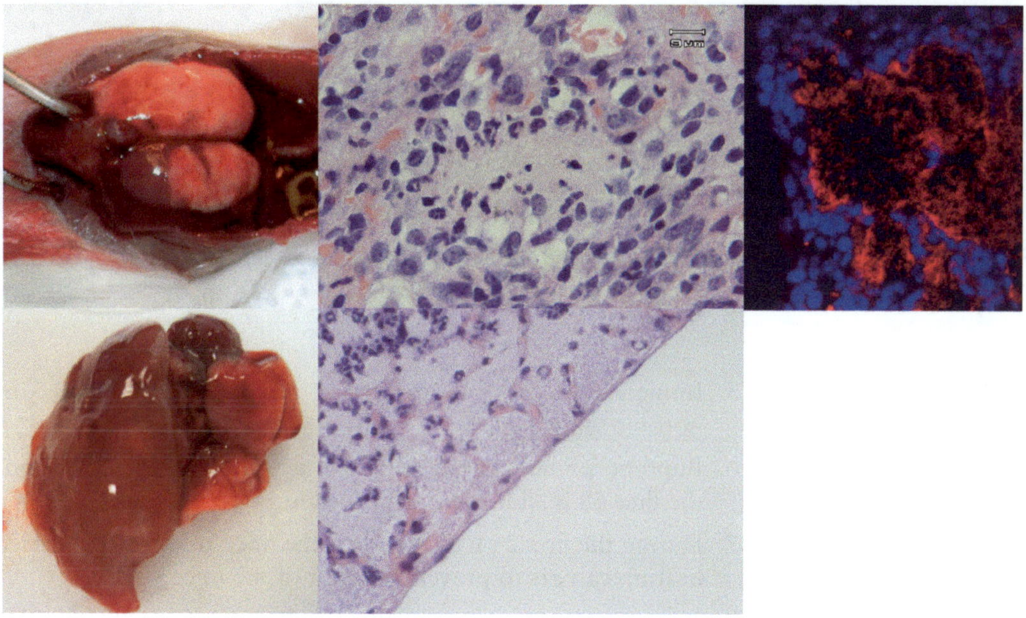

Fig. 2 Macroscopic and microscopic appearances of chronic *P. aeruginosa* lung infection in mice

incubation is presented as log CFU per lung by multiplication with dilution factor.

11. Affected lung parts from the remaining number of mice are fixed in a 4 % W/V formaldehyde solution. For evaluation *see* **Note 3** (Fig. 2).
12. The lung homogenates are centrifuged at 4,400 × *g* for 10 min and the supernatants isolated and kept at −70 °C until cytokine analysis.

3.2 Chronic Wound Infection Model

1. The animals are anesthetized subcutaneously (sc) with a 0.25 ml 1:1 mixture of etomidate and midazolam (10 ml/kg body weight). BALB/c mice are relatively susceptible in this model compared to the more resistant C3H/HeN mouse strain [11].
2. The mice are placed on a sledge and covered with a fire blanket with a window (1.7 × 2.6 cm) corresponding to 6 % of the total body area (*see* **Note 4**).
3. Above the fire blanket a metal plate with a window (1.7 × 2.6 cm) is placed.
4. The sledge with the mouse, the fire blanket, and the metal plate are moved into a stream of hot air with a temperature of 330 °C delivered by a hot air blower for 5 s. This procedure results in a third-degree burn confirmed by histological examination.

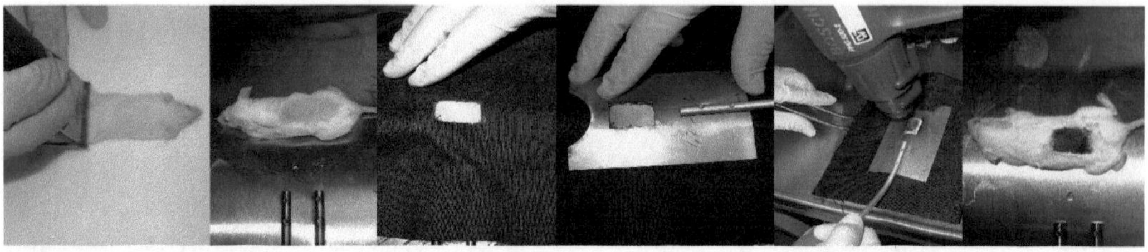

Fig. 3 Burn wound infliction. The structural damage enabling establishing a biofilm infection

5. Immediately after the procedure, the mice are given 0.5 ml isotonic saline s.c.
6. Buprenorphine is given every 8 h as pain relief therapy during the first 48 h after the burn procedure.
7. During the first 24 h of recovery, the mice are kept at electrical heating carpets to prevent cooling of the mice.
8. Mice are injected s.c. with 100 μl challenge solution (10^7 CFU/ml) or PAO1-free alginate beads beneath the burn wound 2–4 days after infliction of thermal lesion (Fig. 3).
9. Mice are sacrificed by intraperitoneal injection of a pentobarbiturate overdose.
10. Quantitative bacteriology: Swabs from beneath the eschars, in both uninfected and infected groups of mice, are carried out immediately after sacrificing the mice and are plated onto modified Conradi-Drigalski medium, selective for Gram-negative rods. Growth of bacteria is noted and presented as +/− presence of *P. aeruginosa*. Additionally, wounds are placed in 4 ml sterile isotonic saline and kept on ice until they are homogenized at 14,000 × *g* for 45 s. Serial dilutions of wound homogenates are prepared in saline, and aliquots of 0.1 ml are spread onto modified Conradi-Drigalski medium. Homogenates are centrifuged at 5,000 × *g* for 15 min, and the supernatants are isolated and kept at −80 °C until cytokine analysis. Counting of the number of CFUs is carried out the following day, and the actual number of CFU in the samples is calculated by multiplication with the dilution factor.
11. Histopathology: Wound tissue is carefully removed in toto by scalpel and fixed in formaldehyde; histological slides are prepared by H&E staining. The degree and type of inflammation are estimated as described previously [12] by a trained pathologist and are scored by index as no inflammation (−) to severe inflammation (+++) and acute (dominated by polymorphonuclear neutrophils, PMN), chronic (dominated by mononuclear cells, MN), or neither type dominating (PMN/MN), respectively (Fig. 4).

a

Group	Day 1	Day 4	Day 7
Alginate − PAO1	0/6	0/6	0/5
Alginate + PAO1	6/6	5/5	6/6

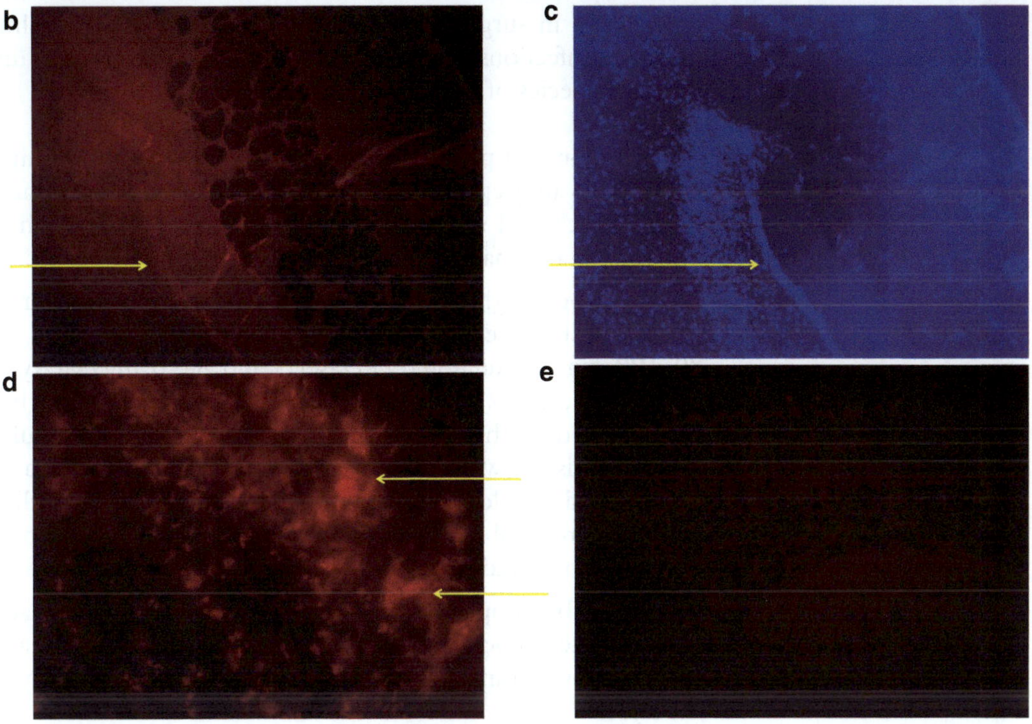

Fig. 4 *Pseudomonas aeruginosa* biofilm-like structures in chronic wounds

3.3 Chronic Osteomyelitis Model

3.3.1 Infection Preparation

Pathogenic challenge will be initiated via a contaminated 0.25 mm diameter stainless steel insect pin that will be generated as follows:

1. The pins will be autoclaved and stored in 70 % ethanol. Pins are then dried.
2. For infections where an early biofilm is first grown on the pins prior to implantation, the dried pins are incubated in a 1:100 dilution of overnight TSB culture of *S. aureus* in fresh Luria broth for 2–24 h at 37 °C. The inoculating dose of bacteria has been determined to be 1×10^6 CFU of *S. aureus* per pin. The pins will be air dried for 5 min before trans-tibial surgical implantation. This procedure is often required to ensure infection with strains other than *S. aureus*, including *Streptococcus pyogenes*, *Staphylococcus epidermidis*, *Acinetobacter baumannii*, *Klebsiella pneumoniae*, and *Pseudomonas aeruginosa*.
3. The second pathogenic challenge mechanism mimics clinical cases of intra/perioperative contamination. In this procedure, a sterile stainless steel insect pin (sterilized as previously indicated) will be used for the trans-tibial implantation first followed by a

S. aureus inoculum that is equivalent to $10 \times 10^{2-3}$ CFUs that is directly applied to the pin as a 1 μl microbial suspension dose in PBS post-implantation. This second challenge mechanism is a better representation of bacterial colonization and infection progression in surgical patients that develop chronic biofilm-mediated infections but is often not a high enough inoculum for other species of bacteria.

3.3.2 Infection Procedure

1. The surgeon will prepare for surgery by thoroughly washing hands with an aseptic surgical scrub as well as wearing individually packaged sterile surgical gloves, the appropriate hair protection, and mask.
2. Mice will be weighed prior to initiating surgical procedures. Mice will be anesthetized by intraperitoneal injection of 80–100 mg/kg ketamine (Ketaset®) and 8–10 mg/kg xylazine using a 25–27 gauge needle. After anesthetization has been verified by the lack of response to toe pinch or stimulation of vibrissae (whiskers), mice will be shaved in the area to be operated on (lower left leg) using battery-operated clippers. The area will be surgically prepared using three alternating scrubs of Betadine and 70 % alcohol.
3. For trans-tibial implantation, a small 0.5 cm incision is made with a sterile scalpel in the left lower leg (medial tibia) of each mouse, about 5 mm above the ankle and just below (1–2 mm) the knee.
4. For infection with an early biofilm-coated pin, one bacterial-coated pin following incubation in bacterial culture for 2–24 h or sterile pin controls will be placed transcortically through the left tibia via medial to lateral implantation. The pin will be bent at both ends for stability and cut adjacent to the skin on both ends, which will allow it to be covered by the skin and eliminate the risk of additional environmental exposure. The ends of the pin will be cut as blunt as possible, to allow for easy placement of skin and muscle back over the ends on the implanted pin and minimal risk of environmental exposure after surgery.
5. For infections mimicking intra/perioperative infection, the sterile pins will be implanted, bent, and cut followed by a 1 μl dose applied directly to the pin post-implantation.
6. Surgical incisions will be closed using sterile, size 3-0 to 5-0 monofilament sutures.
7. During the procedure mice will be kept warm via the use of a heat lamp. A thermometer will be placed at the level of the mouse to ensure that the temperature stays between 80 and 85 °F. The lamp position will be adjusted accordingly to maintain proper temperature.

8. Postoperative monitoring (particularly the respiratory rate) will take place every 2–5 min during the procedure, then every 10–15 min during recovery (until righting response has returned). During recovery from anesthesia, mice will be warmed with supplemental heat from an incandescent heat lamp. Mice will also be treated with buprenorphine (0.05–0.1 mg/kg subcutaneously twice daily, using a 25–27 gauge needle) for 48 h to treat postoperative pain, with the first dose being given at the time of ketamine/xylazine administration to ensure that analgesia is in place at the onset of surgical stimulation.

3.3.3 Evaluation

1. Animals will be anesthetized and euthanized by exsanguination. Mice will be anesthetized by an intraperitoneal injection of 100 mg/kg ketamine and 10 mg/kg xylazine. Exsanguination will be performed via cardiac puncture, utilizing a 1.0 ml syringe and 27-gauge needle. Cervical dislocation will be performed after exsanguinations in order to confirm that the mice are deceased.
2. Following euthanization, we will harvest and process the spleens, draining lymph nodes, and tibias/pin implants. All soft tissue is removed from the tibia and the bone is weighed. PBS is added to the bone in a 3:1 ration (ml PBS: g bone) with protease inhibitor cocktail as described in Materials for cytokine analysis.
3. In order to get representative colony-forming unit counts from biofilms, it is important to break up the biofilm conglomerates by homogenization [16, 17]. Using a Kinematica Polytron P1200E handheld homogenizer at maximum speed (30,000 × *g*), first, disinfect with 70 % ethanol and then wash twice in sterile PBS, followed by homogenization for 1 min on ice. Immediately serially dilute and plate for CFU determination and store on ice. Once serial dilutions have been accomplished, spin down bone fragments and slurry and harvest supernatant if cytokine levels are to be determined.

4 Notes

1. The alginate enters through a central needle. The exit nozzle, which is centrally in line with the axis of the needle, has been countersunk externally. The countersunk leads to the aerodynamical effect so that the jet has a smaller diameter when passing the nozzle than before at the needle. The needle is enclosed in a pressure chamber with an exit through the orifice. The size of the drop is determined by the nozzle size, the product flow rate, and the pressure inside the chamber. The product

flow rate is controlled by a syringe pump to be connected to the product nozzle. The pressure chamber is controlled by the pressure controlling unit. The pressure set point is fixed with a potentiometer. For the small beads we use the 0.250 mm nozzle, an alginate flow rate 20 ml/h, and the airflow 105 mBar. For the large beads the 0.500 mm nozzle, alginate flow rate 60 mL/h, and airflow 35 mBar are used. Diameters for large beads are 136 μm (range 74–205 μm) and 40 μm (range 15–85 μm) for small beads.

2. *P. aeruginosa* PAO 579 (IATS O:2/5) was kindly provided by J. R. W. Gowan, Department of Bacteriology, Medical School, University of Edinburgh, UK. In other experiments clinical isolates have been used [18, 19].
3. The fixed lungs are embedded in paraffin wax and cut into 5 μm thick sections, followed by hematoxylin and eosin staining. The entire lung slides and is scanned at a low magnitude, and, from an average evaluation of a minimum of five representative areas at higher magnitude (400× or 1,000×), the type and degree of lung inflammation is estimated. The type of inflammation is categorized as acute type (>90 % PMNs), chronic type (>90 % mononuclear cells (MNs)), both types present, neither dominating (PMN/MN) or no inflammation (NI). The degree of inflammation is scored on a scale from 0 to 3 +, where 0 means no inflammation, + means mild focal inflammation, ++ mean moderate to severe focal inflammation, and +++ means severe inflammation to necrosis, or severe inflammation throughout the lung. Finally, the localization of the inflammation in the airway lumen or parenchyma is noted.

   To confirm the nature of the biofilm-like structures in the airways and infected tibias, deparaffinized tissue sections are analyzed by fluorescence in situ hybridization (FISH) using peptide nucleic acid (PNA) probes. A mixture of Texas Red-labeled, *P. aeruginosa*-specific PNA probe and fluorescein isothiocyanate (FITC)-labeled, universal bacterium PNA probe in hybridization solution is added to each section and hybridized in a PNA FISH Workstation at 55 °C for 90 min covered by a lid. The slides are washed for 30 min at 55 °C in wash solution. VECTASHIELD mounting media with 4′,6-diamidino-2-phenylindole (DAPI) is applied, and a cover slip is added to each slide. Slides are read using a fluorescence microscope equipped with a FITC, a Texas Red, and a DAPI filter.
4. The burn surface is calculated by using the Meeh formula: $A = KW^{2/3}$ where A = body surface area, K = 9, and W = weight in grams.

Acknowledgements

Mark Shirtliff contributed fully as a co-author for this chapter and should be regarded as such.

References

1. Høiby N, Ciofu O, Johansen HK et al (2011) The clinical impact of bacterial biofilms. Int J Oral Sci 3:55–65
2. Lebeaux D, Chauhan A, Rendueles O, Beloin C (2013) From in vitro to in vivo models of bacterial biofilm-related infections. Pathogens 2:288–356
3. Hall-Stoodley L, Stoodley P, Kathju S et al (2012) Towards diagnostic guidelines for biofilm-associated infections. FEMS Immunol Med Microbiol 65:127–145
4. Moser C, Jensen PO, Kobayashi O et al (2002) Improved outcome of chronic *Pseudomonas aeruginosa* lung infection is associated with induction of a Th1-dominated cytokine response. Clin Exp Immunol 127:206–213
5. Moser C, Hougen HP, Song Z et al (1999) Early immune response in susceptible and resistant mice strains with chronic *Pseudomonas aeruginosa* lung infection determines the type of T-helper cell response. APMIS 107: 1093–1100
6. Prabhakara R, Harro JM, Leid JG et al (2011) Suppression of the inflammatory immune response prevents the development of chronic biofilm infection due to methicillin-resistant *Staphylococcus aureus*. Infect Immun 79:5010–5018
7. Trøstrup H, Thomsen K, Christophersen LJ et al (2013) *Pseudomonas aeruginosa* biofilm aggravates skin inflammatory response in BALB/c mice in a novel chronic wound model. Wound Repair Regen 21:292–299
8. Brady RA, Leid JG, Calhoun JH et al (2008) Osteomyelitis and the role of biofilms in chronic infection. FEMS Immunol Med Microbiol 52:13–22
9. Christensen LD, Moser C, Jensen PO et al (2007) Impact of *Pseudomonas aeruginosa* quorum sensing on biofilm persistence in an in vivo intraperitoneal foreign body infection model. Microbiology 153:2312–2320
10. Pedersen SS, Shand GH, Hansen BL, Hansen GN (1990) Induction of experimental chronic *Pseudomonas aeruginosa* lung infection with *P. aeruginosa* entrapped in alginate microspheres. APMIS 98:203–211
11. Moser C, Johansen HK, Song Z et al (1997) Chronic *Pseudomonas aeruginosa* lung infection is more severe in Th2 responding BALB/c mice compared to Th1 responding C3H/HeN mice. APMIS 105:838–842
12. Calum H, Moser C, Jensen PO et al (2009) Thermal injury induces impaired function in polymorphonuclear neutrophil granulocytes and reduced control of burn wound infection. Clin Exp Immunol 156:102–110
13. Li D, Gromov K, Soballe K et al (2008) Quantitative mouse model of implant-associated osteomyelitis and the kinetics of microbial growth, osteolysis, and humoral immunity. J Orthop Res 26:96–105
14. Prabhakara R, Harro JM, Leid JG et al (2011) Murine immune response to a chronic *Staphylococcus aureus* biofilm infection. Infect Immun 79:1789–1796
15. Harro JM, Daugherty S, Bruno VM et al (2013) Draft genome sequence of the methicillin-resistant *Staphylococcus aureus* isolate MRSA-M2. Genome Announc 1:e00037-12
16. Fux CA, Wilson S, Stoodley P (2004) Detachment characteristics and oxacillin resistance of *Staphylococcus aureus* biofilm emboli in an in vitro catheter infection model. J Bacteriol 186:4486–4491
17. Brady RA, O'May GA, Leid JG et al (2011) Resolution of *Staphylococcus aureus* biofilm infection using vaccination and antibiotic treatment. Infect Immun 79:1797–1803
18. Moser C, Van Gennip M, Bjarnsholt T et al (2009) Novel experimental *Pseudomonas aeruginosa* lung infection model mimicking long-term host-pathogen interactions in cystic fibrosis. APMIS 117:95–107
19. van Gennip M, Moser C, Christensen LD et al (2009) Augmented effect of early antibiotic treatment in mice with experimental lung infections due to sequentially adapted mucoid strains of *Pseudomonas aeruginosa*. J Antimicrob Chemother 64:1241–1250

Chapter 10

Animal Models to Investigate Fungal Biofilm Formation

Jyotsna Chandra, Eric Pearlman, and Mahmoud A. Ghannoum

Abstract

Microbial biofilms play an essential role in several infectious diseases and are defined as extensive communities of sessile organisms irreversibly associated with a surface, encased within a polysaccharide-rich extracellular matrix (ECM), and exhibiting enhanced resistance to antimicrobial drugs. Forming a biofilm provides the microbes protection from environmental stresses due to contaminants, nutritional depletion, or imbalances, but is dangerous to human health due to their inherent robustness and elevated resistance.

The use of indwelling medical devices (e.g., central venous catheters, CVCs) in current therapeutic practice is associated with 80–90 % of hospital-acquired bloodstream and deep tissue infections. Most cases of catheter-related bloodstream infections (CRBSIs) involve colonization of microorganisms on catheter surfaces where they form a biofilm. Additionally, *Fusarium solani* and *F. oxysporum* were the causative organisms of the 2005/2006 outbreak of contact lens-associated fungal keratitis in the United States, Europe, the UK, and Singapore, and these infections involved formation of biofilms on contact lens. Fungal biofilm formation is studied using a number of techniques, involving the use of a wide variety of substrates and growth conditions. In vitro techniques involving the use of confocal scanning laser/scanning electron microscopy, metabolic activity assay, dry weight measurements, and antifungal susceptibility assays are increasingly used by investigators to quantify and evaluate biofilm morphology. However, there are not many in vivo models used to validate biofilm-associated infections. In this protocol, we describe a clinically relevant rabbit model of *C. albicans* biofilm-associated catheter infection to evaluate the morphology, topography, and architecture of fungal biofilms. We also describe a murine model of contact lens-associated *Fusarium* keratitis.

Evaluation of the formation of fungal biofilms on catheters in vivo, their analysis using scanning electron microscopy (SEM) and quantitative catheter culture (QCC), and treatment of biofilms using antimicrobial lock therapy can be completed in ~20–25 days using the described methods. The rabbit model has utility in evaluating the efficacy of lock solutions. In addition, the murine model of contact lens-associated *Fusarium* keratitis enables characterizing/comparing the formation of *Fusarium* biofilms on contact lenses in vitro and determining their role in vivo.

Key words Animal model, Intravascular devices, Catheters, Biofilms, Fungus

Gianfranco Donelli (ed.), *Microbial Biofilms: Methods and Protocols*, Methods in Molecular Biology, vol. 1147, DOI 10.1007/978-1-4939-0467-9_10, © Springer Science+Business Media New York 2014

1 Introduction

Most of the nosocomial infections are device-related, particularly central venous catheter (CVC)-related, with up to 25–30 % resulting in infection-related mortality. *Candida* species are the fourth most common cause of these nosocomial bloodstream infections [1]. Most cases of catheter-related bloodstream infections (CRBSIs) involve the colonization of microorganisms on catheter surfaces where they eventually become embedded in a biofilm [2–5]. Additionally, *Fusarium solani* and *F. oxysporum* were the causative organisms of 2005/2006 outbreak of contact lens-associated fungal keratitis in the United States, Europe, the UK, and Singapore and involved biofilm formation on contact lens [11, 18 and 19]. Other organisms most commonly associated with CVC biofilms are *Staphylococcus aureus*, coagulase-negative staphylococci, aerobic gram-negative bacilli, and *C. albicans* [6]. Treatment guidelines for infections related to intravascular catheters suggest removal of the affected devices [6, 7]. However, removal of CVCs is not always easy or feasible (e.g., for patients with coagulopathy or limited vascular access) and is associated with healthcare expenses and complications due to catheter removal. Therefore, biofilm-associated CRBSIs represent an important problem, underscoring the need to find novel clinically relevant ways to prevent and treat such infections.

In vitro fungal biofilm formation is studied using a number of techniques, which may help in our understanding of their biology as well as identify potential antibiofilm agents. However, the clinical relevance of biofilms formed on catheters has not been extensively studied in vivo. Our CVC-associated *C. albicans* rabbit model was the first clinically relevant animal model that evaluated the effectiveness of an antifungal agent (liposomal amphotericin B [lipoAmB]) as a lock therapy [8, 10] for the treatment of biofilms formed intraluminally on catheters. In this study, we used quantitative catheter culture (QCC) and scanning electron microscopy (SEM) to study the antibiofilm effect of antifungals. QCC results showed zero colony-forming units (CFUs) for the catheters treated with liposomal amphotericin B compared to untreated controls and fluconazole-treated groups. SEM revealed abundant biofilm in the control and fluconazole groups, while inserted catheters from the liposomal amphotericin B treated group were cleared [8, 10]. Quantification of in vivo catheter biofilms is commonly performed using QCC method. In our studies, we optimized and used the QCC assay to quantify biofilms formed on catheters in vivo, as described below.

In this protocol, we describe our rabbit model and provide detailed step-by-step procedure for in vivo catheter placement, inoculation of catheter to form biofilm, catheter removal, quantitative catheter culture of biofilms formed, and SEM of catheter segments to evaluate biofilm surface topography [8, 10]. In addition, we provide a description of how this model can be used to evaluate the efficacy of antifungal lock therapy. Evaluation of the formation of fungal biofilms on catheters in vivo, their analysis using SEM and QCC, and treatment of biofilms using antimicrobial lock therapy can be completed using the described methods in ~20–25 days.

In addition, we describe a murine model of contact lens-associated *Fusarium* keratitis, which characterizes biofilm formation on contact lenses in vitro and examines the effect of biofilm on contact lens-associated *Fusarium* keratitis in vivo [11]. The methods described here are adapted from protocols described by Schinabeck et al. [8], Chandra et al. [10], and Sun et al. [11] and can be completed in a typical laboratory setting.

2 Materials

2.1 Reagents for Rabbit Model (Adapted from [8, 10])

2.1.1 General Reagents

1. Phosphate-buffered saline 1× without calcium and magnesium.
2. Milli-Q water. If not available, deionized or double-distilled water can also be used.

2.1.2 Reagents for Catheter Placement, Inoculation, and Removal in Rabbits

1. Female New Zealand White rabbits weighing 2.5–3.0 kg.
2. *Candida albicans* M61.
3. Ketaset (Ketamine HCl Injection 100 mg/mL).
4. AnaSed Injection® (xylazine 100 mg/mL).
5. 1-mL syringe with 27½-gauge needle.
6. 3-mL syringe.
7. 22-gauge 1-in. needles.
8. 22-gauge 1½-in. needles.
9. 25-gauge 5/8-in. needles.
10. 70 % ethyl alcohol.
11. Silastic tubing (0.04-in. internal diameter and 0.085-in. external diameter).
12. Polyethylene cuff.
13. 0.9 % Sodium Chloride Injection.
14. Heparin.
15. Heparinized saline: 1 mL of heparin; 1,000 U/mL is drawn up and injected into a vial containing 9 mL of 0.9 % Sodium Chloride Inj.
16. Betadine scrub.
17. Betadine solution.

2.1.3 Reagents for Rabbit Euthanasia

1. Euthasol.

2.1.4 Quantitative Catheter Culture Reagents

1. Sabouraud dextrose.
2. Sabouraud dextrose agar (SDA) (*see* **Note 1**).
3. Sterile Petri dishes.
4. 0.125 g/L chloramphenicol.
5. 0.03 g/L gentamicin sulfate salt.

2.1.5 Scanning Electron Microscopy Reagents

1. 8 % glutaraldehyde solution. Mix 10 mL of 8 % glutaraldehyde with 10 mL of sterile Milli-Q water (*see* **Note 2**).
2. Sodium cacodylate (0.2 M). Dissolve 21.4 g sodium cacodylate in 500 mL Milli-Q water. Adjust pH to 7.4 with HCl (6 N) and autoclave the solution for 15 min. Sodium cacodylate wash buffer (0.1 M) (*see* **Note 2**).
3. 1 % (aqueous) uranyl acetate. Dissolve 1 g in 100 mL Milli-Q water and filter through filter-sterilizing assembly using 0.22-μm filter in dark (*see* **Note 3**).
4. 1 % (aqueous) tannic acid. Dissolve 1 g tannic acid in 100 mL Milli-Q water and filter-sterilize using a filtration assembly fitted with a 0.22-μm-diameter (pore size) filter (*see* **Note 4**).
5. Osmium tetroxide 4 % aqueous solution (OsO_4). 1 % OsO_4 solution in 0.1 M sodium cacodylate. Mix 10 mL of 4 % OsO_4 with 10 mL of sterile Milli-Q water and 10 mL of 0.2 M sodium cacodylate. Final working solution is 1 % OsO_4 in 0.1 M sodium cacodylate (*see* **Note 2**).
6. Ethyl alcohol 200 proof, absolute. Ethanol working solutions. Prepare 25 %, 50 %, 75 %, 95 %, and 100 % ethanol in autoclaved Milli-Q water by making up 25 mL, 50 mL, 75 mL, and 95 mL ethanol to 100 mL, respectively; 100 % is used pure with no water added (*see* **Note 5**).

2.1.6 Antifungal Lock Experiment and Catheter Treatment Reagents

1. Liposomal amphotericin B. Reconstitute liposomal amphotericin B in sterile water and dilute with 5 % sterile dextrose.
2. Fluconazole. Reconstitute and dilute with sterile water.
3. Blood cultures from the catheter are sent to the Microbiology Department and processed with the BacT/ALERT® 3D Microbial Detection System to confirm the presence of yeast.

2.2 Equipment

1. Ultrasonicator.
2. Autoclave.
3. Spectrophotometer.
4. Lazy-L Spreaders.
5. Cell scrapers.

6. Scalpels.
7. Tweezers.
8. Filter sterilization assembly.
9. Glass microanalysis filter holders.
10. Durapore 0.22-μm membrane filters.
11. 50-mL polypropylene centrifuge tubes.
12. BD Falcon tissue culture plates (12-well; 6.0 mL).
13. Fisherbrand* Petri Dishes with Clear Lids (100 mm × 15 mm).
14. Fisherbrand* Media-Miser* Dishes (60-mm × 15-mm small Petri dishes).
15. Microcentrifuge tubes.
16. Rocker.
17. Incubator.
18. Scanning electron microscopy (SEM) specimen mount stubs.
19. Pelco tabs, SEM adhesive tape.
20. Glass slides.
21. Glass cover slips.
22. Parafilm.
23. Water purification apparatus.
24. Scanning electron microscope.
25. Sputter coating machine.

2.3 Reagents and Equipments for Fusarium Keratitis Murine Model (Adapted from [11, 16])

1. *Fusarium oxysporum* FOCS3-a strain (MRL8996). Other isolates can be used, but the commercially available strain from CDC does not form biofilm [16, 17].
2. Lotrafilcon A contact lenses although *Fusarium* also form biofilm on other lenses [17].
3. Tetrazolium XTT for live organisms.
4. Concanavalin Alexa Fluor 488 conjugate.
5. C57BL/6 mice (6–8 weeks old).
6. Mixer Mill MM300.
7. 2,2,2-Tribromoethanol : 1 mL of tert-amyl alcohol added to 1 g of 2,2,2-tribromoethanol and then added 48 mL of 0.9 % sodium chloride.
8. 10 % buffered formalin phosphate.
9. Paraffin.
10. 1 % periodic acid.
11. Schiff's reagent.
12. Gill's hematoxylin.

13. Sandwich ELISAs.
14. Microplate reader.
15. Algerbrush to abrade the corneal epithelium.

3 Methods

3.1 Rabbit Model

3.1.1 Preparation of the Rabbit for Surgery and Operative Procedures for Catheter Insertion (Timing: Day 8, After Acclimatizing Rabbits 0–7 Days)

1. Acclimatize rabbits for 7 days prior to surgery (*see* **Note 6**).
2. Anesthetize rabbits intramuscularly with ketamine 70 mg/kg and xylazine 7 mg/kg (*see* **Note 7**).
3. Shave the right cervical, shoulder, and scapular regions with electric clippers.
4. Wash the shaved area with Betadine soap followed by isopropyl alcohol and Betadine solution.
5. Place the rabbit on the surgery table in dorsal recumbency, with the scrubbed area covered in gauze.
6. Make a "pillow" of sterile gauze and place under the neck for support.
7. Tape forearms and jaw loosely to the table.
8. Cover the animal with sterile drape leaving only the incision site open. Make 1–2-cm incision in the right anterolateral cervical region exposing the external jugular vein.
9. Free a segment of the external jugular vein, just distal to the bifurcation of the internal and external maxillary veins from subcutaneous fat and place an 18″ × ¼″ Penrose drain under the vein (Fig. 1a).
10. Place two segments of 3-0 Vicryl suture proximally and distally to the Penrose drain.
11. Flush the catheter with sterile saline and clamp it with a hemostat.
12. Using a #11 scalpel blade, make an incision in the isolated segment of vein which controls the bleeding with upward traction of the Penrose drain (*see* **Note 8**).
13. Insert the catheter into the vein caudally 4 cm, up to the cuff (Fig. 1b) placing the catheter tip in the right anterior vena cava as demonstrated in the venogram in Fig. 1g (*see* **Note 9**).
14. Tie the proximal and distal ligatures and withdraw blood to test catheter patency (Fig. 1c) (*see* **Note 10**).
15. Flush the catheter with heparinized saline.
16. Create a subcutaneous tunnel by passing a hemostat cephalad through a 1.0-cm incision in the intracapsular region to the external jugular vein incision site.

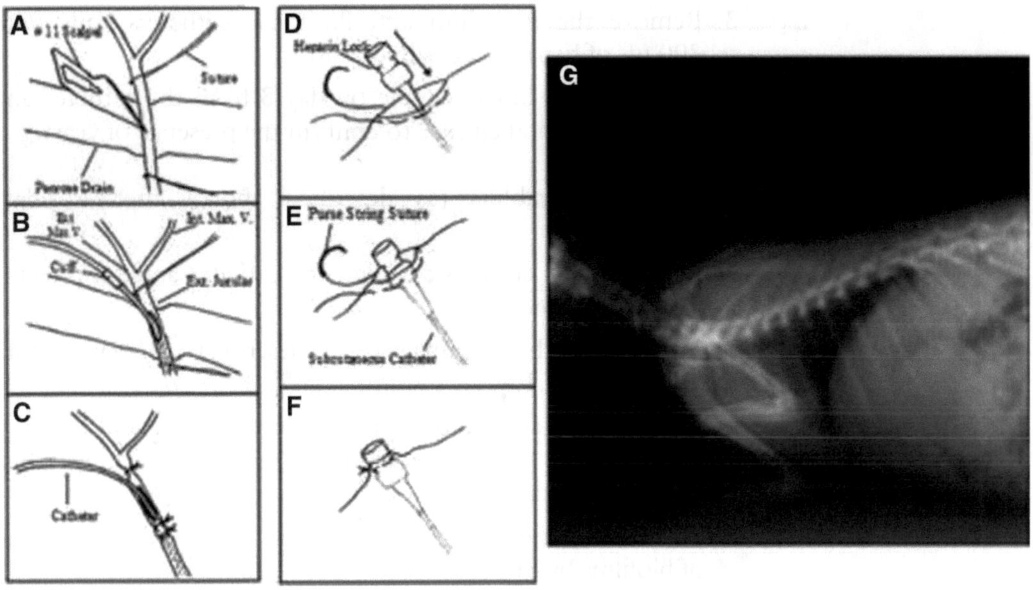

Fig. 1 Surgical placement of the intravenous catheter. (**a–c**) Catheter insertion into the external jugular vein, (**d–f**) attachment of the heparin lock device to skin, and (**g**) postoperative venogram of catheter placement. Adapted from [8]

17. Use the hemostat to pull the catheter through the subcutaneous tunnel, cut the excess catheter, and place an 18-gauge Luer Stub Adapter and sterile heparin lock device on the proximal end.
18. Test the catheter patency again by withdrawing blood and flushing with heparinized saline.
19. Bury the Luer hub in the subcutaneous tract and use a purse-string suture to attach the heparin lock device flush to the skin (Fig. 1d–f).

3.1.2 Postsurgical Procedures

1. Dress the rabbit in an orthopedic stockinette sleeve with arm holes cut.
2. Wrap the rabbit loosely in blankets, place in a heated recovery cage, and monitor every 30 min.
3. When the rabbit is able to sit sternally, return it to its cage.

3.1.3 Catheter Inoculation (Timing: Day 8)

1. *Candida albicans* isolate M61, obtained from a central venous catheter tip of patient with catheter-associated candidiasis, is used for inoculating catheters. From the frozen stock, subculture *C. albicans* cells and place appropriate volume of *Candida* cell suspension in saline to obtain a cell density of 10^7 cells in 300 μL inoculum. This is the primary inoculum used for inoculating catheters.
2. Lock lumen of catheters with 300 μL of 1×10^7 CFU of *C. albicans* and allow it to dwell for 24 h.

3. Remove the inoculum and flush the catheters daily with 300 μL of heparinized saline (100 U).
4. Obtain 5 mL blood sample on day 3 from the catheter and submit for blood culture to confirm the presence of yeast.

3.1.4 Catheter Removal (Timing: Day 18)

1. Anesthetize rabbits (as described above) after 11 days postinfection.
2. Obtain blood cultures through the catheter and peripherally via cardiac puncture.
3. Euthanize the animals as described below under Subheading 3.1.8.
4. Remove the catheter by using sterile technique and divide it into proximal (subcutaneous tunnel) and distal (intravenous) 4-cm segments.
5. Divide each segment in half (2 cm) for QCC and examination of biofilms by SEM.

3.1.5 Quantitative Catheter Culture (Timing: Day 18–20)

1. Cut each 2-cm segment (proximal and distal) in half lengthwise and place in 10 mL sterile saline.
2. Sonicate specimens at 40,000 Hz using ultrasonicator for 12 min at 4-min intervals and then vortex for 15 s.
3. Serially dilute specimens with saline prior to plating 1 mL aliquots into SDA supplemented with chloramphenicol and gentamicin. Do this step in triplicate plates.
4. Incubate the plates for 48 h at 37 °C.
5. Count the colony-forming units (CFUs).

3.1.6 Scanning Electron Microscopy (Adapted from [5]) (Timing: Day 20–25)

1. Sample fixation. Add 2 mL of 2 % glutaraldehyde to wells of 12-well tissue culture plates containing the remaining 2-cm catheter segments (proximal and distal) cut in half lengthwise.
2. Cover plates with aluminum foil.
3. Incubate at 4 °C for 2 h (*see* **Note 11**).
4. Using a glass pipette, aspirate the glutaraldehyde solution.
5. Add 2 mL of 0.1 M sodium cacodylate buffer and let stand for 10 min. Remove the buffer by aspiration using a pipette.
6. Repeat above step twice (total three washes over 30 min).
7. Incubate washed catheter segments with 2 mL of 1 % OsO_4 for 1 h at 4 °C (*see* **Note 12**).
8. Repeat washing steps with sodium cacodylate buffer as described above (for three total washes).
9. Add 2 mL of sterile Milli-Q water, let stand for 5 min, and aspirate the liquid.
10. Repeat the above step.

11. To the wells containing the catheter segments above, add 2 mL of 1 % tannic acid and incubate for 30 min in the dark at room temperature.
12. Add 2 mL of sterile Milli-Q water, let stand for 10 min, and aspirate the liquid.
13. Repeat the above wash step twice.
14. Add 2 mL of 1 % uranyl acetate and place the plate in the dark for 1 h (*see* **Note 13**)
15. Add 2 mL of sterile Milli-Q water, let stand for 5 min, and aspirate the liquid.
16. Repeat the above wash step once.
17. Dehydrate the sample by adding 25 %, 50 %, 75 %, and 95 % ethanol for 15 min and 100 % ethanol for 15 min twice. Ethanol must be removed step by step (*see* **Note 14**).
18. Place round black SEM adhesive tape on specimen mounting stubs.
19. Use sterile tweezers to remove the substrate (catheter segments) containing dehydrated cells from 12-well plates (from step above) and place on mounting stubs prepared above (*see* **Note 15**).
20. Press gently to ensure that the biofilm-containing substrate is firmly stuck to the adhesive tape.
21. Place the stub in stub holder and then transfer to a vacuum desiccator (*see* **Note 16**).
22. Seal the desiccator under vacuum and store at room temperature for 24 h (*see* **Note 17**).
23. Sputter coat the stubs with Au/Pd (60/40) for 30 s using the sputter coating machine (*see* **Note 18**).
24. Transfer the stub to the sample holding platform in the scanning electron microscope (*see* **Note 19**).
25. Examine the biofilms under the electron microscope (Fig. 2) following the unit-specific instructions (*see* **Note 20**).
26. Scan through the specimen surface and select representative areas to record.
27. Acquire images at different magnifications (*see* **Note 21**).

3.1.7 Antifungal Lock and Catheter Treatment (Timing: Day 11–18)

In this protocol, lock antifungal solutions for 8 h per day for 7 days were used. In other experiments, 4-h periods can also be used, depending on the drug the time for lock treatment may vary.

1. Liposomal amphotericin B and fluconazole are used in this protocol.
2. Flush catheter with 300 μL of heparinized saline (100 U) on a daily basis for the first 3 days of the study.

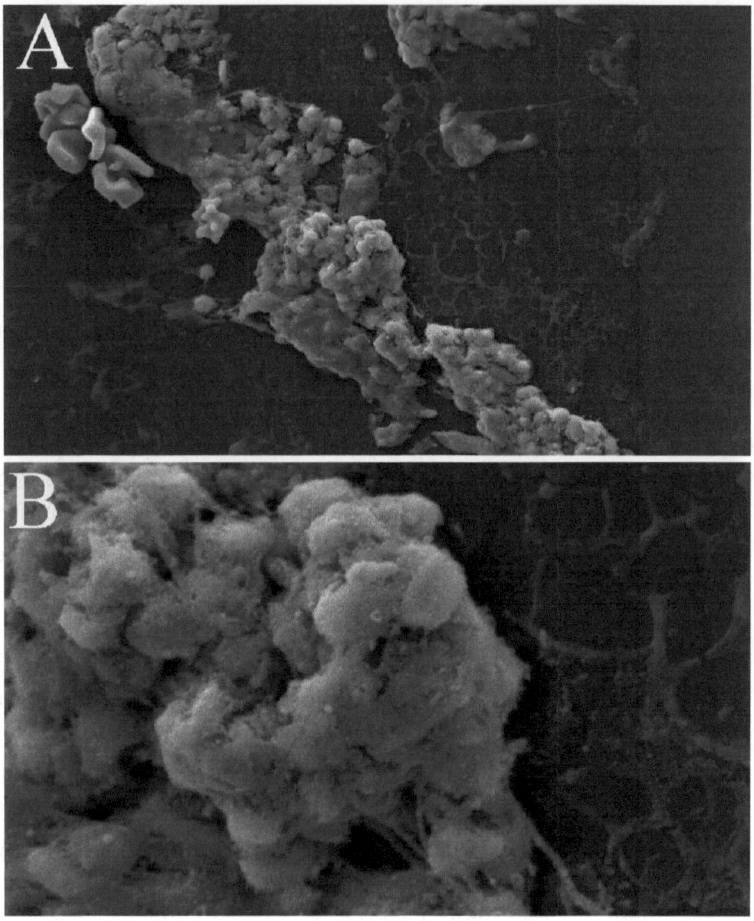

Fig. 2 Mature in vivo *C. albicans* biofilm formation during rabbit model development. Scanning electron micrographs of *C. albicans* biofilms adherent to the intraluminal surface of catheters showing no difference in biofilm architecture at 7 days postinfection (magnification, 6,500×) (**a**) and 3 days postinfection (magnification, 2,500×) (**b**) are shown. Adapted from [8]

3. On day 3 postinfection, draw 5 mL blood through each catheter for culture.
4. To determine the efficacy of antifungal lock therapy, randomize the rabbits into three groups, each consisting of seven animals.
5. Flush group I catheters daily with 300 μL of heparinized saline (100 U).
6. Lock catheter lumens of animals in group II with 300 μL solution containing (1) 3 mg of liposomal amphotericin B, (2) 100 U of heparin, and (3) 5 % dextrose solution (since dextrose solution is used for reconstituting lipoAmB, we use it as a blank).

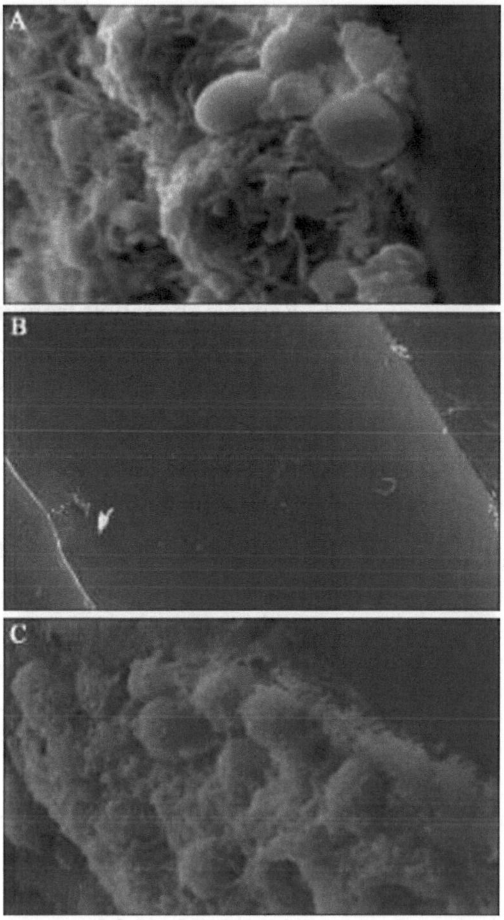

Fig. 3 Effectiveness of antifungal lock therapy. Scanning electron micrographs of intraluminal catheter surfaces following 7 days of therapy with heparinized saline (magnification, 5,000×) (**a**), liposomal amphotericin B (magnification, 121×) (**b**), and fluconazole (magnification, 3,500×) (**c**) are shown. Adapted from [8]

7. Lock catheter lumens of rabbits in group III with 300 μL solution containing (1) 3 mg of fluconazole, (2) 100 U of heparin, and (3) sterile normal saline (*see* **Note 22**).
8. Lock antifungal solutions for 8 h per day for 7 days (*see* **Note 23**).
9. Upon completion of each daily treatment, remove antifungal lock solution and flush catheters with 300 μL of heparinized saline.
10. After 7 days of antifungal lock therapy, anesthetize animals and obtain blood through the catheter and via a cardiac puncture.
11. Euthanize animals as described below.
12. Remove catheters under sterile conditions and perform QCC and SEM (Fig. 3) as described above.

3.1.8 Rabbit Euthanasia (Timing: Day 18)

1. Animals are given a standard dose of 3 mL Euthasol as determined by the Animal Resource Center and IACUC regulations (about 390 mg/kg).
2. When drawing up the working solution of drug, the rim of the plunger in the syringe should be even with the mark for the appropriate amount.
3. After drawing up the solution, affix the appropriate size needle to the syringe.
4. Remove the rabbit from its housing and restrain.
5. Once the rabbit is comfortably restrained, give IM anesthesia as described above.
6. Place rabbit back in its housing until fully anesthetized.
7. Place rabbit on right side lateral recumbency and wet target area with alcohol.
8. Feel for strongest point of heartbeat with thumb.
9. Insert needle into the heart chamber, pulling back on the syringe the whole time.
10. Once blood is observed, stop advancing needle and draw up appropriate amount of blood if sample is required.
11. While leaving needle in place, direct euthanasia needle alongside.
12. Pull back on plunger to check for blood to be sure needle is in the heart.
13. Remove blood collection syringe and inject euthanasia solution (*see* **Note 24**).
14. When dosing is complete, remove the needle and discard the needle and syringe in a biohazard sharp container.
15. Confirm death by verifying absence of respiration, cardiac function, corneal reflex, muscle tone, and mucus membrane color.
16. When finished, place rabbit in double-bagged biohazard bags and follow institutional disposal procedures (*see* **Note 25**).
17. If done in a biosafety cabinet, wipe thoroughly with Clidox-soaked paper towels.
18. Wash hands after handling animals.

3.2 Murine Fusarium Keratitis Model (Adapted from [11, 16])

Following is the summary of methods for murine model of contact lens-associated *Fusarium* keratitis.

3.2.1 Fusarium Culture and Biofilm Formation

1. Culture *Fusarium* at 37 °C in Sabouraud dextrose broth in a shaking incubator.
2. For biofilm development, harvest conidia from a 40-h broth culture by standard filtration methods, count, and standardize to 1×10^6 conidia in PBS.

3. Incubate lotrafilcon A (or other) contact lenses with the above standardized culture for 90 min at 37 °C (adherence phase) in 12-well plates.
4. Wash the lenses gently in 4 mL PBS and then incubate in SDB at 37 °C for 48 h on a rocking platform (biofilm growth phase).

3.2.2 Murine Model of Fusarium Biofilm-Induced Keratitis

1. Anesthetize mice by intraperitoneal injection of 0.4 mL 2,2,2-tribromoethanol.
2. Subject the corneal epithelium to either three parallel scratches or abrasion of a 1-mm-diameter area of the central corneal epithelium, as described earlier [11] (*see* **Note 26**).
3. Place a 2-mm-diameter punch from the contact lenses with formed biofilm on the abraded corneal surface. Mice remain anesthetized on a heating pad.
4. After 2 h (when the mice recover from anesthesia), remove the contact lenses.
5. At each time point (24 and 48 h) thereafter, examine corneal opacification using a dissecting microscope (Fig. 4).
6. Euthanize the mice, remove, and homogenize the eyes either to determine the number of CFUs or to process the eyes for histology.

3.2.3 Quantification of Fungi in Infected Eyes

1. Homogenize whole eyes under sterile conditions in 1 mL PBS at 33 Hz for 4 min.
2. Perform a series of log dilutions in replicate and plate onto Sabouraud's agar.
3. Incubate the plates at 37 °C for 40 h; determine the number of CFUs (in the lowest dilution) by direct counting. Note: The lower limit of detection is ten organisms.

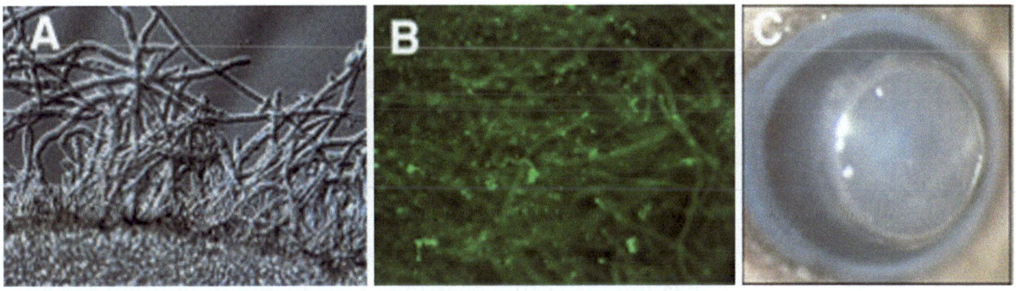

Fig. 4 Murine model of *Fusarium* biofilm keratitis. *Fusarium* conidia were incubated 90 min with lotrafilcon A contact lenses, washed, and incubated an additional 48 h. (**a**) *Fusarium* hyphae associated with the contact lens (DIC microscopy). (**b**) After incubation with FITC-ConA, showing hyphae and extracellular matrix. (**c**) Murine model: 2-mm-diameter punch of a contact lens with attached *Fusarium* biofilm on the corneal surface of a C57BL/6 mouse. Original magnification: (**a**, **b**) 400× and (**c**) 20×. Adapted from [11]

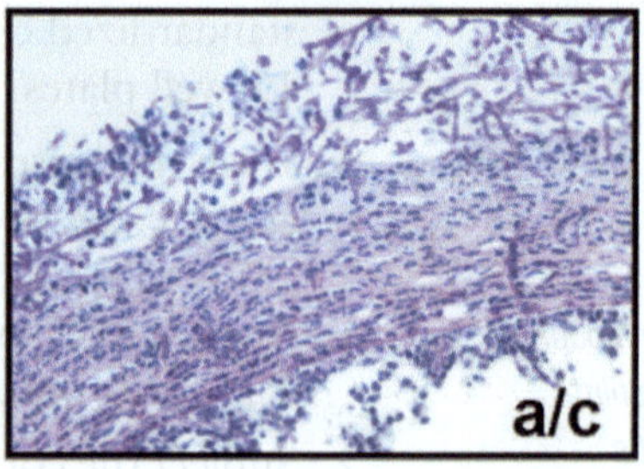

Fig. 5 Histopathology of *Fusarium* keratitis. 24 h after adding the lens, mice are euthanized, and eyes are processed for histology. Note the hyphae penetrating through the cornea and into the anterior chamber (a/c) and the pronounced cellular infiltrate in the cornea. Original magnification is 400×. Adapted from [11]

3.2.4 Histologic Preparations

1. For histologic analysis (Fig. 5), fix the eyes in 10 % buffered formalin phosphate and embed in paraffin.
2. Stain 5-µm sections by 1 % periodic acid solution for 30 min, rinse with water, placed in Schiff's reagent for 30 min, wash well with tap water for 10 min, counterstain with hematoxylin, and examine by light microscopy.

3.2.5 Detection of Cytokines in the Cornea

1. Excise mouse corneas using a 2-mm trephine and homogenize corneal tissue using Mixer Mill MM300 for 4 min at 30 cycles per second.
2. After centrifugation, determine cytokine levels in soluble corneal extracts by sandwich ELISA according to the manufacturer's directions and as described earlier [11]. Briefly, coat a 96-well microplate with the diluted capture antibody for overnight at room temperature; wash three times by 0.05 % Tween 20 in PBS; block plate by 5 % BSA/PBS at room temperature for 2 h; wash as same above; add samples or standard for 2 h at room temperature; wash as same; add the diluted detection antibody for 2 h at room temperature; wash as same; add the working dilution (1:200) of streptavidin-HRP for 20 min; wash as same; add substrate solution for 20 min. Add stop solution [11].
3. Measure the absorption at 450 nm on a microplate reader.

4 Notes

1. Avoid overheating SD agar plates during autoclaving, as overheating may cause charring of the medium. Pouring the liquefied agar into Petri dishes slowly will help avoid bubble formation in plates.
2. Glutaraldehyde (irritant, allergen, carcinogen), sodium cacodylate, osmium tetroxide, and uranyl acetate are highly toxic. Handle them with care.

3. Avoid light exposure and immediately transfer to 4 °C. Highly toxic. This reagent is radioactive and should be stored covered by lead sheathing. Lead and uranyl are toxic; avoid contact with skin and eye.
4. Avoid light exposure and store at room temperature after preparation. Toxic (irritant).
5. Keep away from flame. The solutions should be stored in tightly capped bottles to prevent evaporation.
6. Rabbits can be unpredictable as they are usually timid and are easily upset. Rabbits can be lifted firmly by the scruff with the head pointing away from your body and the other hand used to support the rabbit's hind end. Rabbits can be cradled to your body. If the rabbit struggles, place it on a solid surface and calm it by shielding the rabbit's eyes by tucking its head into the crook of your arm; this should help keep the animal calm. When finished with procedures, keep a firm grip on the scruff and gently release the rabbit in to the correctly marked cage.
7. There can be anesthetic complications including decreased respiration and heart rate which may lead to death. Proper animal restraining techniques are important for proper dosing and for minimizing the stress on the animal. Before attempting injections, one should be competent in rabbit handling.
8. To avoid ripping the vein into two pieces, when incising the vein, do not transect more than 1/3 of the way through. Care should be taken to avoid surgical complications, i.e., excessive bleeding, tearing blood vessels, nerve damage, and death.
9. Caution should be taken when moving the rabbit to lateral recumbency to avoid dislodging the catheter.
10. To avoid needlesticks, never recap needles.
11. These samples can be stored in 2 % glutaraldehyde overnight if sealed (with parafilm) and kept at 4 °C in the dark.
12. Samples can be stored for 3–4 h.
13. Samples can be stored overnight at 4 °C, sealed (with parafilm) and in the dark (cover with aluminum foil).
14. Samples can be stored for up to 4 weeks in 100 % ethanol at 4 °C, sealed with parafilm (to ensure the ethanol is not evaporated). To protect from light, cover with aluminum foil.
15. Be careful to place the catheters on the stubs, making sure that catheters with *C. albicans* biofilms remain exposed at the top.
16. Use appropriate tweezers to handle the stubs since they can easily slip off and fall.
17. Proper vacuum desiccation is critical to ensure complete evaporation of ethanol and to prevent rehydration of the sample. Presence of even the smallest amount of water due to rehydration from the atmosphere will cause interference with subsequent steps.

18. Care should be taken not to exceed the sputter coating step for more than 30 s, since exceeding this time may impart excess charge on the samples and result in burnt areas when observed under the microscope.
19. Care should be taken while placing the stubs in the holding platform to avoid dropping them.
20. We observe samples under ESEM (model XL3C Philips microscope) using voltage of 15 kV and spot size of 4. These settings may need to be optimized for different microscopes.
21. When recording images at high magnifications (e.g., 5,000×), it is important to acquire them quickly since the intense electron source at these magnifications can result in overexposure ("burning") and poor-quality images.
22. The concentrations of drugs used in the study were guided by dosages in prior case reports. It is necessary to check the dosage from published reports to minimize the number of animals used and to prevent harm by excessive dose concentrations. The fluconazole used in this study required a 10-min incubation in a hot water bath to dissolve completely.

    In this protocol, lock antifungal solutions for 8 h per day for 7 days were used. In other experiments, 4-h periods can also be used, depending on the drug the time for lock treatment may vary.
23. Lock treatment times may vary by drug.
24. Do not remove syringe until heart stops beating in case additional euthanasia solution is needed.
25. Ketamine and Euthasol are drugs under control of the Drug Enforcement Agency (DEA), and all usage must be recorded in a drug logbook. Controlled drugs must be stored in a double-locked cabinet in a secure area. Be careful to avoid accidental needlesticks; never recap needles.
26. Care should be taken while making scratches or abrasions to avoid perforating the cornea.

Acknowledgments

Studies by our group were supported by funds from NIH to MAG (RO1DE17846) and the Oral HIV/AIDS Research Alliance (OHARA/NIDCR), grant number BRS-ACURE-S-11-000049-110229, PKM (R21EY021303 and R21AI074077), and the CWRU/UH Center for AIDS Research (CFAR, NIH grant number P30 AI036219). Studies were also supported by NIH funds to EP (R01EY18612 and P30EY11373).

References

1. Wisplinghoff H et al (2004) Nosocomial bloodstream infections in US hospitals: analysis of 24,179 cases from a prospective nationwide surveillance study. Clin Infect Dis 39:309–317, PMID: 15306996
2. Passerini L, Lam K, Costerton JW, King EG (1992) Biofilms on indwelling vascular catheters. Crit Care Med 20:665–673, PMID: 1572192
3. Chandra J et al (2001) Antifungal resistance of candidal biofilms formed on denture acrylic *in vitro*. J Dent Res 80:903–908, PMID: 11379893
4. Chandra J et al (2001) Biofilm formation by the fungal pathogen *Candida albicans* – development, architecture and drug resistance. J Bacteriol 183:5385–5394, PMID: 11514524
5. Chandra J, Mukherjee PK, Ghannoum MA (2008) In vitro growth and analysis of Candida biofilms. Nat Protoc 3:1909–1924, PMID: 19180075
6. Mermel LA et al (2001) Guidelines for the management of intravascular catheter-related infections. Clin Infect Dis 32:1249–1272, PMID: 11303260
7. Mermel LA et al (2009) Clinical practice guidelines for the diagnosis and management of intravascular catheter-related infection: 2009 update by the Infectious Diseases Society of America. Clin Infect Dis 49:1–45, PMID: 19489710
8. Schinabeck MK et al (2004) Rabbit model of Candida albicans biofilm infection: liposomal amphotericin B antifungal lock therapy. Antimicrob Agents Chemother 48:1727–1732, PMID: 15105127
9. Andes D et al (2004) Development and characterization of an *in vivo* central venous catheter *Candida albicans* biofilm model. Infect Immun 72:6023–6603, PMID: 15385506
10. Chandra J, Long L, Ghannoum MA, Mukherjee PK (2011) A rabbit model for evaluation of catheter-associated fungal biofilms. Virulence 2(5):466–474, PMID: 21921676
11. Sun Y, Chandra J, Mukherjee P, Szczotka-Flynn L, Ghannoum MA, Pearlman E (2010) A murine model of contact lens-associated fusarium keratitis. Invest Ophthalmol Vis Sci 51(3):1511–1516
12. Donlan RM (2002) Biofilms: microbial life on surfaces. Emerg Infect Dis 8:881–890
13. Mohammadi Z, Palazzi F, Giardino L, Shalavi S (2013) Microbial biofilms in endodontic infections: an update review. Biomed J 36: 59–70
14. Kalathil S, Khan MM, Lee J, Cho MH (2013) Production of bioelectricity, bio-hydrogen, high value chemicals and bioinspired nanomaterials by electrochemically active biofilms. Biotechnol Adv 31:915–924
15. Nicastri E, Petrosiillo N, Viale P, Ippolito G (2001) Catheter-related bloodstream infections in HIV-infected patients. Ann N Y Acad Sci 946:274–290
16. Mukherjee PK, Chandra J, Yu C, Sun Y, Pearlman E, Ghannoum MA (2012) Characterization of fusarium keratitis outbreak isolates: contribution of biofilms to antimicrobial resistance and pathogenesis. Invest Ophthalmol Vis Sci 53(8):4450–4457
17. Imamura Y, Chandra J, Mukherjee PK, Lattif AA, Szczotka-Flynn LB, Pearlman E, Lass JH, O'Donnell K, Ghannoum MA (2008) Fusarium and Candida albicans biofilms on soft contact lenses: model development, influence of lens type, and susceptibility to lens care solutions. Antimicrob Agents Chemother 52(1):171–182
18. Khor WB, Aung T, Saw SM, Wong TY, Tambyah PA, Tan AL, Beuerman R, Lim L, Chan WK, Heng WJ, Lim J, Loh RS, Lee SB, Tan DT (2006) An outbreak of Fusarium keratitis associated with contact lens wear in Singapore. JAMA 295:2867–2873
19. Gaujoux T, Chatel MA, Chaumeil C, Laroche L, Borderie VM (2008) Outbreak of contact lens-related Fusarium keratitis in France. Cornea 27:1018–1021

Chapter 11

Nonmammalian Model Systems to Investigate Fungal Biofilms

Marios Arvanitis, Beth Burgwyn Fuchs, and Eleftherios Mylonakis

Abstract

Medical advances have resulted in an increase in the number of patients in immunocompromised states, vulnerable to infection, or individuals fitted with medical devices that form niches for microbial infections. These infections are difficult to treat and have significant morbidity and mortality rates. An important factor in the pathogenesis of fungal diseases is the development of biofilm-forming communities, enabling the invasion of host tissues and resistance to antimicrobial compounds. To investigate the genetic requirements for filamentation and seek compounds that inhibit the process, invertebrate hosts are employed as models of in vivo infection. The purpose of our review is to highlight methods that can be utilized to investigate fungal filamentation, an important step in the development of biofilms, in the invertebrate hosts *Galleria mellonella, Caenorhabditis elegans*, and *Drosophila melanogaster*.

Key words Biofilm, *Caenorhabditis elegans*, *Drosophila melanogaster*, Filamentation, Fungal infection, *Galleria mellonella*

1 Introduction

Although microbial biofilms have been extensively studied, our knowledge on fungal biofilms is still rudimentary [1, 2]. Fungal biofilm formation has been described in *Candida* species as a series of sequential steps for the sake of simplicity, although one should be aware that all these steps may occur concurrently during in vivo biofilm development [3]. First, fungal cells adhere to a surface (adherence step). Attached cells aggregate and form filaments (initiation step), followed by extracellular polymeric substance (EPS) production and acquisition of polymicrobial resistance (maturation step). Finally, fungal cells are released from the biofilm into the surrounding area (dispersal step). Of note, with the exception of fungi such as *Cryptococcus neoformans* [4] and *Pneumocystis* species [5], most fungal biofilms studied to date share a common characteristic, the requirement for filamentation for their development [6].

Gianfranco Donelli (ed.), *Microbial Biofilms: Methods and Protocols*, Methods in Molecular Biology, vol. 1147, DOI 10.1007/978-1-4939-0467-9_11, © Springer Science+Business Media New York 2014

The cells within biofilms form a community. They develop ways to communicate with each other, through a method called quorum sensing [7], detecting nutrients and controlling population size. As a biofilm, the cells generate physical barriers, thus mediating protection from the surrounding environment. Matrix formation promotes the physical exclusion of compounds. Those compounds that are able to penetrate into the biofilm can be ejected via efflux pumps. In the case of *C. albicans* biofilms, persister cells evolve, forming a small population of the cells within the biofilm that are dormant and tolerant to stress [8, 9]. The arsenal of defenses makes biofilm infections notorious for antibiotic resistance and thus difficult to treat [10].

It is now realized that biofilms play a role in the majority of clinically documented infections [10]. Further, with the ever-growing development of the field of transplant medicine and with the increase in life expectancy of people in immunocompromised states, such as those with AIDS, cancer patients, and transplant recipients [11, 12], systemic fungal infections are becoming a common encounter in clinical practice [13, 14]. The fact that these infections are difficult to treat and are usually accompanied by high mortality rates [13, 15] of 35–65 % for systemic infections underscores the need for the better understanding of every aspect of the pathophysiology of these diseases, including the function and creation of fungal biofilms.

Mice are the most common experimental hosts used in the study of fungal infections due to the striking similarity of their immune system to that of humans. However, their use has important limitations, such as lengthy reproduction time, housing expenses, and ethical considerations. All the above observations stress the need for the use of alternative hosts in experimental mycology. The perfect host in that regard would be an easy to use and to obtain host, genetically tractable and affected by fungi at mammalian temperatures. Moreover, there should be no ethical constraints associated with its use. Finally, the virulence traits necessary for fungal infection in this host should reflect the ones involved in the pathogenesis of human fungal infections. Of course, the perfect host does not exist. However, many nonmammalian model hosts, like the nematode *Caenorhabditis elegans* [16], the fruit fly *Drosophila melanogaster* [17], the larvae of the greater wax moth *Galleria mellonella* [17], and the plant *Arabidopsis thaliana* [18], share enough of the above characteristics to make them favorable screening models in experimental mycology and thus have been used to date in several studies of fungal diseases.

The purpose of our review is to describe the ways in which these nonmammalian models can be used to study fungal biofilms. It is important to note, however, that although many researchers have studied bacterial biofilms in invertebrates [19, 20], to our knowledge, no experiment to date has directly observed and studied fungal biofilms in vivo in these hosts. Nevertheless, given the fact that, as was already mentioned, filamentation plays a crucial role in most of the fungal biofilms, we will endeavor to describe the materials and methods used to study fungal filamentation in the hosts *G. mellonella* [21, 22], *C. elegans* [23–25], and *D. melanogaster* [26, 27].

2 Materials

2.1 Fungal Filamentation in Galleria mellonella

1. *G. mellonella* larvae at sixth instar state are obtained from the supplier (we use Vanderhorst Wholesale, St. Mary's, OH) and stored at room temperature for up to 1 week.
2. Fungal strains. The most commonly used strains in these experiments are *Candida* spp.
3. YPD broth (1 % yeast extract, 2 % dextrose, 2 % peptone): To prepare 1 L solution, dissolve 10 g of yeast extract and 20 g of peptone in 800 mL of ddH_2O, autoclave, and add 200 mL of 10 % dextrose that has been filter sterilized.
4. Phosphate-buffered saline (PBS): To prepare a 1 L sock solution of 10× PBS, dissolve 80 g NaCl, 2 g KCl, 14.4 g $Na_2HPO_4 \cdot 2H_2O$, and 2.4 g KH_2PO_4 in 800 mL of distilled water. After complete mixing, bring the solution volume up to 1 L. Autoclave to sterilize. Use 1× PBS in your experiments by diluting the stock solution tenfold and filter sterilizing.
5. Insect physiologic saline (IPS): 150 mM NaCl, 5 mM KCl, 100 mM Tris–HCl, pH 6.9 with 10 mM EDTA, and 30 mM $C_6H_7NaO_7$ (sodium citrate).
6. Grace's insect medium.
7. 95 % ethanol.
8. Petri dishes.
9. Kimwipes.
10. Hamilton syringes (26 gauge, 10 μL capacity).
11. 37 °C incubator.
12. Tissue processor.
13. Mesh biopsy collection bags (*see* **Note 1**).
14. Glass chamber slides.
15. Periodic acid-Schiff (PAS) stain.
16. FITC (fluorescein isothiocyanate) stain.
17. Microscope (at least 60× magnification).
18. Confocal laser microscope.
19. Centrifuge.

20. Eppendorf tubes.
21. Plastic bins (*see* **Note 2**).
22. Scalpel.

2.2 Fungal Filamentation in *Caenorhabditis elegans*

1. *C. elegans* strains can be acquired from the *Caenorhabditis* Genetics Center (CGC).
2. Fungal strains.
3. 5 mg/mL cholesterol dissolved in ethanol.
4. 1 M $KHPO_4$ buffer: For 1 L dissolve 108.3 g KH_2PO_4 and 35.6 g K_2HPO_4 in 800 mL ddH_2O, then adjust to pH 6.0 with KOH. Bring the volume up to 1 L with ddH_2O. Filter sterilize.
5. Nematode growth medium (NGM) agar: For 1 L medium, dissolve 3 g NaCl, 2.5 g peptone, and 17 g agar in 972 mL ddH_2O. Autoclave and then add, under sterile conditions, 1 mL cholesterol 5 mg/mL stock solution, 1 mL 1 M $CaCl_2$, 1 mL 1 M $MgSO_4$, 25 mL 1 M KPO_4 buffer, and 1 mL of streptomycin 100 mg/mL stock (for a final concentration of 100 μg/mL) (*see* **Note 3**). Pour the medium onto plates and store at 4 °C (*see* **Note 4**).
6. *E. coli* OP50 strain.
7. M9 solution: Add 3 g KH_2PO_4, 6 g Na_2HPO_4, and 5 g NaCl in 1 L of water and autoclave. After cooling down add 1 mL sterilized 1 M $MgSO_4$.
8. Bleach.
9. 5 M NaOH.
10. Microscope.
11. Rotisserie.
12. Table top centrifuge.
13. YPD broth plus antibiotics (kanamycin 45 μg/mL, streptomycin 100 μg/mL, and ampicillin 100 μg/mL).
14. Brain-heart infusion (BHI) broth: Add 37 g BHI powder in 1 L ddH_2O and autoclave.
15. BHI agar plus antibiotics: Add 37 g BHI powder and 12.5 g agar in 1 L distilled water and autoclave. After cooling add 1 mL kanamycin 45 mg/mL (for a final concentration of 45 μg/mL).
16. Roller drum.
17. 6-well plates.
18. 3 % glutaraldehyde.
19. Millonig's buffer: In 500 mL distilled water, add 1.8 g NaH_2PO_4, 23.25 g Na_2HPO_4, and 5 g NaCl. Adjust pH to 7.3 and then add ddH_2O to bring the volume up to 1 L.

20. Ethanol.
21. 100 % propylene oxide.
22. LX-112 resin.
23. BEEM capsules.
24. Leica Ultracut-R microtome.
25. Mesh copper grids (EMS).
26. Diamond knife.
27. 2 % uranyl acetate.
28. Reynold's lead citrate.
29. JEOL 1200 transmission electron microscope.
30. Gatan CCD camera.
31. 15 °C incubator, 25 °C incubator.

2.3 Fungal Filamentation in Drosophila melanogaster

1. Fungal strains.
2. Adult fly lines: $Tl^{r632}/Tl^{I\text{-}RXA}$ Toll-deficient flies are used in the experiment, because wild type strains are resistant to fungal infections [28]. $Tl^{I\text{-}RXA}$ is a null allele and Tl^{r632} is a temperature-sensitive allele which is deactivated in 29 °C.
3. Fly food (*see* **Note 5**).
4. Fly-food vials.
5. Stereoscopic microscope equipped with a controllable CO_2-flow fly pad.
6. 29 °C incubator.
7. Tungsten stainless steel needle held in a pin vise.
8. Bunsen burner.
9. Sterile disposable petri dishes (100 × 15 mm).
10. 10 % vol/vol formaldehyde.
11. Ethanol.
12. Methyl benzoate.
13. Paraffin.
14. Microtome.

3 Methods

3.1 Observation of Filamentation in Galleria mellonella

1. Preparation of the inoculum: Fungal strains are grown overnight in the appropriate media and growth conditions. We hereby present the example of *C. albicans* since it is the fungus most commonly used in such experiments. *C. albicans* cells are grown overnight in YPD at 30 °C with agitation. The fungal cells are collected with centrifugation and washed

twice with PBS. Then, cells are counted with a hemocytometer and used immediately at the appropriate concentration suspended in PBS.

2. *G. mellonella* larvae are selected for uniformity in size and appearance. The larvae should be between 250 and 350 mg and light in color (*see* **Note 6**).
3. Clean the Hamilton syringes: Syringes are first cleaned with 10 % bleach three times, then with 100 % ethanol, ddH_2O, and 1× PBS sequentially. Clean the syringes again after six injections or between fungal strains.
4. Larvae inoculation: *G. mellonella* larvae are inoculated by direct injection into the hemocoel. A volume of 10 μL of inoculum is delivered into their last left proleg using a clean Hamilton syringe (*see* **step 3**) (Fig. 1a). Immobilize the larva using a three-finger pincer grasp and insert the needle into the injection site (*see* **Note 7**) (Fig. 1a). For *C. albicans* infections, the inoculum concentration is 10^6 cells/larvae for a lethal infection (*see* **Note 8**).
5. Following the inoculation place the larva onto a kimwipe to let it recover and to observe the amount of bleeding, hemolymph leakage from the wound. A small spot on the paper due to larval bleeding is normal, but if the larva continues to bleed, discard it (*see* **Notes 9** and **10**).
6. After all the larvae within the infection group are inoculated, put them into a petri dish and incubate them at 37 °C. The incubation temperature can be adjusted depending on the experiment (*see* **Notes 11** and **12**). Depending on the infecting pathogen, the larvae will melanize, appearing dark in color (Fig. 1b). Melanization is observed during an infection with *C. albicans* but not with *Cryptococcus neoformans.*
7. Release of the internal organs: At appropriate time points three larvae per group are cut open from the upper part of the body to the lower and then squeezed to release their internal structures (Fig. 1c). At first, the fluid that comes out is the hemolymph, which is discarded, and then the larvae are held over an Eppendorf tube containing 1 mL of formalin, and the internal organs are released into the tube. The collected internal structures are stored at 4 °C overnight and the internal organs are subsequently put into mesh biopsy collection bags which are placed in an automated tissue processor (Autotechnicon Mono). The tissue is serially dehydrated with 50, 70, and 90 % ethanol for 1 h each and then with 100 % ethanol for 3 h. The tissue is then transferred to xylene for 3 h and mounted in paraffin. Paraffin blocks are allowed to harden overnight at −20 °C [21].
8. Tissue visualization for filamentation: Thin sections of paraffin-embedded tissue are cut, stained with periodic acid-Schiff (PAS), and observed under the microscope (Fig. 1d) (*see* **Note 13**).

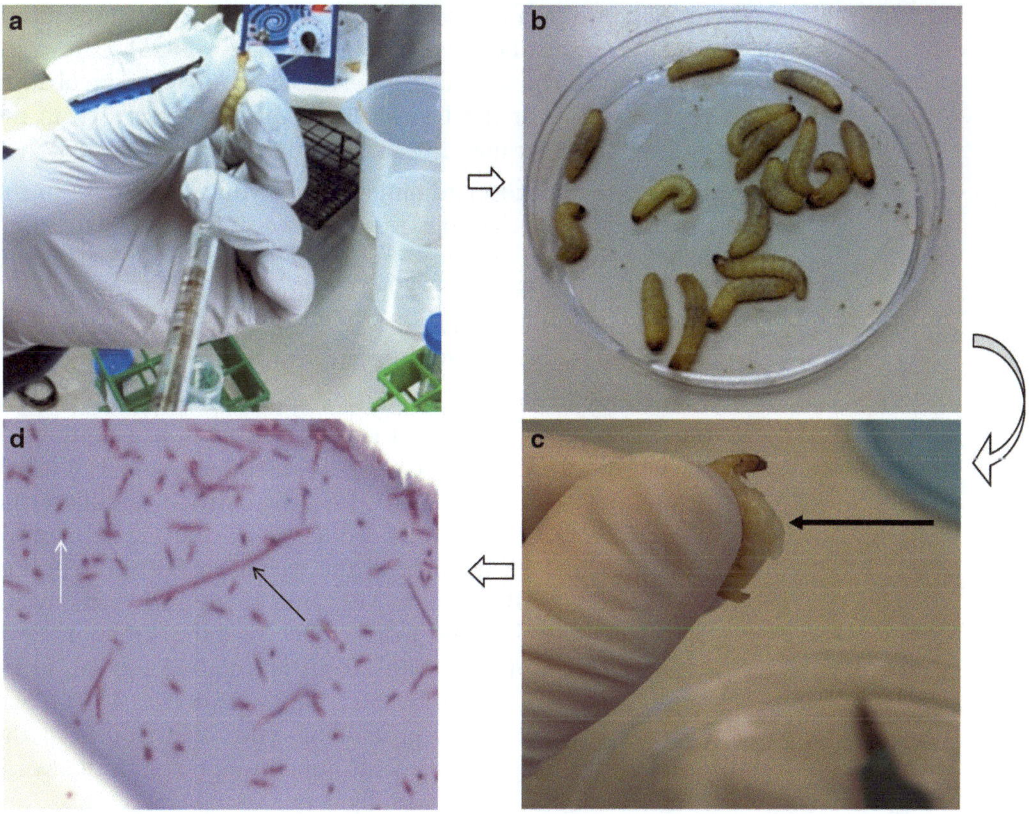

Fig. 1 Fungal filamentation in *G. mellonella*. (**a**) Inject the larvae on their last left proleg using Hamilton syringes. (**b**) Infected larvae become melanized. (**c**) Cut the larva open with a scalpel and squeeze to release the internal structures (*arrow*). (**d**) Fixed tissue is sectioned and stained with PAS to observe fungal cells (*white arrow*) and filaments (*black arrow*)

3.2 Observation of Filamentation in *Caenorhabditis elegans*

1. Use nematodes at the L4 stage which you can synchronize as follows: Cut a ~1 cm × ~1 cm square of NGM agar containing recently hatched nematodes from stock plates (stored at 15 °C) with any sterilized sharp tool. Transfer the agar square to a new NGM plate with *E. coli* and place the upper surface of the square directly on to the surface of the new plate. Incubate worms at 15 °C for 3–4 days until they become gravid. After that, wash the plate with 6 mL M9 buffer and pour the M9 (containing the gravid worms) into a 15 mL centrifuge tube. Add M9 buffer to a total volume of 15 mL. Centrifuge for 30 s at 1,500 × *g*. Remove the supernatant, leaving the worms in 500 μL buffer. Add 400 μL 100 % bleach plus 100 μL 5 M NaOH to the tube and mix gently until the cuticles of half of the worms have ruptured (*see* **Note 14**). Monitor worms under the dissecting microscope during this step to avoid breaking the nematode eggs. Quench the reaction by adding M9, and subsequently, wash the eggs three times with M9. After the final wash, remove the supernatant and add 5 mL M9. Leave the tube at room

temperature on a rotisserie for at least 18 h to allow the eggs to hatch. Transfer the hatched worms to fresh NGM plates with *E. coli* and incubate at 25 °C for 60 h. After the incubation period, worms at the L4 to early adult stages are used in the proceeding experiments to infect with fungi (*see* **Note 15**).

2. Prepare the appropriate fungal strains you want to use in the experiment. For example, for *Candida* strains, inoculate 2 mL of YPD and let them grow at 30 °C, overnight. Prepare lawns by spreading 200 μL of each culture on 100 × 15 mm plates that contain solid BHI media with 45 μg/mL kanamycin (*see* **Note 16**). Incubate the plates at 30 °C for 48 h.
3. Nematode inoculation: Transfer approximately 100 worms onto each lawn and allow them to feed for 4 h. Gently wash the worms off the plates using M9 buffer (*see* **Note 17**). Wash nematodes four times with M9 using gravity to collect the worms. Remove the liquid from the tube using aspiration. After the final wash, suspend the worms in 500 μL M9 and 500 μL 80 % BHI, 20 % M9, and 45 μg/mL kanamycin. Transfer 70–80 infected worms into wells that contain 2 mL liquid medium of 80 % M9, 20 % BHI, and 45 μg/mL kanamycin (*see* **Note 18**). After 60 h of incubation at 25 °C, filaments can be observed protruding from 40 to 60 % of the nematodes. Images of filaments puncturing through the worm cuticle can be observed with light microscopy (Fig. 2).
4. *C. elegans* filamentation staining: At appropriate time intervals stain 20–40 of the worms in 200 μL of 10 μM FUN-1 and 25 μg/mL Concanavalin A-Alexa Fluor 488 for 45 min. Take pictures with a confocal laser microscope (*see* **Notes 19–22**).

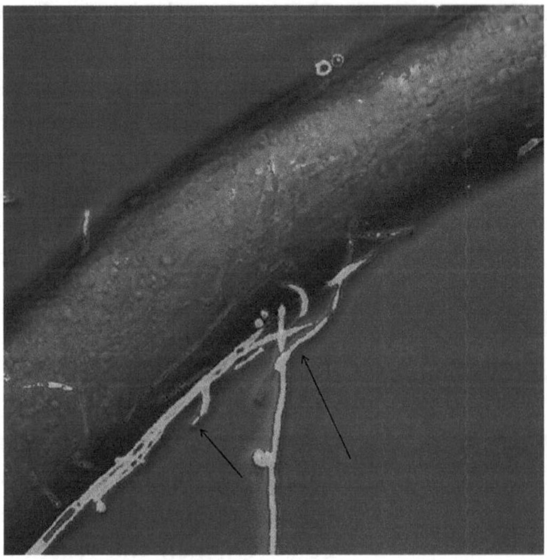

Fig. 2 *C. elegans* infected with GFP-expressing *C. albicans*. *Arrows* point to fungal filaments protruding out of the nematode's cuticle

3.3 Observation of Filamentation in Drosophila melanogaster

1. Prepare working suspensions of fungal cells: Grow the fungus you want to use on appropriate solid media. Pour 0.5 mL sterile water onto the plates and spread to create a cell suspension. Dilute the cell suspension to an appropriate concentration (e.g., 10^8 spores/mL for *zygomyces*).
2. Preparation of the mutant flies: Cross flies carrying a null allele of Toll ($Tl^{I\text{-}RXA}$) with flies that are carrying a thermosensitive allele of Toll (Tl^{r632}) [26, 28, 29] (*see* **Note 23**).
3. Inoculation of the flies: Anesthetize the flies by placing them on the CO_2-flow fly pad. Dip a tungsten needle into the fungal cell suspension after sterilizing it and insert it into the dorsolateral aspect of the fly thorax. Return injected flies to the fly-food vial and observe them for 3 h. If they die within this time period, it means they sustained an injection injury and should not be used in the experiment. Maintain infected flies at 29 °C (*see* **Notes 24** and **25**).
4. Observation for filamentation: On day 1 after infection, fix the flies with 10 % vol/vol formaldehyde, process them and embed them in paraffin (*see* **Note 26**). Cut tissue sections using a microtome, stain them with the appropriate stain (e.g., Crystal violet for *zygomyces* or Grocott-Gomori methenamine-silver nitrate (GMS) stain for *Candida*) and observe under the microscope [26, 30] (Table 1).

Table 1
Differences in the methods used to study fungal filamentation among different hosts

	Galleria mellonella	Caenorhabditis elegans	Drosophila melanogaster
Breeding and synchronization of the host	Larvae are acquired directly from the vendor and can be used within 7 days	Nematodes can be synchronized as described in the text	Cross flies carrying the thermosensitive allele with the ones carrying the null allele of Toll to produce the heterozygotes needed for the experiments
Inoculation method	Direct injection	Nematodes are left to feed on fungal lawns	Injection, rolling on fungal cultures or feeding on fungal lawns
Common incubation temperatures	25–37 °C	15–25 °C	<22–27 °C
Observation of fungal filaments	Observe thin sections of paraffin-embedded tissue from the *G. mellonella* internal organs stained with *PAS*	Observe the worm directly either without stain or with FUN-1 and Concanavalin A-Alexa Fluor 488 stains under a confocal laser microscope	Stain thin sections of whole flies embedded in paraffin and observe under the microscope

4 Notes

1. We have found these bags will keep the larvae internal structures together so they can be fixed in an automated processor.
2. By depositing the *G. mellonella* acquired from the vendor into a plastic bin, the larvae are easier to inspect and sort so they can be selected for uniformity in color and size.
3. Streptomycin does not prevent OP50 *E. coli* growth but does inhibit contamination.
4. Using a consistent volume of media for each plate will reduce the need to refocus the microscope between plates when inspecting the nematodes. For example, 20 mL of media can be used for 15×100 mm plates.
5. Fly food is obtained in powder form and dehydrated from a biological supply company. However, since it is the only source of water for flies, it should be rehydrated properly for best results.
6. Some of the larvae from the vendor shipment can be dark in color due to a melanization effect. Consistent results are achieved when the light-colored larvae are utilized for the experiments.
7. The insertion of the needle should be easy. If you feel tension when trying to insert the needle, adjust the position slightly.
8. We make a 10^8 cells/mL stock solution in PBS and inject 10 μL for a final concentration of 10^6 cells/larva.
9. There are two reasons why you should discard these larvae. First, they may be dead soon due to uncontrollable bleeding, and second, even if they survive, it is impossible to predict what amount of the inoculum remained inside the larva, leading to experimental error.
10. Shortly after injecting the larvae with *C. albicans*, the *G. mellonella* will begin to melanize, starting at the injection site.
11. Keep in mind that differences in temperature alter the *G. mellonella* immune response [31, 32], so the temperature at which the experiment was conducted should be clearly stated in order to have reproducible results.
12. Two control groups should be included for each experiment, uninjected larvae and *G. mellonella* injected with PBS. Both are incubated for the allotted time of the experiment. The uninjected control group will ensure the quality of the larvae lot throughout the course of the experiment. Those injected with PBS will enable the monitor of any effects from the trauma of injecting the larvae.

13. FITC: An alternative technique can be used to observe filamentation of fungal cells during their interaction with *G. mellonella* hemocytes in vitro. This technique utilizes the stain fluorescein isothiocyanate (FITC) to allow for better visualization of fungal cells. Specifically, culture fungal cells overnight, wash them with PBS and then add 0.1 mg/mL FITC and incubate for 30 min in the dark at room temperature. Following the incubation, wash three times with PBS containing 1.5 % fetal bovine serum (FBS). Collect healthy *G. mellonella* hemolymph by cutting the larvae at the base of the right last true leg and let it bleed in tubes containing cold, sterile IPS. Centrifuge the collected hemolymph at 700 × *g* for 5 min at 4 °C and then pass the supernatant through 0.45 mm filters to separate the cells from the hemolymph. Suspend FITC-labeled fungal cells in 10 % cell-free hemolymph in IPS at 37 °C for 60 min with agitation for opsonization. Take the *G. mellonella* hemocytes and suspend 5×10^4 of them in 500 μL of Grace's medium on glass chamber slides. Incubate for 2 h at 37 °C. Remove the medium and add 100 μL of opsonized, FITC-labeled fungal cells and incubate at 37 °C for another 2 h. Wash the chamber slides with PBS containing 100 mg/mL $CaCl_2$, 100 mg/mL $MgCl_2$, and 5 mM glucose; fix overnight with 4 % paraformaldehyde; and observe for filamentation using a microscope.
14. The worms will be a "V" shape as they break open.
15. We favor the use of the mutant strain *glp-4;sek-1*, because of the adverse effects that can occur with the N2 wild-type strain, such as matricidal death due to premature hatching of the eggs within the nematode. *glp 4* mutants are unable to produce progeny at 25 °C, but they are also much more resistant to *C. albicans* killing than the wild-type strain [33]. *SEK-1* encodes for a mitogen-activated protein kinase within the *C. elegans* orthologue of the human p38 kinase cascade, which plays an important role to the nematode's immune response to pathogens [34]. Therefore, the *glp-4;sek-1* strain is susceptible to C. *albicans* and does not exhibit significantly different survival compared to the N2 strain [23].
16. Take care that the lawn does not extend to the wall of the plates. Food too close to the edge of the plate will entice worms to crawl onto the plastic where they will dry out.
17. By washing gently and disposing as little of the fungal cells as possible, fewer of the fungi will transfer into the liquid portion of the assay.
18. The mixture of BHI, M9, and kanamycin is made fresh the day of the experiment.
19. FUN-1 is a fluorescent yellow stain that is absorbed by metabolically active fungal cells and fluoresces red when illuminated

at 480 nm. Concanavalin A-Alexa Fluor 488 is a stain that fluoresces green at 512 nm and is used to stain polysaccharides and filaments.

20. Alternatively you can watch the worms without special staining with a microscope with 40× magnification and observe for penetrative filamentation (defined as any breach in the worm cuticle by filamentous cells).
21. Worms are transferred to an agarose pad on a glass slide. The pad is made of 2 % agarose.
22. Electron microscopy of infected nematodes: This is an additional method that can be used and allows for a more detailed visualization of host-pathogen interactions and was first described by Cruz et al. [24]. After incubation in liquid medium (*see* Subheading 3.2, **step 3**) for 4 days, fix specimens in 3 % glutaraldehyde overnight. Rinse in Millonig's buffer for 5 min and layer the sample with 2 % OsO_4 for 60 min at 4 °C. Rinse the samples with deionized water and dehydrate them sequentially using 20, 70, and 95 % ethanol (10 min each), 100 % ethanol (10 min, repeat three times), and 100 % propylene oxide (10 min, repeat three times). Permeate the samples with 50 % LX-112 resin and 50 % propylene oxide for 3 h followed by 100 % LX-112 for 2 h before embedding in 100 % LX-112 in BEEM capsules for overnight polymerization at 70 °C. Cut 500 nm thick sections using a glass knife and a Leica Ultracut-R microtome. Heat fix to glass slides and stain for 20 s with Toluidine Blue to select the most appropriate areas for imaging. Trim the blocks and cut 120 nm thin sections using the same ultramicrotome and a diamond knife (Diatome-U.S.), place on 100 and 150 mesh copper grids (EMS), and stain for 15 min with 2 % uranyl acetate. Rinse with ddH_2O and further stain for 5 min with Reynold's lead citrate. Image the grids using a JEOL 1200 transmission electron microscope at 60 kV and capture with a 2 k×168 2 k Gatan CCD camera.
23. Fly strains can be acquired from many different sources. A commonly used source is the Bloomington stock center at Indiana University.
24. 29 °C is the temperature at which the flies are most sensitive to fungal infection due to a temperature-sensitive loss-of-function allele.
25. There are other ways to inoculate the flies, for example, by letting them feed on fungal cell lawns or by rolling them on fungal cultures [27], but the injection method allows for a more accurate quantification of the inoculum.
26. The fixation and processing can be performed using standard histologic procedures [35]. Specifically, dip the flies into a

10 % vol/vol solution of formaldehyde and fix them at 4 °C overnight. Then dehydrate the sample by submerging into 40 % (twice), 70, and 100 % (twice) ethanol sequentially at room temperature for 10 min in each. Then incubate the tissue in methyl benzoate and methyl benzoate/paraffin (1:1 mixture) for 30 min in each and subsequently submerge it in two changes of melted paraffin at 60–65 °C for 60 min in each change. Finally, embed the sample in a third change of melted paraffin and leave it at room temperature overnight to become hard.

References

1. Costerton JW, Geesey GG, Cheng KJ (1978) How bacteria stick. Sci Am 238:86–95
2. Douglas LJ (2003) *Candida* biofilms and their role in infection. Trends Microbiol 11:30–36
3. Finkel JS, Mitchell AP (2011) Genetic control of *Candida albicans* biofilm development. Nat Rev Microbiol 9:109–118
4. Martinez LR, Casadevall A (2007) *Cryptococcus neoformans* biofilm formation depends on surface support and carbon source and reduces fungal cell susceptibility to heat, cold, and UV light. Appl Environ Microbiol 73:4592–4601
5. Cushion MT, Collins MS, Linke MJ (2009) Biofilm formation by *Pneumocystis spp.* Eukaryot Cell 8:197–206
6. Fanning S, Mitchell AP (2012) Fungal biofilms. PLoS Pathog 8:e1002585
7. Romling U, Balsalobre C (2012) Biofilm infections, their resilience to therapy and innovative treatment strategies. J Intern Med 272:541–561
8. Nobile CJ, Mitchell AP (2007) Microbial biofilms: *e pluribus unum*. Curr Biol 17: R349–R353
9. Lewis K (2008) Multidrug tolerance of biofilms and persister cells. Curr Top Microbiol Immunol 322:107–131
10. Hall-Stoodley L, Stoodley P (2009) Evolving concepts in biofilm infections. Cell Microbiol 11:1034–1043
11. Weber R, Ruppik M, Rickenbach M et al (2013) Decreasing mortality and changing patterns of causes of death in the Swiss HIV Cohort Study. HIV Med 14:195–207
12. Chen Y, Wang H, Kantarjian H et al (2013) Trends in chronic myeloid leukemia incidence and survival in the United States from 1975–2009. Leuk Lymphoma 54:1411–1417
13. Tabah A, Koulenti D, Laupland K et al (2012) Characteristics and determinants of outcome of hospital-acquired bloodstream infections in intensive care units: the EUROBACT International Cohort Study. Intensive Care Med 38:1930–1945
14. Pappas PG, Alexander BD, Andes DR et al (2010) Invasive fungal infections among organ transplant recipients: results of the Transplant-Associated Infection Surveillance Network (TRANSNET). Clin Infect Dis 50:1101–1111
15. Sipsas NV, Kontoyiannis DP (2012) Invasive fungal infections in patients with cancer in the intensive care unit. Int J Antimicrob Agents 39:464–471
16. Anastassopoulou CG, Fuchs BB, Mylonakis E (2011) *Caenorhabditis elegans*-based model systems for antifungal drug discovery. Curr Pharm Des 17:1225–1233
17. Lionakis MS (2011) Drosophila and Galleria insect model hosts: new tools for the study of fungal virulence, pharmacology and immunology. Virulence 2:521–527
18. Prithiviraj B, Bais HP, Jha AK et al (2005) *Staphylococcus aureus* pathogenicity on *Arabidopsis thaliana* is mediated either by a direct effect of salicylic acid on the pathogen or by SA-dependent, NPR1-independent host responses. Plant J 42:417–432
19. Mulcahy H, Sibley CD, Surette MG et al (2011) *Drosophila melanogaster* as an animal model for the study of *Pseudomonas aeruginosa* biofilm infections *in vivo*. PLoS Pathog 7:e1002299
20. Atkinson S, Goldstone RJ, Joshua GW et al (2011) Biofilm development on *Caenorhabditis elegans* by *Yersinia* is facilitated by quorum sensing-dependent repression of type III secretion. PLoS Pathog 7:e1001250

21. Fuchs BB, O'Brien E, Khoury JB et al (2010) Methods for using *Galleria mellonella* as a model host to study fungal pathogenesis. Virulence 1:475–482
22. Fuchs BB, Eby J, Nobile CJ et al (2010) Role of filamentation in Galleria mellonella killing by Candida albicans. Microbes Infect 12:488–496
23. Breger J, Fuchs BB, Aperis G et al (2007) Antifungal chemical compounds identified using a C. elegans pathogenicity assay. PLoS Pathog 3:e18
24. Cruz MR, Graham CE, Gagliano BC et al (2013) *Enterococcus faecalis* inhibits hyphal morphogenesis and virulence of *Candida albicans*. Infect Immun 81:189–200
25. Muhammed M, Coleman JJ, Mylonakis E (2012) Caenorhabditis elegans: a nematode infection model for pathogenic fungi. Methods Mol Biol 845:447–454
26. Chamilos G, Lewis RE, Hu J et al (2008) *Drosophila melanogaster* as a model host to dissect the immunopathogenesis of zygomycosis. Proc Natl Acad Sci U S A 105:9367–9372
27. Lionakis MS, Kontoyiannis DP (2012) Drosophila melanogaster as a model organism for invasive aspergillosis. Methods Mol Biol 845:455–468
28. Lemaitre B, Nicolas E, Michaut L et al (1996) The dorsoventral regulatory gene cassette spatzle/Toll/cactus controls the potent antifungal response in Drosophila adults. Cell 86:973–983
29. Lionakis MS, Lewis RE, May GS et al (2005) *Toll*-deficient *Drosophila* flies as a fast, high-throughput model for the study of antifungal drug efficacy against invasive aspergillosis and *Aspergillus* virulence. J Infect Dis 191: 1188–1195
30. Chamilos G, Lionakis MS, Lewis RE et al (2006) *Drosophila melanogaster* as a facile model for large-scale studies of virulence mechanisms and antifungal drug efficacy in *Candida* species. J Infect Dis 193:1014–1022
31. Mowlds P, Kavanagh K (2008) Effect of pre-incubation temperature on susceptibility of *Galleria mellonella* larvae to infection by *Candida albicans*. Mycopathologia 165:5–12
32. Wojda I, Kowalski P, Jakubowicz T (2009) Humoral immune response of *Galleria mellonella* larvae after infection by *Beauveria bassiana* under optimal and heat-shock conditions. J Insect Physiol 55:525–531
33. Beanan MJ, Strome S (1992) Characterization of a germ-line proliferation mutation in *C. elegans*. Development 116:755–766
34. Kim DH, Feinbaum R, Alloing G et al (2002) A conserved p38 MAP kinase pathway in *Caenorhabditis elegans* innate immunity. Science 297:623–626
35. Kucherenko MM, Marrone AK, Rishko VM et al (2010) Paraffin-embedded and frozen sections of *Drosophila* adult muscles. J Vis Exp 46:2438

Part II

Investigations on Anti-biofilm Compounds and Strategies

Chapter 12

Microbiological Methods for Target-Oriented Screening of Biofilm Inhibitors

Livia Leoni and Paolo Landini

Abstract

The ability of many pathogenic bacteria to grow as a biofilm results in lower susceptibility to antibiotic treatments and to the host immune response, thus leading to the development of chronic infections. The understanding that biofilms can play an important role in bacterial virulence has prompted the search for inhibitors of biofilm development and of biofilm-related cellular processes. In this report, we present two examples of target-based microbiological screenings for antimicrobials endowed with anti-biofilm activity, aimed respectively at the inhibition of the signal molecule cyclic di-GMP and of quorum sensing.

Key words Biofilm, Antimicrobials, Signal molecules, Quorum sensing, Acyl-homoserine lactone, Elastase, c-di-GMP, Diguanylate cyclase

1 Introduction

Biofilm and planktonic cells differ significantly in their physiology, gene expression pattern, and morphology. Bacteria growing in biofilms are less sensitive to treatments with antimicrobial agents compared to planktonic cells [1–4]. Although the molecular mechanisms of tolerance to antibiotics are not yet fully understood, it appears that "persister cells," i.e., bacterial cells with a dormant metabolic state that provides insensitivity to most antibiotics, play an important role in decreased antibiotic sensitivity (reviewed in ref. 5). Due to their lack of sensitivity to antibiotics, there is a strong need for novel approaches aimed to the inhibition of cellular processes linked to biofilm formation, maintenance, and dispersal, which have therefore become important targets for the discovery of novel chemical inhibitors. Such inhibitors may be used either alone or in combination with conventional antimicrobial agents in anti-infective therapies. Since biofilm formation responds to signal molecules that can trigger the transition from single cell to microbial biofilms, strategies aimed at the development of anti-biofilm agents

Gianfranco Donelli (ed.), *Microbial Biofilms: Methods and Protocols*, Methods in Molecular Biology, vol. 1147,
DOI 10.1007/978-1-4939-0467-9_12, © Springer Science+Business Media New York 2014

often target either signal molecule biosynthesis or their transduction pathways.

Quorum sensing (QS) is a cell-cell communication process that plays a key role in virulence and biofilm formation in bacteria; hence, QS inhibitors (QSI) are considered promising anti-biofilm agents [6–8]. The most widespread QS signals in proteobacteria are *N*-acyl-L-homoserine lactones (acyl-HSL), and the mechanism of acyl-HSL-dependent QS in the human pathogen *Pseudomonas aeruginosa* has been widely studied (reviewed in ref. 9). *P. aeruginosa* has two interconnected acyl-HSL QS systems, based on the production of *N*-3-oxododecanoyl-homoserine lactone ($3OC_{12}$-HSL) and *N*-butanoyl-homoserine lactone (C4-HSL) as signal molecules. $3OC_{12}$-HSL is synthesized by the LasI synthase, and, when it reaches a threshold concentration in the bacterial cell, it can be bound by transcriptional regulator LasR. The LasR/$3OC_{12}$-HSL complex activates transcription of hundreds of genes, including genes involved in the production of virulence factors, in the expression of efflux pumps involved in antibiotic resistance [6, 10].

Although QS is not strictly necessary for biofilm formation in *P. aeruginosa*, it plays a crucial role in shaping the properties of both biofilm matrix and cells. *P. aeruginosa* mutants impaired in QS typically produce biofilms more susceptible to antibiotics and to the host immune system, and QS is therefore considered a promising target for novel antimicrobial agents [6, 8]. Indeed, several compounds inhibiting *P. aeruginosa* acyl-HSL QS have been discovered by library screening of synthetic and natural compounds. Some of these compounds are able to prevent *P. aeruginosa* infection in animal models and to increase *P. aeruginosa* biofilms susceptibility to both antibiotics and the host immune system [6, 8, 11]. Recently, 2-aminobenzimidazole derivatives were discovered that are capable of dispersing *P. aeruginosa* biofilms [12].

Another class of signal molecules instrumental in biofilm regulation is modified nucleotides, such as cyclic di-GMP monophosphate (c-di-GMP) and, possibly, cyclic di-adenosine monophosphate (c-di-AMP); the involvement of the latter in biofilm formation, however, has not yet been characterized. Accumulation of c-di-GMP promotes biofilm formation and maintenance via a variety of different mechanisms resulting in increased production of cell adhesion factors, including allosteric activation of enzyme activity and acting as a ligand for regulatory proteins [13–15]. Intracellular levels of c-di-GMP are determined by two classes of enzymes with opposite activities: diguanylate cyclases, which synthesize c-di-GMP, and c-di-GMP phosphodiesterases that hydrolyze it into the inactive diguanylate phosphate (pGpG) form (reviewed in ref. 16). Genes involved in c-di-GMP biosynthesis and turnover are conserved in all *Eubacteria*, while absent in animal species [17], thus making enzymes involved in c-di-GMP biosynthesis an interesting target for anti-biofilm agents; indeed, diguanylate cyclases have been targeted for development of specific inhibitors both via chemical design of substrate analogues [18] and via high-throughput screenings [19].

In this chapter, we describe screening strategies for inhibitors of either QS or c-di-GMP biosynthesis. Both strategies are defined by a primary assay, suitable for high-throughput screening of chemical compounds, and a secondary assay, used to confirm the activity and target specificity of the hit compounds in the primary assays. Bacterial strains used as biosensors in the screening assays presented are available from the authors upon request (L.L. for the *P. aeruginosa* QS biosensors and P.L. for *Escherichia coli* strains overproducing c-di-GMP).

As a general remark, it should be pointed out that, while we chose to illustrate target-oriented screening strategies linked to the production of signal molecules in Gram-negative bacteria (viz., acyl-HSL and c-di-GMP), some of the techniques presented here can be applied to a variety of different biological systems. For instance, the crystal violet assay is commonly used for biofilm determination by any bacterium, either Gram-positive or Gram-negative, and even for eukaryotic microorganisms. However, in the method presented here, it is the utilization of a specific indicator strain (the AM70/pTOPOAdrA strain, see below) that confers the crystal violet assay specificity for inhibitors of c-di-GMP biosynthesis. Likewise, reporter-based assays modeled on the one described here can be devised for different QS systems, both in Gram-negative and Gram-positive bacteria.

2 Materials

- Bacterial strains: *P. aeruginosa* PA14 [20] and PA14-R3 biosensor [21], *E. coli* AM70/pTOPOAdrA [22].
- Growth media: Luria-Bertani broth (LB: 10 g/L NaCl; 10 g/L tryptone; 5 g/L yeast extract); Luria-Bertani agar (LA, as LB plus 15 g/L agar); M9Glu/sup medium (82 mM Na_2HPO_4, 24 mM KH_2PO_4, 130 mM NaCl, 19 mM NH_4Cl, 1 mM $MgSO_4$, 0.1 mM $CaCl_2$, 0.4 % glucose, 0.25 g/L tryptone, 0.125 g/L yeast extract). When necessary, ampicillin is supplied at 100 μg/mL.
- Phosphate buffer saline (PBS): 137 mM NaCl; 2.7 mM KCl; 10 mM Na_2HPO_4; 2 mM KH_2PO_4, pH 7.4.
- MOPS buffer: 83.7 g/L 3-(*N*-morpholino) propanesulfonic acid (MOPS); 13.6 g/L sodium acetate, trihydrate; 3.7 g/L ethylenediaminetetraacetic acid (EDTA), disodium salt.
- Crystal violet, 1 % aqueous solution.
- Black, clear-bottom 96-wells microtiter plates.

- Wallac 1420 Victor³V multilabel plate reader.
- Polystyrene 96-well microtiter plates.
- 12.5 cm Supelcosil LC-18-DB, 3 μm particle size, reversed-phase HPLC column.
- HPLC buffers: buffer A (0.1 M KH_2PO_4 pH 6); buffer B (0.1 M KH_2PO_4 pH 6, 20 % CH_3OH).
- Elastin-Congo red.
- Protease Buffer (100 mM Tris-HCl; 1 mM $CaCl_2$, pH 7.5).

3 Methods

3.1 Rationale of the Primary Screening System for QSI

The screening system for inhibitors of *P. aeruginosa* QS [21] is based on a biosensor strain (PA14-R3) able to detect the QS signal molecule $3OC_{12}$-HSL. The PA14-R3 strain carries a nonfunctional allele of the *lasI* gene and is thus unable to synthesize $3OC_{12}$-HSL; however, it can respond to exogenous $3OC_{12}$-HSL provided either through supply of the purified molecule or by co-cultivation with a wild-type *P. aeruginosa* proficient in $3OC_{12}$-HSL production, such as the PA14 strain. The screening system is detailed in Fig. 1. The $3OC_{12}$-HSL signal synthesized by the wild-type PA14 diffuses into the PA14-R3 biosensor and induces bioluminescence emission. The addition of a molecule with inhibitory activity towards any process related to the $3OC_{12}$-HSL-dependent QS system, namely, $3OC_{12}$-HSL synthesis, transport, and perception, will reduce light emission by the biosensor with respect to a control coculture grown in the absence of any chemical compound.

3.2 QSI Screening

1. Grow *P. aeruginosa* PA14 wild type and the $3OC_{12}$-HSL reporter strain PA14-R3 overnight at 37 °C on LA plates.
2. Scrape bacteria from LA plate surfaces and dilute in 1 mL of LB to an absorbance at 600 nm (A_{600nm}) of 0.09 and 0.03 for PA14-R3 and PA14, respectively (3/1 reporter/wild-type ratio).
3. Aliquot 100 μL per well of the co-culture in a 96-well microtiter black plates with clear bottom.
4. As untreated control, add 100 μL of LB in six wells containing the co-culture.
5. Set up serial dilutions of the compounds to 2× the final concentrations to be tested. For example, chemical compounds to be used in the screening assays are dissolved in 10 mg/mL DMSO and then diluted to 200, 20, and 2 μg/mL in LB medium (to obtain 100, 10, and 1 μg/mL final concentrations in the assay). Aliquot 100 μL of each compound in the microtiter wells containing the coculture.

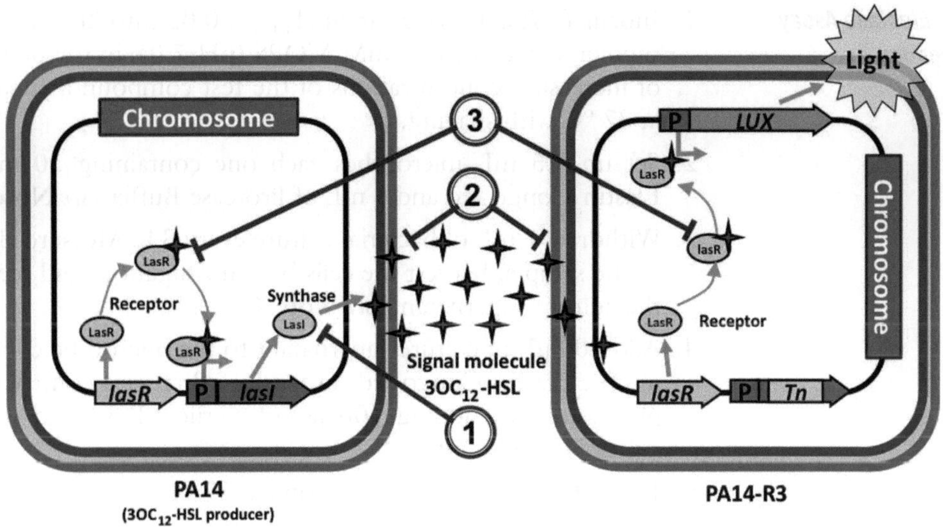

Fig. 1 Schematic representation of the PA14/PA14-R3 cocultivation screening system. The wild-type PA14 produces the $3OC_{12}$-HSL signal that in turn activates PA14-R3 bioluminescence emission. PA14-R3 is a PA14 derivative in which a transcriptional fusion between the LasR-dependent *rsaL* promoter (P) and the *luxCDABE* operon (LUX) was chromosomally integrated at the *attB* neutral site of the chromosome. In addition, the *lasI* gene encoding $3OC_{12}$-HSL synthase was inactivated by transposon insertion (Tn) in the PA14-R3 strain. Inactivation of the *lasI* gene prevents self-activation of the reporter system, thus allowing the screening system to identify also inhibitors of acyl-HSL transport and uptake. Molecules interfering with synthesis of the signal molecule by LasI in the PA14 strain (reaction marked as (*1*) in the figure), transport of the signal molecule across the cell envelope (*2*), and activity of the signal molecule receptor LasR (*3*) will all cause a reduction of luminescence in comparison to the untreated control. *Tn*: *MrT7 transposon* (Reproduced from ref. 21 with permission)

6. Incubate the microtiter plate at 37 °C for 4 h with gentle shaking.
7. Measure A_{600nm} and light counts per second (LCPS), simultaneously (*see* **Note 1**).
8. Put back the microtiter plate at 37 °C and incubate for further 2 h, then repeat the above measurement (*see* **Note 2**).

3.3 Secondary Assay: Elastase Production

Elastase is a *P. aeruginosa* secreted protease that mainly targets mammalian elastin and plays a key role in virulence [4]. Transcription of the elastase gene, *lasB*, is strongly activated by QS [9]. Therefore, a convenient method to validate the QS inhibitory activity of a hit compound is to measure its effect on elastase production, in comparison with an untreated control.

The elastase assay is based on the Elastin-Congo red reagent, a water-insoluble powder in which elastin is bound to the Congo red dye. Hydrolysis of elastin causes the release of the dye in the aqueous phase. The amount of released (soluble) dye is proportional to the level of hydrolyzed elastin and, ultimately, to the level of elastase present in a *P. aeruginosa* culture supernatant. The method described here is modified from reference [23].

3.4 Elastase Assay Procedure

1. Inoculate *P. aeruginosa* at an $A_{600nm} \approx 0.05$ into 20 mL of LB supplemented with 50 mM MOPS (pH 7.0), in the presence of increasing concentrations of the test compound. Incubate at 37 °C with shaking.
2. Set up 1.5 mL microtubes each one containing 20 mg of Elastin-Congo red and 1 mL of Protease Buffer (*see* **Note 3**).
3. Withdraw 1 mL of bacterial culture every 3 h. Measure A_{600nm} of the sample, harvest the cells by centrifugation, and recover the culture supernatant (*see* **Note 4**).
4. Add 100 μL of culture supernatant to the microtube containing the Elastin-Congo red suspension. Prepare a control sample (blank) by adding 100 μL of sterile LB instead of the culture supernatant.
5. Incubate 2 h with gentle shacking at 37 °C.
6. Centrifuge 3 min at 12,000 ×*g* at room temperature.
7. Read optical absorbance at 495 nm of the clear supernatants, using as blank the control sample (see above). Normalize with respect to the A_{600nm} of the corresponding culture.

3.5 Rationale of the Primary Screening System for Inhibitors of c-di-GMP Biosynthesis

The AM70 strain is a derivative of the standard *E. coli* laboratory strain MG1655 carrying a deletion of the *csgA* gene [22]. The *csgA* gene encodes the main subunit of curli fibers, the most important adhesion factor in *E. coli*, and its deletion results in the inability to form biofilm on microtiter plates. Transformation of AM70 with a multicopy plasmid carrying the *adrA* gene (pTOPOAdrA), encoding a diguanylate cyclase, stimulates cellulose production through c-di-GMP biosynthesis. Cellulose acts as an adhesion factor and promotes cell adhesion to microtiter plates, which can be detected by crystal violet staining. Chemical compounds interfering with c-di-GMP biosynthesis, such as the sulfonamide antibiotic sulfathiazole, completely inhibit biofilm formation and prevent cell adhesion to microtiter plates (Fig. 2).

3.6 Screening for Inhibitors of c-di-GMP Biosynthesis by Crystal Violet Assay

1. Grow the AM70/pTOPOAdrA strain overnight (*see* **Note 5**) in M9Glu/sup medium supplemented with ampicillin. Dilute the culture to $A_{600nm} = 0.02$ in fresh medium with no antibiotic and distribute 100 μL in each well of a microtiter plate.
2. Set up serial dilutions of the compounds to 2× the final concentrations to be tested. For example, chemical compounds to be used in the screening assays are dissolved in 10 mg/mL 50 % DMSO and then diluted to 200, 20, and 2 μg/mL in M9Glu/sup medium (to obtain 100, 10, and 1 μg/mL final concentrations in the assay). DMSO at final concentrations of 0.5 % or lower had no effect on biofilm formation. Aliquot 100 μL of each compound to be tested to the microtiter wells inoculated with the AM70/pTOPOAdrA culture.
3. Incubate for 20 h at 30 °C (*see* **Note 6**).

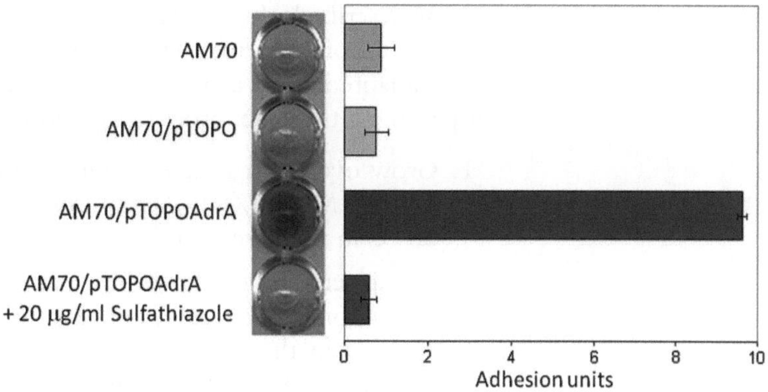

Fig. 2 c-di-GMP-dependent biofilm formation in *E. coli* AM70. Biofilm was detected with the crystal violet staining method described in the text. *Left panel*: biofilm visualization after crystal violet staining (**step 8** of the procedure Subheading 3.6); *right panel*: quantification of biofilm cells as adhesion units (**steps 9–11** of the same procedure). Biofilm formation only takes place when the diguanylate cyclase AdrA is overexpressed from the multicopy plasmid pTOPOAdrA, and it is inhibited by the sulfonamide antibiotic sulfathiazole [22]

4. Determine the A_{600nm} of the liquid culture. This can be automatically done by a plate reader spectrophotometer (for HTS) or by collecting 100 μL and adding them to 900 μL of fresh medium in a 1-mL cuvette (this value is indicated as "A_{600nm} culture" and will be used to quantify the amount of biofilm formed).
5. Remove the liquid culture and wash the microtiter plate wells with 200 μL PBS.
6. Air-dry the microtiter well for 10–20 min.
7. Add 200 μL of 1 % crystal violet and incubate at room temperature.
8. Remove crystal violet solution, wash with water, and drain excess water by placing microtiter plates upside down on blotting paper. At this stage, biofilm cells are stained by crystal violet, allowing visualization (*see* Fig. 2). For quantitative evaluation of biofilm, proceed as follows.
9. Resuspend the stained biofilm cells in 200 μL ethanol and incubate for at least 20 min; if needed, scrape the biofilm layer to facilitate solubilization (*see* **Note 7**).
10. Measure the A_{590nm} of the solubilized crystal violet (CV_{A590nm}).
11. Quantitative determination of biofilm is obtained as "adhesion units," i.e., the value obtained by the ratio CV_{A590nm}/A_{600nm} culture (values obtained at **step 10** and **step 4** of the procedure).

Compounds tested were considered active when leading to a more than fourfold reduction in the adhesion units value.

As described, the assay aims at the discovery of molecules able to inhibit biofilm formation. The crystal violet method can also be easily adapted to screen for compounds able to promote biofilm dispersal. In this case, proceed as follows:

1. Grow overnight bacterial cultures diluted to $A_{600nm} = 0.02$ in 190 μL M9Glu/sup medium in microtiter plates for 20 h at 30 °C in the absence of any additional compounds.
2. Set up serial dilutions of the compounds to 20× the final concentrations to be tested. For example, chemical compounds to be used in the screening assays are dissolved in 10 mg/mL 50 % DMSO and then diluted to 2,000, 200, and 20 μg/mL in M9Glu/sup medium (to obtain 100, 10, and 1 μg/mL final concentrations in the assay). DMSO at final concentrations of 0.5 % or lower had no effect on biofilm dispersal. Aliquot 10 μL of each compound to be tested to the microtiter wells inoculated with the AM70/pTOPOAdrA culture.
3. Incubate for 2 h at 30 °C.

Proceed from **step 4** onwards as described for inhibitors of biofilm formation.

3.7 Secondary Screening for Inhibitors of c-di-GMP Biosynthesis: HPLC Determination

The method described here has a detection limit of about 0.3 pmol/mg bacterial cells (dry weight). In the AdrA-overexpressing strain, c-di-GMP concentrations range in the orders of 300–500 pmol/mg ([22]; Fig. 3). The HPLC separation protocol was originally described in reference [24] and adopted to c-di-GMP determination in *Escherichia coli* and *Borrelia burgdorferi* nucleotide extracts, as described [22, 25].

1. Grow overnight cultures of AM70/pTOPOAdrA (usually 20 mL culture).
2. Harvest bacterial cells by centrifugation (*see* **Note 8**). Resuspend the pellet in PBS buffer and transfer the sample in a *previously weighed* eppendorf tube. Centrifuge again to collect cells and discard the supernatant. Weigh the eppendorf tube and annotate the weight of the pellet (dry weight). Store the pellet at −20 °C (*see* **Note 9**).
3. Thaw the pellet and resuspend it in cold 0.4 M $HClO_4$ at a ratio of 0.4 mL/50 mg bacterial cells (dry weight).
4. Disintegrate bacterial cells by sonication (*see* **Note 10**). Remove cell debris and precipitated proteins by centrifugation (12,000 × *g*, 10 min, 4 °C).
5. Neutralize supernatant with 1 M K_2CO_3 (*see* **Note 11**), keep on ice for 10 min, and centrifuge at 12,000 × *g* for 3 min. Freeze supernatant at −20 °C (*see* **Note 12**). Thaw the supernatant, centrifuge at 12,000 × *g* for 3 min, and inject into the HPLC column.

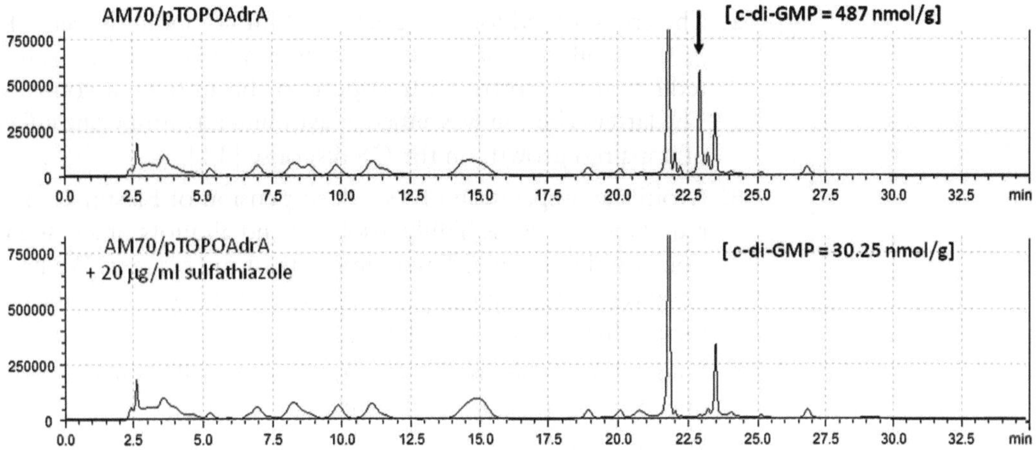

Fig. 3 HPLC determination of c-di-GMP intracellular concentrations. HPLC chromatograms in the AM70 *E. coli* strain overexpressing the diguanylate cyclase AdrA (AM70/pTOPOAdrA), in the presence (chromatogram shown below) and in the absence (chromatogram shown above) of the antibiotic sulfathiazole. The peak corresponding to c-di-GMP, with a retention time of 22.6 min, is marked by an *arrow*; the peak with a retention time of 21.8 min corresponds to NAD, as determined with a purified standard, and its intracellular concentration is not affected by exposure to sulfathiazole

6. Separation of nucleotides is carried out on a HPLC system equipped with a diode-array detector using a 12.5 cm Supelcosil LC-18-DB, 3 μm particle size, reversed-phase column, at a temperature of 18 °C. Elution conditions are 9 min at 100 % buffer A, 6 min up to 12 % buffer B, 2.5 min up to 45 % buffer B, and 2.5 min up to 100 % buffer B, holding at 100 % buffer B for 5.5 min and returning to 100 % buffer A in 5 min. Flow rate is maintained at 1.3 mL/min. In these conditions, purity index of c-di-GMP peak is 0.96. Its identity as genuine c-di-GMP is determined by coelution and identical UV absorption spectra with a c-di-GMP standard (*see* **Note 13**). c-di-GMP concentrations are calculated from its extinction coefficient (ε) of 23,700 at 254 nm [26].

The result of a c-di-GMP determination in bacterial cells grown either in the presence or in the absence of the sulfonamide antibiotic sulfathiazole is shown in Fig. 3.

4 Notes

1. For the Wallac 1420 Victor³V multilabel plate reader (Perkin Elmer), relevant parameters for bioluminescence measurement were as follows: emission aperture, large; counting time, 1 s. Relevant parameters for absorbance measurement were as follows: filter 595/60; excitation aperture, normal; reading time, 0.1 s.

2. The criteria used for the selection of hit compounds were (1) ≥50 % inhibition of bioluminescence emission and (2) ≤20 % reduction of growth with respect to the untreated controls. The latter criterion was aimed at avoiding any unspecific effect of impaired growth on the QS response [11].
3. Avoid the preparation of a stock suspension of Elastin-Congo red. This powder is highly insoluble and aliquots of a suspension could contain different amounts of the reagent. We have observed that aliquoting the powder in each sample microtube enhances assay reliability.
4. Under these conditions, *P. aeruginosa* usually starts to produce fairly detectable levels of elastase at around $A_{600} \approx 3.0$.
5. *E. coli* strains harboring plasmids carrying diguanylate cyclase-encoding genes, such as pTOPOAdrA, tend to lose their phenotype with time. It is best to transform the AM70 strain shortly before utilization and streak one transformant colony on LA plates. Stored at 4 °C, the colonies retain their phenotype for a few weeks.
6. Incubation at temperature of 30 °C or lower is crucial for production of cellulose, i.e., the main adhesion factor produced in response to c-di-GMP synthesis by the AdrA protein [27]. AdrA stimulates biofilm formation even at 37 °C but through an unknown mechanism.
7. Resuspension of the biofilm can be problematic and can be helped by vigorous pipetting. Utilization of 10 % acetone in ethanol allows quicker resuspension of the biofilm (But make sure that your plasticware is resistant to solvents!).
8. This step is just for harvesting cells, so speed and times for centrifugation can vary. Our standard conditions are $6,000 \times g$ for 10 min at 4 °C.
9. It is suggested to freeze the pellet, since freezing and thawing the cells will facilitate their lysis in the sonication step. Frozen pellets can be stored for several days.
10. The efficiency of sonicators can vary greatly. Sonication must be carried out on ice, giving pulses of no longer than 30 s, with intervals of 30 s, to avoid heating of the sample. The procedure is repeated until significant decrease of turbidity in the bacterial suspension can be detected by eye.
11. The final pH should be 6.0. This value is obtained by adding 32 μL 1 M K_2CO_3 to 150 μL of the acid supernatant.
12. It is suggested to freeze the neutralized supernatant before HPLC analysis to facilitate removal of CO_2 formed during the reaction between $HClO_4$ and K_2CO_3.
13. Using these separation conditions, the c-di-GMP peak is eluted as a unique peak with a slightly higher retention time than

NAD, a nucleotide with a high intracellular concentration. NAD intracellular concentration can thus be used as a control for specific inhibition of c-di-GMP biosynthesis (see also Fig. 3).

Acknowledgments

We wish to thank Nadia Raffaelli for the protocol of the HPLC method for c-di-GMP determination and for critical reading. Work in P.L.'s laboratory has been supported by the Italian Foundation for Research on Cystic Fibrosis (project FFC#13/2009, with the contribution of Delegazione Novara and Delegazione Cosenza-2) and by the CHEM-PROFARMA-NET Research Program (Project RBPR05NWWC_004) from the Italian Ministry for University and Research.

Funding to L.L. has been provided by the Italian Foundation for Research on Cystic Fibrosis (projects FFC#14/2010 and FFC#13/2011, with the contribution of delegazione FFC di Vittoria-Ragusa, Delegazione FFC del Lago di Garda con Gruppi di Sostegno di Chivasso, dell'Isola Bergamasca e della Valpolicella, Delegazione FFC di Verbania e Antonio Guadagnin & Figlio).

References

1. Costerton JW, Lewandowski Z, Caldwell DE et al (1995) Microbial biofilms. Annu Rev Microbiol 49:711–745
2. Anderl JN, Franklin MJ, Stewart PS (2000) Role of antibiotic penetration limitation in *Klebsiella pneumoniae* biofilm resistance to ampicillin and ciprofloxacin. Antimicrob Agents Chemother 44:1818–1824
3. Jabra-Rizk MA, Meiller TF, James CE, Shirtliff ME (2006) Effect of farnesol on *Staphylococcus aureus* biofilm formation and antimicrobial susceptibility. Antimicrob Agents Chemother 50:1463–1469
4. Driscoll JA, Brody SL, Kollef MH (2007) The epidemiology, pathogenesis and treatment of *Pseudomonas aeruginosa* infections. Drugs 67:351–368
5. Lewis K (2008) Multidrug tolerance of biofilms and persister cells. Curr Top Microbiol Immunol 322:107–131
6. Bjarnsholt T, Tolker-Nielsen T, Høiby N, Givskov M (2010) Interference of *Pseudomonas aeruginosa* signalling and biofilm formation for infection control. Expert Rev Mol Med 12:e11
7. Høiby N, Ciofu O, Johansen HK et al (2011) The clinical impact of bacterial biofilms. Int J Oral Sci 3:55–65
8. Galloway WR, Hodgkinson JT, Bowden S et al (2012) Applications of small molecule activators and inhibitors of quorum sensing in Gram-negative bacteria. Trends Microbiol 20:449–458
9. Williams P, Winzer K, Chan WC, Cámara M (2007) Look who's talking: communication and quorum sensing in the bacterial world. Philos Trans R Soc Lond B Biol Sci 362: 1119–1134
10. Schuster M, Greenberg EP (2006) A network of networks: quorum-sensing gene regulation in *Pseudomonas aeruginosa*. Int J Med Microbiol 296:73–81
11. Imperi F, Massai F, Ramachandran Pillai C et al (2013) New life for an old drug: the anthelmintic drug niclosamide inhibits *Pseudomonas aeruginosa* quorum sensing. Antimicrob Agents Chemother 57:996–1005
12. Frei R, Breitbach AS, Blackwell HE (2012) 2-Aminobenzimidazole derivatives strongly inhibit and disperse *Pseudomonas aeruginosa* biofilms. Angew Chem Int Ed Engl 51:5226–5229
13. Kader A, Simm R, Gerstel U et al (2006) Hierarchical involvement of various GGDEF domain proteins in rdar morphotype development of *Salmonella enterica* serovar Typhimurium. Mol Microbiol 60:602–616

14. Fazli M, O'Connell A, Nilsson M et al (2011) The CRP/FNR family protein Bcam1349 is a c-di-GMP effector that regulates biofilm formation in the respiratory pathogen *Burkholderia cenocepacia*. Mol Microbiol 82:327–341
15. Romling U, Galperin MY, Gomelsky M (2013) Cyclic di-GMP: the first 25 years of a universal bacterial second messenger. Microbiol Mol Biol Rev 77:1–52
16. Tamayo R, Pratt JT, Camilli A (2007) Roles of cyclic diguanylate in the regulation of bacterial pathogenesis. Annu Rev Microbiol 61: 131–148
17. Galperin MY (2004) Bacterial signal transduction network in a genomic perspective. Environ Microbiol 6:552–567
18. Spangler C, Kaever V, Seifert R (2011) Interaction of the diguanylate cyclase YdeH of Escherichia coli with 2′, (3′)-substituted purine and pyrimidine nucleotides. J Pharmacol Exp Ther 336:234–241
19. Sambanthamoorthy K, Sloup RE, Parashar V et al (2012) Identification of small molecules that antagonize diguanylate cyclase enzymes to inhibit biofilm formation. Antimicrob Agents Chemother 56:5202–5211
20. Rahme LG, Stevens EJ, Wolfort SF et al (1995) Common virulence factors for bacterial pathogenicity in plants and animals. Science 268:1899–1902
21. Massai F, Imperi F, Quattrucci S et al (2011) A multitask biosensor for micro-volumetric detection of N-3-oxo-dodecanoyl-homoserine lactone quorum sensing signal. Biosens Bioelectron 26:3444–3449
22. Antoniani D, Bocci P, Maciag A et al (2010) Monitoring of diguanylate cyclase activity and of cyclic-di-GMP biosynthesis by whole-cell assays suitable for high-throughput screening of biofilm inhibitors. Appl Microbiol Biotechnol 85:1095–1104
23. Ohman DE, Burns RP, Iglewski BH (1980) Corneal infections in mice with toxin A and elastase mutants of *Pseudomonas aeruginosa*. J Infect Dis 142:547–555
24. Stocchi V, Cucchiarini L, Magnani M et al (1985) Simultaneous extraction and reverse-phase high-performance liquid chromatographic determination of adenine and pyridine nucleotides in human red blood cells. Anal Biochem 146:118–124
25. Kostick JL, Szkotnicki LT, Rogers EA et al (2011) The diguanylate cyclase, Rrp1, regulates critical steps in the enzootic cycle of the Lyme disease spirochetes. Mol Microbiol 81:219–223
26. Hayakawa Y, Reiko N, Hirata A et al (2003) A facile synthesis of cyclic bis(3′ → 5′)diguanylic acid. Tetrahedron 59:6465–6471
27. Zogaj X, Nimtz M, Rohde M et al (2001) The multicellular morphotypes of *Salmonella typhimurium* and *Escherichia coli* produce cellulose as the second component of the extracellular matrix. Mol Microbiol 39:1452–1463

Chapter 13

In Vitro Screening of Antifungal Compounds Able to Counteract Biofilm Development

Marion Girardot and Christine Imbert

Abstract

Fungi are able to grow as a single-species or a more complex biofilm attached to inert surfaces (catheters…) or tissues (lung…). This last form is a microbial niche which must be considered as a major risk factor of developing a human fungal infection. Nowadays, only a few therapeutic agents have been shown to be active against fungal biofilms in vitro and/or in vivo. So there is a real need to find new anti-biofilm molecules. Here we describe in detail some rapid, 96-well microtiter plate-based methods, for the screening of compounds with anti-biofilm activity against *Candida* spp. yeasts. Two approaches will be considered: prophylactic or curative effects of the tested compounds by producing biofilms on two supports – polystyrene well surfaces and catheter sections.

Key words Biofilms, *Candida* spp., Antifungal compounds, Catheters, Microtiter plates, Fungal infections, Colorimetric assay, XTT

1 Introduction

Biofilms are consortia of microbial cells aggregated, attached to a surface or an interface and embedded in a self-produced extracellular matrix [1, 2]. In the human body, fungi can grow as a single-species or a more complex biofilm that would host few fungal and/or bacterial species. Fungal biofilms can develop on both superficial and deeper sites, be attached to inert surfaces (catheters, prosthesis…) or tissues (sinus, lung…). Both yeasts and filamentous fungi are able to form biofilms both in vitro and in vivo; currently, *Candida* spp. and *Aspergillus* spp. biofilms are the most studied and the better understood fungal biofilms.

A biofilm acts as a microbial niche that is able to cause infections; so the biofilm must be considered as a major risk factor of developing a fungal infection. It has been widely shown that microbes growing as a biofilm (sessile microbes) have a different phenotype compared to those of planktonic microorganisms;

Gianfranco Donelli (ed.), *Microbial Biofilms: Methods and Protocols*, Methods in Molecular Biology, vol. 1147, DOI 10.1007/978-1-4939-0467-9_13, © Springer Science+Business Media New York 2014

importantly, the clinically most relevant consequence is their lack of sensibility to a wide range of antimicrobial agents [3–7].

Up to now, the in vitro efficiency of antifungal agents against a fungal species has been mainly evaluated measuring the minimal inhibitory concentration (MIC), studying more or less fungal strains. However, in vitro and in vivo data do not necessarily agree: an antifungal agent sometimes displays a very low MIC towards a strain suggesting an in vitro sensitivity; however, this molecule may fail to fight the fungal disease at the patient level. This lack of correlation may be explained, at least for a part, by the different status of microbial cells in patients (biofilm) versus for MIC tests (planktonic).

So, the knowledge of the biofilm lifestyle and the significant available current data focusing on fungal infections associated with the development of a biofilm should encourage microbiologists to test the activity of antifungal compounds on both planktonic (MIC) and sessile microorganisms. Moreover, the recently published ESCMID guidelines for the diagnosis and management of *Candida* diseases highlight specifically the case of catheter-related bloodstream infections which require a cautious choice of an antifungal agent known to be active against *Candida* spp. biofilms and also additional measures, such as removal of the device when possible [8, 9].

So, nowadays, the anti-biofilm activity should be taken in account to evaluate the interest of antifungal compounds.

Only a few agents have been shown to be active against fungal biofilms in vitro and/or in vivo. The most effective available agents against *Candida* biofilms would be echinocandins and amphotericin B lipid formulations although azole compounds would be poorly efficient; however, beyond the biofilm status, it is essential to consider the fungal genus and species as the active agents against *Aspergillus* biofilms are quite opposite to those active on *Candida* biofilms [3, 10–16].

So, there is a need to find new anti-biofilm molecules, ideally with new mechanisms of action to expand this very short list of available conventional antifungal agents able to counteract fungal biofilms.

Here we describe in detail some rapid, reproducible, accurate, and inexpensive 96-well microtiter plate-based methods for the screening of compounds with anti-biofilm activity against *Candida* spp. yeasts. Depending on the required level of screening, we suggest to produce biofilms directly either onto the polystyrene well surface or on catheter sections. The method using wells as substrates is faster but less effective; it is ideal for a broad screening. Biofilms produced onto silicone catheters are certainly closer to those that develop in vivo in animal models or in patients; however, their production needs more steps making it more time-consuming. In addition, some steps require more precision using silicone catheters, and approximate realization may cause a significant variability in results. It has been demonstrated that the kinetic as well as the architecture of the biofilm may be different depending on the substrate's nature [5, 17, 18]. So, the method with silicone sections permits a more relevant screening.

Moreover, different approaches can be considered to test the candidate anti-biofilm compounds depending on the desired target of action resulting in prophylactic or curative interests.

In the prophylactic approach, these compounds may be able to prevent the adhesion phase, which is the early phase during the biofilm growth process [5]. Two strategies would be used to test their anti-adherent activity: in one case, the culture medium would be enriched with the tested molecule to evaluate the ability of yeasts to adhere to polystyrene surfaces in presence of this molecule and to form a biofilm. Otherwise, polystyrene surfaces would be pretreated by the tested molecule, to obtain an active noncovalent covering layer, and the adhesion of *Candida* spp. to these pretreated surfaces and their ability to form biofilms would be then evaluated.

In the curative approach, the anti-biofilm compounds may be screened for their ability to act on later phases of the biofilm growth, such as intermediate or mature phases.

So here are described two methods to test in microplates the anti-biofilm activity of molecules using two materials, polystyrene (well surfaces) and silicone (catheter sections), and corresponding to prophylactic or curative approaches.

2 Materials

1. *C. albicans* strains isolated from patients (*see* **Notes 1–3**).
2. Presterilized, polystyrene, flat-bottomed, untreated 96-well microtiter plates.
3. Catheters in silicone 100 % (medical grade; 2 mm and 3.2 mm in inside and outside diameters, respectively).
4. Sabouraud agar slants for yeasts' subcultures.
5. Yeast Nitrogen Base medium supplemented with 30 mM of glucose for biofilm development and treatment; YNB-Glc.
6. Tetrazolium salt 2,3-bis[2-methyloxy-4-nitro-5-sulfophenyl]-2H-tetrazolium-5-carboxanilide (XTT) salt.
7. Menadione.
8. Sterile phosphate-buffered saline (PBS), pH: 7.2 stored at 4 °C until use.
9. Neat fetal bovine serum stored at −20 °C until use.
10. Dimethyl sulfoxide.
11. Ringer's lactate.
12. Acetone.
13. Saliva stored at −20 °C until use.

14. DL-Dithiothreitol.
15. Distilled water.
16. Ethanol 95 %.
17. Microtiter plate reader with 490 (or 450) nm optical filter.
18. Inverted microscope.
19. Multichannel pipet.
20. Fluconazole.
21. Amphotericin B.
22. Other studied anti-biofilm compounds.

3 Methods

Carry out all procedures at room temperature unless otherwise specified. Microplates will be always incubated covered with their lids. This section describes the following steps: (1) compounds and yeasts preparation, (2) methods for antifungal tests on polystyrene and then silicone surfaces, and (3) method for biofilm quantification and the reporting results.

3.1 Preparation of Compounds for Anti-biofilm Tests

3.1.1 Tested Compounds

1. Tested concentrations: the tested concentrations against biofilms will be in the range of planktonic MIC to 1,000 × planktonic MIC.
2. Depending on the solubility of the tested compounds, different solvents can be used to prepare stock solutions:
   - for polar compounds, preferred solvent would be distillated water.
   - DMSO will be used for the nonpolar compounds (*see* **Note 4**).
   - Tween 80 can be added as a surfactant to allow the dispersal of oil components (except mineral oil). We advise 0.1 % (v/v) (*see* **Note 5**).

The toxicity towards the cell model of other solvents should be pretested before using it. The solubilization of the compound in the solvent can be increased by shaking and sonication.

1. Highly concentrated (we advice ≥10 × MIC) stock solutions of the studied compound can be aliquoted into small volumes and stored at −80 °C.
2. The compound solutions at working concentrations can be stored at 4 °C up for 24 h in closed tubes.
3. Immediately before use, the compound in solution is homogenized by shaking.

3.1.2 Antifungal Agents (Positive Control)

1. The drugs used as control are solubilized following manufacturer's instructions.
2. Final working concentrations of each drug (in the range of 20–1,000 × the planktonic MIC) are prepared in YNB-Glc medium.
3. If needed, these antifungal solutions at concentration ≥20× MIC can be aliquoted into smaller volumes (in YNB-Glc medium) and stored at −80 °C until required.

3.2 Yeasts Preparation

1. Yeast isolates are stored either as glycerol stocks at −80 °C or on Sabouraud dextrose agar slants. All strains are maintained on slopes of Sabouraud dextrose agar and subcultured monthly.
2. 2 days before experiments start, pre-cultivate the yeasts on Sabouraud agar slants at 37 °C for 48 h.
3. Transfer a loopful of this culture to 25 mL of YNB-Glc in a sterile Erlenmeyer flask of 100 mL; incubate without shaking at 37 °C for 16 h (*see* **Notes 6–8**).
4. Transfer the obtained blastospores suspension in a 50 mL plastic tube; harvest blastospores and wash twice in sterile PBS; for each washing, centrifuge the suspension at 4,000 × *g* for 10 min at 25 °C, remove the supernatant, and add 25 mL of PBS (*see* **Notes 9** and **10**).
5. Add 10 mL of YNB-Glc to the washed yeasts after a last time removing of the supernatant; resuspend them.
6. Determine the concentration of the obtained fungal suspension using Kova cells and microscope (*see* **Note 11**).
7. Adjust this final suspension at a cell density of 5 × 106 blastospores/mL in YNB-Glc medium (*see* **Notes 12** and **13**).

3.3 Tests Using Polystyrene Well Surfaces

Here, we will discuss (1) the evaluation of the prophylactic effect of a candidate compound against *Candida* biofilms with two different methods, in presence or not of tested compound during the development of the biofilm, and (2) the evaluation of the curative effect of a candidate compound against a preformed biofilm.

3.3.1 Evaluation of the Prophylactic Effect

Evaluation of the Anti-adherent and Anti-biofilm Activities of Experimental Medium Supplemented with Tested Compound

1. Before starting the experiment, calculate the exact number of requested 96-well microplates depending on the number of strains, tested molecules, and conditions (concentrations). Prepare as many microplates as necessary (*see* **Notes 14–16**).
2. Add 200 μL of FBS to each test well.
3. Incubate the microplates overnight at 4 °C without shaking (*see* **Notes 17** and **18**).
4. Remove FBS in each well using vacuum aspiration (*see* **Note 19**).

5. Wash each well to remove FBS surplus: add 200 μL PBS in wells and then remove it using vacuum aspiration (*see* **Note 19**).
6. Add 196 μL of YNB-Glc in each well of the first column of microplates.
7. Add 100 μL of YNB-Glc in wells of other columns (columns 2–11).
8. Add 4 μL of the stock solution (the highest working concentration), solvent control, or antifungal agent control (positive control) in three successive wells of the first column of microplates (*see* **Notes 20–22**).
9. A serial twofold dilution is prepared: after mixing with a multichannel pipet the solutions in the wells of the first column, remove 100 μL of this first column and add it in the second column. After mixing with the pipet the solutions in the wells of the second column, remove 100 μL of this column and add it in the third column. Repeat this operation until column 11 after which the final 100 μL are discarded.
10. Add 100 μL of the yeast suspension (5×10^6 cells/mL) in each well.
11. Incubate the microplates without shaking for 2 h or/and 22 h at 37 °C to permit adhesion and biofilm formation, respectively (*see* **Note 23**).

Evaluation of the Anti-adherent and Anti-biofilm Activities Induced by Pretreatment of Polystyrene Well Surfaces by Tested Compound

1. Prepare as many microplates as necessary (*see* **Notes 14–16**).
2. Add 196 μL of FBS in each well of the first column of a microplate and 100 μL of FBS in wells of the other columns (columns 2–12) (*see* **Note 18**).
3. Add 4 μL of the studied stock solution, solvent control, or antifungal agent (positive control) in the wells of the first column in three successive wells by product (*see* **Notes 20–22**).
4. A serial twofold dilution is prepared: after mixing with a multichannel pipet the solutions in the wells of the first column, remove 100 μL of this column and add it in the second column. After mixing with the pipet the solutions in the wells of the second column, remove 100 μL of this column and add it in the third column. Repeat this operation until column 11 after which the final 100 μL are discarded.
5. Incubate the microplates at 4 °C (overnight) or at 37 °C (1 h) (without shaking) (*see* **Note 23**).
6. Remove noncoated surplus of both FBS and tested compound in each well using vacuum aspiration (*see* **Note 19**).
7. Add 200 μL of the yeast suspension (5×10^6 cells/mL) in each well.
8. Incubate the microplates without shaking for 2 h or/and 22 h at 37 °C to permit adhesion and biofilm formation, respectively (*see* **Note 23**).

3.3.2 Evaluation of the Curative Effect of Tested Compound

1. Prepare as many microplates as necessary (*see* **Notes 14–16**).
2. Add 200 μL of FBS to each test well. Incubate the microplates at 4 °C (overnight) or at room temperature (1 h) (*see* **Note 18**).
3. Remove FBS in each well using vacuum aspiration (*see* **Note 19**).
4. Wash each well to remove FBS surplus: add 200 μL PBS in wells and then remove it using vacuum aspiration (*see* **Note 19**).
5. Add 200 μL of the suspension of blastospores at 5×10^6 blastospores/mL in each well pretreated with FBS and incubate microplates for 1 h at 37 °C (without shaking).
6. Remove carefully nonadherent cells in each well using vacuum aspiration; the culture medium must be totally removed (*see* **Note 19**).
7. Wash twice with 250 μL PBS each well covered with adherent yeasts to remove all the nonadherent yeasts.
8. Add 250 μL of YNB-Glc to each well containing adherent cells.
9. Incubate the microplates without shaking at 37 °C for 2, 24, 48 h, or other point times in YNB-Glc to obtain biofilms with the desired stages of maturation.
10. For the preparation, use new 96-well microplates.
11. Add 196 μL of YNB-Glc in each well of the first column of microplates and 100 μL of YNB-Glc in the wells of other columns.
12. Add 4 μL of the stock solution, solvent control, or antifungal agent (positive control) in three successive wells of the first column of microplates (*see* **Notes 20–22**).
13. A serial twofold dilution is prepared: after mixing with a multichannel pipet the solutions in the wells of the first column, remove 100 μL of this column and add it in the second column. After mixing with the pipet the solutions in the wells of the second column, remove 100 μL of this column and add it in the third column. Repeat this operation until column 11 after which the final 100 μL are discarded.
14. Take now both the microplates with preformed biofilms and those with tested compounds.
15. Transfer 50 μL of each well of the microplates with tested compounds in the corresponding well of the microplates containing biofilms, being careful not to disrupt the biofilms (*see* **Note 24**).
16. Incubate preformed biofilms with tested compounds for 24 h without shaking (*see* **Note 23**).

3.4 Tests Using Silicone Sections

This section describes the procedure of the anti-biofilm test using silicone sections including (1) the preparation of the substrate, (2) the growing of biofilms, (3) the procedure of adding of tested compounds, and (4) the evaluation of the post-anti-biofilm activity of tested compounds.

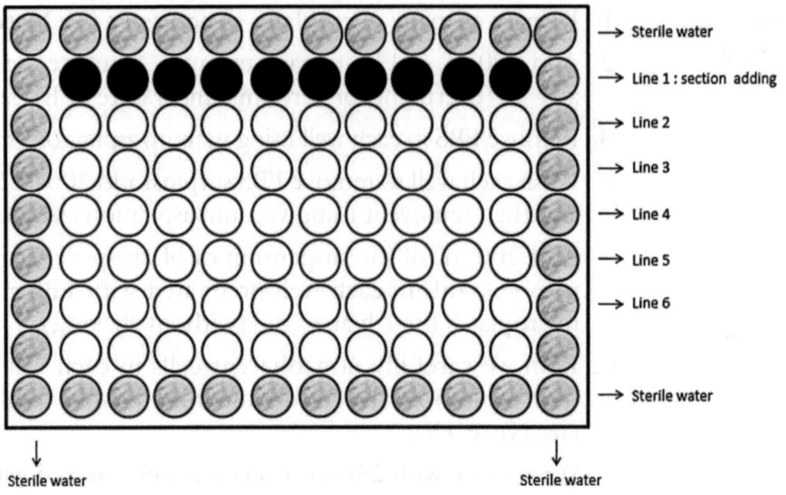

Fig. 1 Typical microplate map showing the arrangement of silicone sections. External wells are filled with sterile water to prevent dehydration losses, especially for the preparation of mature biofilms. At the beginning of the experiment, one silicone section is placed in each well drawn in black

3.4.1 Substrate Preparation

1. Prepare as many microplates as necessary (*see* **Notes 14–16**).
2. Add 200 µL of sterile water in all external wells; these wells are left unchanged for the experiment duration (*see* **Note 25** and Fig. 1).
3. Cut calibrated silicone sections of 5 mm from a silicone tubing using a scalpel and autoclave them (*see* **Notes 26–28**).
4. Put one calibrated section at the bottom of each well of the second line of the microplate; avoid adding sections in the wells of the outermost lines. A microplate map suggesting a useful arrangement of silicone sections is shown in Fig. 1 (*see* **Note 29**).
5. Add 200 µL of FBS to each well containing a silicone section. Incubate microplates overnight (*see* **Notes 17** and **30**).
6. Microplates are incubated at 4 °C (overnight) or at room temperature (1 h) (without shaking).

3.4.2 Development of Biofilm

1. Remove FBS in each well using vacuum aspiration and wash them twice with 200 µL of YNB-Glc (*see* **Note 19**).
2. Silicone sections are incubated in wells for 1 h at 37 °C (without shaking) with 200 µL of the fungal suspension at 5 × 106 yeasts/mL.
3. Move each silicone section with adhered cells in a new well of the microplate (use the third line of the microplate; Fig. 1) without taking off medium; use pliers.
4. Add 250 µL of YNB-Glc to each well containing silicone sections with adhered cells.
5. Incubate for 2, 24, 48 h, or other times in YNB-Glc to obtain biofilms with different stages of maturation.

3.4.3 Testing of Active Compounds Against Preformed Biofilms

1. Add 50 μL of the studied compound at the tested concentration to the ten wells with preformed biofilms of the third line of the microplate; each condition must be replicated ten times (*see* **Note 31**).
2. Incubate the preformed biofilms with the studied compound for 2–24 h (*see* **Notes 32** and **33**).

3.4.4 Evaluation of the Post-anti-biofilm Activity of Tested Compounds

1. To evaluate the post-anti-biofilm effect of a compound, move (using pliers) each section with a treated biofilm to corresponding well on the line below (line 4 of the microplate) and incubate it with 250 μL of YNB-Glc (without adding the studied compound) for 24–72 h (*see* **Note 34**).

3.5 Quantification of the Metabolic Activity of Fungal Cells Within Biofilms

The anti-biofilm effects are monitored using a metabolic assay. The principle is based upon the reduction of XTT tetrazolium salt to tetrazolium formazan, colored product, by mitochondrially active yeasts in the presence of an electron-coupling agent, menadione. This product can be measured spectrophotometrically.

Described here are the steps for the preparation of the XTT-menadione solution and then the colorimetric assay on both polystyrene and silicone surfaces.

3.5.1 Preparation of the XTT-Menadione Solution

1. Prepare XTT sodium salt powder as a 1 mg/mL saturated stock solution in Ringer's lactate (*see* **Note 35**).
2. Filtrate the XTT stock solution using 0.22 μm syringe filter (*see* Notes **36** and **37**).
3. Prepare small volume aliquots and store at −80 °C until use (*see* **Notes 38–41**).
4. Prepare menadione as a 10 mM solution in 100 % acetone (*see* **Note 42**).
5. Prepare small volume aliquots (*see* **Notes 43** and **44**).
6. Store at −80 °C until use (*see* **Notes 45** and **46**).

The working solution of XTT-Menadione corresponds to XTT (300 mg/L) – menadione (0.13 mM) in PBS.

7. Defrost XTT and menadione aliquots.
8. To obtain the XTT-menadione working solution needed to analyze 100 biofilm replicates, add in a tube the following: 600 μL of the filtered solution of XTT at 1 mg/mL plus 50 μL of menadione 10 mM plus 4.35 mL of PBS (*see* **Note 47**).

3.5.2 Colorimetric Assay: Biofilms Formed on Polystyrene Well Surfaces

1. After incubation, remove carefully waste medium in each test well using vacuum aspiration; the culture medium must be totally removed (*see* **Note 19**).
2. Wash twice each well coated with adherent yeasts or biofilms with 200 μL of sterile PBS (*see* **Note 48**).

3. Add 200 µL of PBS to each well containing adherent cells or biofilms (*see* **Note 49**).
4. Add 50 µL of XTT-menadione solution extemporaneously prepared to each well (tests and control wells) of the microplates.
5. Incubate the microplates in the dark (surrounded with aluminum foil) for 3 h at 37 °C without shaking.
6. After a short and gentle agitation, measure absorbance at 492 nm using the microplate reader. Absorbance at 492 nm corresponds to the XTT formazan produced by XTT reduction and indicates the metabolic activity of *Candida* yeasts in the biofilms (*see* **Notes 50–52**).

3.5.3 Colorimetric Assay: Biofilms Formed on Silicone Sections

1. After incubation, move the sections coated with biofilms (either exposed or not exposed to the studied compound) in a new well (using pliers), on the line below of the microplate (line 4, without posttreatment effect investigation, or line 5, with posttreatment effect investigation) (*see* **Note 53**). Add 200 µL of PBS to each well containing sections coated with biofilm.
2. Add 50 µL XTT-menadione solution extemporaneously prepared to each well containing sections coated with biofilms (either exposed or not exposed to the studied compound) (*see* **Note 54**).
3. Incubate the microplates in the dark (surrounded with aluminum foil) for 3 h at 37 °C without shaking.
4. After shaking the plates or mixing with a multichannel pipet the solutions in the wells, remove 200 µL of the reaction medium of each well and add it into the corresponding new wells of the line below (wells of the line 5, without posttreatment effect investigation, or line 6, with posttreatment effect investigation) (*see* **Note 55**).
5. Measure absorbance at 492 nm using the microplate reader (*see* **Notes 50–52**).

3.6 Statistical Analyses

An analysis of variance (ANOVA) and a Scheffe's test are applied to determine statistical differences between the groups ($p < 0.001$) (*see* **Note 56**).

The results can be transcribed in the form of graphs showing the metabolic activity (absorbance at 492 nm) according to the concentrations of tested compounds and controls. Results significantly different from control are marked. Regarding experiments with polystyrene surfaces, the anti-adherent activity of studied compounds can be evaluated measuring the percentage of inhibition of the fungal adherence; the percentage of fungal adherence inhibition is calculated as follows: inhibition $(\%) = 100 \times [1(A_{492\text{nm treated strain}}/A_{492\text{nm untreated strain}})]$ where $A_{492\text{nm}}$ is the absorbance measured at 492 nm. Decreases are calculated as the mean inhibition for each *Candida* species.

In the case of the experiment exposed in Subheading 3.5.2 and **Note 48**, the percentages of adhesion may be calculated as follows: $100 \times (\text{mean } A_{492\text{nm with treatment on sessile yeast inoculum}} / \text{mean } A_{492\text{nm with treatment on total yeast inoculum}})$.

All experiments must be performed at least twice with at least three replicates.

4 Notes

1. For the first screening of an anti-biofilm active component, we advise the use of *C. albicans* species; however, strains of other *Candida* species can be studied, especially *C. glabrata* and *C. parapsilosis* which are often incriminated in infections associated with catheters, both in adult and pediatric patients.
2. We advise to include both clinical and reference (such as ATCC or IHEM) strains.
3. We advise to use clinical strains originally isolated from an infected catheter to evaluate anti-biofilm activity in a model mimicking catheter-associated biofilm.
4. The volume of DMSO in each well should not exceed 2 % of the total volume to avoid the observation of DMSO activity on the cells.
5. For instance, this can be used for essential oils.
6. To make the organization easier, this incubation will start at about 18 h pm and finish at about 8 h am.
7. This incubation step is without shaking to avoid the formation of germ tubes in order to obtain mainly blastospores to mimic the typical early adhesion phase occurring during biofilm formation; however, if we expect to study the activity of a molecule purposely against germ tube adhesion, shaking should be preferred.
8. We advise to use YNB-Glc instead of RPMI to mostly obtain yeast forms and thus avoid hyphae formation.
9. PBS will be out of the fridge early enough to be used at room temperature.
10. It is not essential that the centrifugation is made at exactly 25 °C; yeasts can be centrifuged at room temperature.
11. The Kova cells can be replaced by other hemocytometer such as Malassez hemocytometer; we do not advise the use of McFarland standard because of the variable size of yeasts depending both on the species and the strains.
12. The yeast cell concentration is a key part of the experiment and has been optimized in preliminary tests to form a monolayer covering the whole tested surface.

13. The time between the yeasts preparation and their addition to the tested substrate (polystyrene or silicone surfaces) should be as quick as possible to avoid the modification of the morphology of yeasts which may modify their adhesive behavior.
14. Microplates will be removed from the incubator for each experimental step, so prepare as many culture plates that studied maturation times of biofilms. Multiple wells in plates with replicate biofilms will be prepared for each condition to be tested: each time point, each active molecule concentration including negative controls, etc.
15. Some components, such as essential oils and terpenic derivatives, are volatile. They must be tested in separate microplates to avoid cross-activity between wells that could bias the screening results.
16. Also because of the volatile nature of some molecules, the arrangement of the different tested concentrations of the studied volatile molecules will be cautious, only in decreasing order, and sufficiently spaced apart.
17. The incubation can be shorter: 1 h incubation at 37 °C is enough to obtain FBS coating onto the well surface.
18. FBS can be replaced by saliva (2 %) if we are interested in oral biofilm instead of catheter-associated biofilm. Saliva may be prepared as follows: nonstimulated saliva collected from 12 healthy volunteers will be pooled, denatured with DTT (2.5 mmol/L), shaken in the ice during 10 min, centrifuged for 20 min at 3,000 × *g* and 4 °C, diluted in three volumes of distilled water (25 %), and filtrated. This saliva stock solution will be aliquoted and stored at −80 °C. Before the experiment, saliva will be defrosted and diluted at 2 % in sterile distilled water.
19. The aspiration of the media from the well must be performed carefully: angle the tip for vacuum aspiration on the corner of the well. If vacuum pump is not available, a pipet (mono- or multiple-channel) may be used.
20. This volume is retained when using DMSO as solvent for the stock solution of the tested compound so that DMSO volume does not exceed 2 % of the final volume.
21. In each line, one well is kept untreated as negative control: column 12.
22. Each tested product occupies three lines in microplates.
23. If the tested compound is sensitive to light, the plates will be surrounded by aluminum foil.
24. We advice the use of a multichannel pipet for this step.
25. Some of these experiments may last for several days resulting in potential dehydration losses in wells, especially in the most external wells of the microplates. All external wells of each

microplate are filled with 200 μL of sterile water to prevent dehydration which may result in the non-immersion of some silicone sections in wells and consequently disrupt the value of the anti-biofilm tests.

26. Preparation and autoclaving steps of sections are time-consuming and must be done before the starting day of the experiment.
27. Silicone sections can be put in a glass hemolyze tube to be autoclaved (fill the tube with calibrated sections).
28. Silicone can be replaced with PMMA if the study focuses on dental biofilms.
29. This microplate's organization reduces the risk of handling mistake and makes the experiment steps easier.
30. FBS must completely cover the silicone sections, both inside (luminal part) and outside.
31. Testing a range of increasing or decreasing concentrations of compound is possible, but it should be prepared in new 96-well microplates, and then 50 μL of each concentration is transferred to microplates containing biofilms.
32. The duration of the treatment will be chosen depending on the mimicked model: we advise to treat for 24 h to mimic a systemic therapy and for 2–12 h maximum to mimic a lock therapy.
33. To mimic a lock therapy approach, we advise to test compound at concentrations in the range of 100–1,000 × MIC.
34. The post-anti-biofilm effect may be useful to demonstrate the remanence of the anti-biofilm activity.
35. XTT solution is light sensitive, so it should be covered with aluminum foil during preparation, storage, and utilization.
36. 0.45 μm syringe filter can also be used.
37. A yellow precipitate may be observed on the filter since it is a saturated solution; however, this is normal and has no consequence.
38. Volume in each aliquot must be chosen depending on the number of biofilm replicates prepared in each experiment; for example, in order to analyze 100 biofilm replicates by experiment, we advise the preparation of aliquots of about 600 μL of XTT filtrated stock solution.
39. XTT stock solution can be stored at −20 °C.
40. XTT must be prepared before the day of the beginning of the experiment.
41. It is not recommended to store XTT longer than 1 year because the activity of the reagent may decrease over time.
42. Menadione solution must be prepared in a glass tube because acetone may attack plastic material.

43. The preparation of small volume aliquots of menadione must be fast to minimize acetone evaporation losses.
44. Volumes in each aliquot must be chosen depending on the number of biofilm replicates prepared in each experiment, for example, in order to analyze 100 biofilm replicates by experiment, we advise the preparation of aliquots of 60 μL of menadione solution, and the exact needed volume for 100 biofilm replicates is 50 μL of menadione; however, we advise preparing 60 μL aliquots because acetone pipetting may be difficult and a slightly larger volume in aliquots makes pipetting of the requested volume easier. Besides the evaporation losses of the acetone will be minimized by a slightly larger volume.
45. Menadione solution can be stored at −20 °C.
46. Menadione solution must be prepared before the day of the beginning of the experiment.
47. The XTT-menadione working solution must be prepared immediately before use and cannot be stored more than a few hours surrounded with aluminum foil.
48. To study the activity of a molecule purposely on the adhesion phase, only half of the wells will be washed at the end of the adhesion step while the second half will not be washed: in this way, the metabolic activities corresponding both to the overall population (adherent plus not adherent yeasts; not washed half) and the adherent population (washed half) can be measured and enable calculating percentages of adherence. However, this cannot be achieved for colored tested compounds since the color, which is not removed in the not washed wells, disrupts the absorbance measurement. In the experiment described in Subheading 3.3.1.2, the color compounds are aspirated at the same time that FBS, so for colored compounds, we advice to use this protocol.
49. At this stage, the biofilm can be observed under an inverted microscope (with a 20 or 40× objectives) or may be visible to the naked eye by looking at the underside of the plates. Images can be captured if the microscope is fitted for image acquisition.
50. If the microplate reader is not able to measure absorbance at 492 nm, the metabolic activity can be evaluated as well by absorbance at 450 nm.
51. The fungal metabolic activity results in the appearance in the well of a pink to orange color depending on the intensity of the activity. Thus, a visual inspection of the plates will suggest a distinct efficacious cutoff concentration.
52. XTT-menadione assay has already demonstrated its relevance, but it can be completed by other methods such as the crystal violet method [19].

53. Moving the coupons allows to not take into account the metabolic activity of any biofilm formed on the surface of polystyrene wells.
54. Catheter sections must be totally immersed in the XTT-menadione solution.
55. The presence of catheter sections in the wells disturbs the measure of absorbance.
56. $p < 0.05$ can be tolerated.

References

1. Costerton JW, Stewart PS, Greenberg EP (1999) Bacterial biofilms: a common cause of persistent infections. Science 284:1318–1322
2. Donlan RM (2002) Biofilms: microbial life on surfaces. Emerg Infect Dis 8:881–890
3. Kuhn DM, George T, Chandra J et al (2002) Antifungal susceptibility of Candida biofilms: unique efficacy of amphotericin B lipid formulations and echinocandins. Antimicrob Agents Chemother 46:1773–1780
4. Mukherjee PK, Chandra J (2004) Candida biofilm resistance. Drug Resist Updat 7:301–309
5. Chandra J, Kuhn DM, Mukherjee PK et al (2001) Biofilm formation by the fungal pathogen Candida albicans: development, architecture, and drug resistance. J Bacteriol 183:5385–5394
6. LaFleur MD, Kumamoto CA, Lewis K (2006) Candida albicans biofilms produce antifungal-tolerant persister cells. Antimicrob Agents Chemother 50:3839–3846
7. Uppuluri P, Nett J, Heitman J et al (2008) Synergistic effect of calcineurin inhibitors and fluconazole against Candida albicans biofilms. Antimicrob Agents Chemother 52:1127–1132
8. Cornely OA, Bassetti M, Calandra T et al (2012) ESCMID* guideline for the diagnosis and management of Candida diseases 2012: non-neutropenic adult patients. Clin Microbiol Infect 18:19–37
9. Ullmann AJ, Akova M, Herbrecht R et al (2012) ESCMID* guideline for the diagnosis and management of Candida diseases 2012: adults with haematological malignancies and after haematopoietic stem cell transplantation (HCT). Clin Microbiol Infect 18:53–67
10. Cateau E, Berjeaud JM, Imbert C (2011) Possible role of azole and echinocandin lock solutions in the control of Candida biofilms associated with silicone. Int J Antimicrob Agents 37:380–384
11. Toulet D, Debarre C, Imbert C (2012) Could liposomal amphotericin B (L-AMB) lock solutions be useful to inhibit Candida spp. biofilms on silicone biomaterials? J Antimicrob Chemother 67:430–432
12. Mukherjee PK, Long L, Kim HG et al (2009) Amphotericin B lipid complex is efficacious in the treatment of Candida albicans biofilms using a model of catheter-associated Candida biofilms. Int J Antimicrob Agents 33:149–153
13. Uppuluri P, Srinivasan A, Ramasubramanian A et al (2011) Effects of fluconazole, amphotericin B, and caspofungin on Candida albicans biofilms under conditions of flow and on biofilm dispersion. Antimicrob Agents Chemother 55:3591–3593
14. Seidler M, Salvenmoser S, Müller FMC (2010) Liposomal amphotericin B eradicates Candida albicans biofilm in a continuous catheter flow model. FEMS Yeast Res 10:492–495
15. Liu W, Li L, Sun Y et al (2012) Interaction of the echinocandin caspofungin with amphotericin B or voriconazole against Aspergillus biofilms in vitro. Antimicrob Agents Chemother 56:6414–6416
16. Fiori B, Posteraro B, Torelli R et al (2011) In vitro activities of anidulafungin and other antifungal agents against biofilms formed by clinical isolates of different Candida and Aspergillus species. Antimicrob Agents Chemother 55:3031–3035
17. Donlan RM, Costerton JW (2002) Biofilms: survival mechanisms of clinically relevant microorganisms. Clin Microbiol Rev 15:167–193
18. Hawser SP, Douglas LJ (1994) Biofilm formation by Candida species on the surface of catheter materials in vitro. Infect Immun 62:915–921
19. Jin Y, Yip HK, Samaranayake YH et al (2003) Biofilm-forming ability of Candida albicans is unlikely to contribute to high levels of oral yeast carriage in cases of human immunodeficiency virus infection. J Clin Microbiol 41:2961–2967

Chapter 14

Biofilm Matrix-Degrading Enzymes

Jeffrey B. Kaplan

Abstract

Polymeric substances such as proteins, polysaccharides, and DNA constitute a major component of the biofilm matrix. Enzymes that depolymerize and degrade these components are useful tools for investigating the composition and function of the biofilm matrix. This chapter provides a brief overview of the most commonly used biofilm matrix-degrading enzymes and presents examples of their applications in biofilm research.

Key words Biofilm, Extracellular DNA, Dispersin B, DNase, Matrix, PNAG, Proteinase K

1 Introduction

One of the defining features of bacterial biofilms is the synthesis of an extracellular polymeric matrix. The biofilm matrix is a complex microenvironment that contains proteins, polysaccharides, DNA, RNA, lipids, dissolved nutrients, phages, and host-derived substances [1]. High-molecular-weight components of the matrix such as adhesive pili, capsular polysaccharides, and extracellular DNA (eDNA) are known to mediate many basic biofilm phenotypes including surface attachment, intercellular adhesion, and biocide resistance [1, 2]. Enzymes that degrade biofilm matrix components have been shown to inhibit biofilm formation and promote detachment of established biofilm colonies [3–6]. Once the biofilm colony is enzymatically dispersed, the bacterial cells become sensitive to killing by a variety of antimicrobial agents including antibiotics [7], detergents [8], disinfectants [4], bacteriophages [9], macrophages [10], and predatory bacteria [11]. Some biofilm matrix-degrading enzymes may be useful agents for the treatment and prevention of biofilm-related infections in clinical settings [4].

Biofilm matrix-degrading enzymes have been used extensively as research tools to help identify the structural components of the biofilm matrix [12] and to establish a role for the matrix in cell-to-cell and cell-to-surface adhesion [13] and biocide

Gianfranco Donelli (ed.), *Microbial Biofilms: Methods and Protocols*, Methods in Molecular Biology, vol. 1147,
DOI 10.1007/978-1-4939-0467-9_14,

resistance [14]. This chapter provides a brief overview of the most commonly used biofilm matrix-degrading enzymes and presents examples of their applications in biofilm research.

2 Materials

2.1 Deoxyribonuclease I

Deoxyribonuclease I (DNase I) is a nuclease that cleaves single-stranded and double-stranded DNA preferentially at phosphodiester bonds adjacent to pyrimidine nucleotides [15]. It results in 5′-phosphorylated polynucleotides that are on average four nucleotides in length. Dozens of published studies have shown that DNase I inhibits biofilm formation and disperses preformed biofilms [4]. Most studies utilize bovine pancreatic DNase I obtained from the Sigma Chemical Company (St. Louis, MO, USA). A more highly purified form of the enzyme (recombinant human DNase I; Pulmozyme®) is available from Genentech (South San Francisco, CA, USA) (*see* **Note 1**).

2.2 Restriction Endonucleases

Restriction endonucleases are bacterial enzymes that cleave DNA at or near specific recognition sequences. They result in 5′-phosphorylated polynucleotides, the average length of which varies depending on the frequency of the enzyme recognition sequence in the target DNA. Highly purified restriction endonucleases are available from several suppliers including New England Biolabs (Ipswich, MA, USA), Life Technologies (Grand Island, NY, USA), and Roche Applied Science (Indianapolis, IN, USA).

2.3 Other Deoxyribonucleases

Other deoxyribonucleases that have been used in biofilm research include *Staphylococcus aureus* thermonuclease, recombinant *Bacillus licheniformis* extracellular nuclease NucB [16], and purified human DNase1L2 [17].

2.4 Dispersin B

Dispersin B is a 42-kDa bacterial glycoside hydrolase that depolymerizes poly-*N*-acetylglucosamine (PNAG) surface polysaccharide [18]. PNAG is produced by many bacteria and fungi (*see* **Note 2**). PNAG plays a role in surface attachment, biofilm formation, antimicrobial resistance, and immune evasion [4]. Recombinant dispersin B (>98 % pure) is available from Kane Biotech, Inc. (Winnipeg, Manitoba, Canada) (*see* **Note 3**).

2.5 Other Glycoside Hydrolases

Other commercially available glycoside hydrolases that have been used in biofilm research include bacterial alginate lyase [19]; amylases produced by *Aspergillus oryzae*, *B. subtilis*, sweet potato, and human [11]; and bacterial cellulases and *N*-glycanases [20–22] (*see* **Note 4**).

2.6 Proteases

Proteases are potent antibiofilm enzymes because they degrade proteinaceous adhesins such as pili, fimbriae, and surface adhesins that are required for bacterial cell-to-cell and cell-to-surface interactions. Subtilisins, a group of serine proteases produced by *Bacillus* spp., are commonly used to control biofilms in industry [23]. Other proteases such as proteinase K, trypsin, and serratiopeptidase (a metalloprotease produced by *Serratia marcescens*) are commonly used in biofilm research. Proteases are available from many commercial suppliers including Sigma Chemical Company, Life Technologies, IBI Scientific (Peosta, IA, USA), and Takeda (Osaka, Japan) (*see* **Note 5**).

3 Methods

Matrix-degrading enzymes have been utilized in both static and flow-cell biofilm cultures. Two procedures are used to measure the effects of enzymes on static cultures. In the first procedure, enzymes are added to the culture at the time of inoculation. Control cultures receive no enzyme. After incubation, biofilm is quantitated by staining or microscopy. In the second procedure, biofilms are cultured in the absence of enzymes, rinsed to remove loosely adherent cells, and then treated with enzymes, usually for a relatively short period of time. Control cultures are treated with enzyme buffer alone. Following enzyme treatment, biofilm is quantitated by staining or microscopy. For flow-cell cultures, biofilms are usually seeded in enzyme-free broth and then perfused with either enzyme-free or enzyme-supplemented broth. Alternatively, biofilms are seeded and perfused with enzyme-free broth and then switched to enzyme-supplemented broth after a mature biofilm has formed [24]. Biofilms cultured in flow cells are usually visualized by microscopy. The following sections present examples of research protocols utilizing biofilm matrix-degrading enzymes in static cultures. All of these protocols can readily be modified for use with other biofilm bacteria simply by substituting the appropriate medium and culture conditions (*see* **Note 6**).

3.1 Identifying Biofilm Matrix Components

The following example illustrates how biofilm matrix-degrading enzymes can be used to identify structural and adhesive matrix components in biofilms of the Gram-negative periodontal pathogen *Aggregatibacter actinomycetemcomitans* [14]:

1. Culture bacteria in a 1-ml volume in 16-mm × 100-mm polystyrene test tubes using previously established culture conditions that promote biofilm formation. Incubate tubes at 37 °C for 24 h.
2. Decant supernatants and rinse biofilms twice with phosphate buffered saline (PBS).

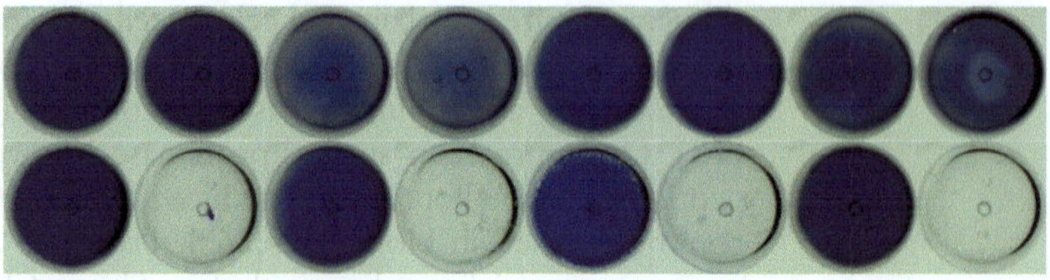

Fig. 1 Detachment of preformed *A. actinomycetemcomitans* strain CU1000 biofilms cultured in polystyrene tubes. Tubes were stained with crystal violet and photographed from the bottom. The *top row* shows 24-h-old biofilms that were treated for 30 min with 100 μg/ml of bovine DNase I, 100 μg/ml of proteinase K, 50 mM EDTA, or 1 % SDS (or the appropriate vehicle alone as a control) prior to staining. The *bottom row* shows biofilms pretreated with 10 μg/ml of dispersin B for 30 min prior to the enzyme and chemical treatments

3. Fill tubes with 1 ml of 10 μg/ml dispersin B in PBS. Fill control tubes with PBS alone. Incubate tubes at 30 °C for 30 min.
4. Decant supernatants and rinse biofilms twice with PBS.
5. Fill tubes with 1 ml of 100 μg/ml bovine DNase I in 40 mM Tris [pH 8.0], 1 mM $CaCl_2$, 10 mM $MgCl_2$; 100 μg/ml proteinase K in 50 mM Tris [pH 8.0]; 50 mM EDTA [pH 8.0]; or 1 % SDS in water. Fill control tubes with the corresponding buffer or water alone. Incubate tubes at 30 °C for 30 min.
6. Decant supernatants and rinse biofilms twice with PBS.
7. Stain biofilms with crystal violet.

Figure 1 shows that dispersin B alone did not detach the biofilms and even resulted in a slight increase in crystal violet binding. This may result from a thinning and weakening of the biofilm matrix which enables more crystal violet dye to enter the biofilm. DNase I and proteinase K did not detach mock-treated biofilms but efficiently detached biofilms pretreated with dispersin B. These results suggest that eDNA, PNAG, and proteinaceous adhesins all contribute the stability of the *A. actinomycetemcomitans* biofilm matrix and that these three polymeric matrix components interact with one other. EDTA efficiently detached dispersin B-treated biofilms but not mock-treated biofilms, suggesting that divalent cations also contribute to matrix stability. Dispersin B sensitized *A. actinomycetemcomitans* biofilms to detachment by SDS, a detergent that denatures proteins, which supports the hypothesis that PNAG interacts with proteinaceous adhesins [14].

3.2 Estimating the Size of eDNA in the Biofilm Matrix

Numerous studies have shown that eDNA functions as a biofilm matrix adhesin in many bacterial biofilms [4]. Although several studies have demonstrated that eDNA consists primarily of genomic DNA released by lysed cells, few studies have investigated the structure of eDNA in the matrix. The following protocol demonstrates how restriction endonucleases can be used to estimate the size of eDNA fragments in the biofilm matrix of *S. aureus* strain SH1000 [12]:

1. Culture bacteria in a 200-μl volume in 96-well microtiter plates under conditions that promote biofilm formation.
2. Aspirate broth and rinse wells twice with PBS.
3. Fill wells with restriction endonuclease buffer (10 mM Tris [pH 7.9], 10 mM $MgCl_2$, 50 mM NaCl, 1 mM dithiothreitol) containing 100 U/ml of restriction endonuclease. Control wells are filled with enzyme buffer alone. Use approximately 10 different restriction endonucleases that yield digestion products of various lengths ranging from ca. 0.4 to 40 kb (*see* **Note 7**). Incubate the plate at 37 °C for 1 h.
4. Rinse biofilms twice with PBS.
5. Quantitate biofilm using crystal violet staining.
6. Calculate percent biofilm detachment ($1 - A_{595[\text{buffer+enzyme}]}/A_{595[\text{buffer alone}]} \times 100$) for each enzyme. Graph percent biofilm detachment versus average restriction fragment size (size of genome in bp/number of enzyme recognition sequences).

Figure 2 shows the results from the restriction endonuclease detachment assay. The amount of biofilm detachment depended on the frequency of the enzyme recognition sequence in the *S. aureus* genome. Enzymes that produced restriction fragments with an average size of <10 kb caused efficient detachment, whereas enzymes that produced restriction fragments with an average size of 11–24 kb caused partial detachment. These findings suggest that the fraction of *S. aureus* eDNA that mediates intercellular adhesion is composed primarily of genomic DNA and that eDNA fragments >11 kb can function as intercellular adhesins.

3.3 Measuring Resistance to Antimicrobial Enzymes in Planktonic Cells

Several studies have shown that DNases can sensitize biofilm cells to killing by antibiotics, antiseptics, and disinfectants [4]. The following protocol illustrates the use of DNase I to demonstrate that eDNA protects planktonic cells from killing by the antimicrobial enzyme lysostaphin (*see* **Note 8**). This protocol was carried out using *S. aureus* strain SH1000:

1. Spread a 100-μl aliquot of an overnight broth culture (>10^8 CFU) onto a blood agar plate and incubate at 37 °C for 24 h.
2. Scrape the cell paste from the agar using a plastic inoculating loop or cell scraper and transfer the cells to a tube containing 3 ml of PBS supplemented with 2 mM $CaCl_2$.
3. Resuspend the cells by vortex agitation and adjust the A_{650} value to approximately 1.5.

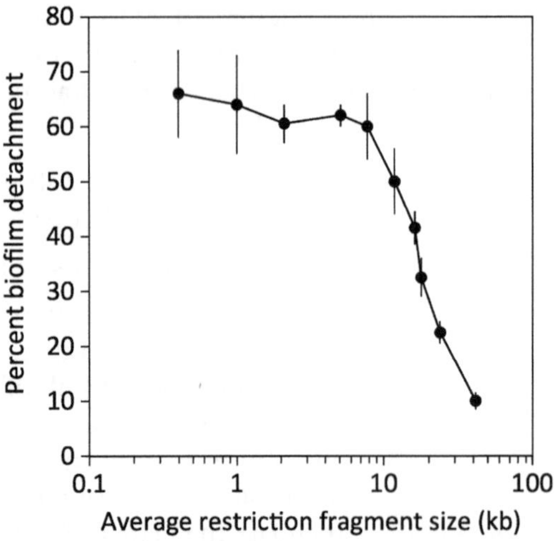

Fig. 2 Detachment of *S. aureus* strain SH1000 biofilms by restriction endonucleases. *S. aureus* strain SH1000 biofilms grown in microtiter plates were treated for 1 h with 100 U/ml of various restriction endonucleases and then rinsed and stained with crystal violet. Percent detachment was calculated as $1 - (A_{595[\text{buffer + enzyme}]}/A_{595[\text{buffer alone}]}) \times 100$. *Values* show means and ranges for duplicate wells. The restriction endonucleases used (and average restriction fragment lengths in kb) were HinfI (0.4), FokI (1.0), HaeIII (2.1), AlwNI (5.1), ApaLI (7.8), NcoI (11.8), KpnI (16.2), AvaI (17.9), BamHI (23.9), and SstI (41.5)

4. Transfer 500-μl aliquots of cells into four 1.5-ml microcentrifuge tubes.
5. Add 5 μl of 1 mg/ml Pulmozyme® to the first tube, 5 μl of 1 mg/ml lysostaphin to the second tube, and 5 μl of 1 mg/ml Pulmozyme® and 5 μl of 1 mg/ml lysostaphin to the third tube. The fourth tube receives no enzymes and serves as a negative control. Mix tubes briefly by vortex agitation (*see* **Note 9**).
6. Transfer 200-μl aliquots from each tube to two wells of a flat-bottom polystyrene microtiter plate.
7. Measure the A_{650} of the wells every 30 min for 5 h.
8. Graph average A_{650} values versus time for each experimental condition.

Lysostaphin is a 27 kDa endopeptidase that cleaves the cross-linking pentaglycine bridges in staphylococcal cell walls [25]. Lysostaphin treatment causes cell lysis and a concomitant decrease in the turbidity of the culture.

Figure 3 shows that Pulmozyme® alone did not cause lysis of *S. aureus* cells, whereas lysostaphin cause a nearly linear decrease in the turbidity of the culture over time. Cultures treated with a combination of Pulmozyme® and lysostaphin exhibited more rapid cell lysis than cultures treated with lysostaphin alone, suggesting that eDNA protects planktonic cells or small clusters of cells from lysis and killing by lysostaphin.

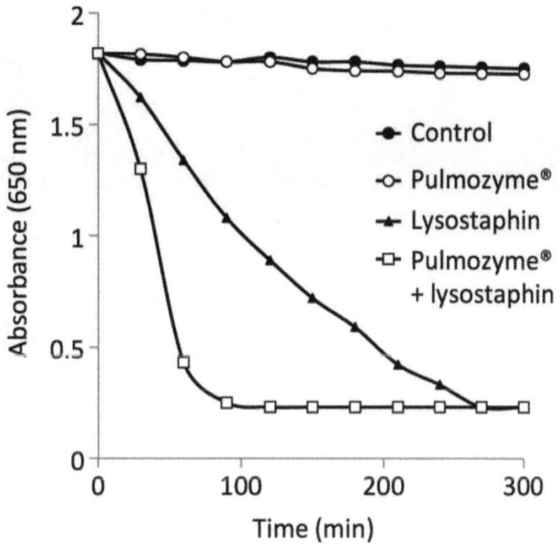

Fig. 3 Killing of *S. aureus* strain SH1000 planktonic cells by lysostaphin. Cell suspensions were treated with Pulmozyme® (10 μg/ml), lysostaphin (10 μg/ml), or Pulmozyme® + lysostaphin (10 μg/ml each), and the A_{650} of the suspension was measured after increasing amounts of time. Cell lysis caused by lysostaphin results in a decrease in the turbidity of the cell suspension

3.4 Extracting eDNA from the Biofilm Matrix

It is often necessary to purify eDNA from the biofilm matrix in order to investigate its structure and composition. However, a large portion of eDNA may be bound to polysaccharides and proteins in the matrix, reducing the yield from conventional extraction procedures. The following example illustrates how pretreating biofilms with *N*-glycanase, dispersin B, and proteinase K can increase the yield of eDNA. This example was carried out using biofilms of *Acinetobacter* sp. strain AC811 [22]:

1. Dilute an overnight culture 1:100 in fresh LB broth and transfer 3-ml aliquots to the wells of a 6-well polystyrene microtiter plate. Incubate at room temperature for 4 days.
2. Rinse biofilms twice with PBS.
3. Harvest biofilm cells by scraping into 1 ml of PBS. Pool cells from multiple wells.
4. Homogenize cells using a tissue homogenizer.
5. Supplement cell suspensions with 1/100 vol of 1 mg/ml *N*-glycanase, 2 mg/ml dispersin B, or 0.5 mg/ml proteinase K. Incubate at 37 °C for 30 min. For combinations of enzymes, treat cells with *N*-glycanase or dispersin B (or both) at 37 °C for 30 min, followed by treatment with proteinase K at 37 °C for another 30 min.
6. Filter cells through a 0.2-μm pore size filter.
7. Purify eDNA by cetyltrimethylammonium bromide (CTAB) precipitation or other suitable method.

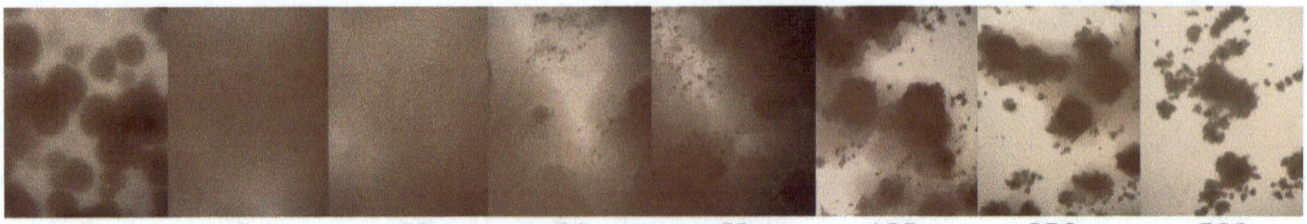

Fig. 4 Biofilm formation by *S. aureus* strain SH1000 in polystyrene microtiter plate wells in the presence of increasing concentrations of proteinase K. Wells were photographed under an inverted microscope. Imaged areas = 3 × 4 mm

Pretreatment of *Acinetobacter* sp. biofilm cells with *N*-glycanase, dispersin B, or proteinase K increased the yield of eDNA by >30-fold when compared to the yield obtained with the control treatment [22]. Matrix-degrading enzymes are superior to conventional methods for eDNA extraction (EDTA, SDS, NaOH) because they do not cause cell lysis.

3.5 Effects of Proteinase K on Biofilm Formation

Proteases such as proteinase K are often efficient biofilm dispersing agents. This protocol measures the effect of proteinase K on biofilm formation by *S. aureus* strain SH1000:

1. Prepare a 5 mg/ml solution of proteinase K in sterile water. Make six serial twofold dilutions of the solution, yielding 2.5, 1.25, 0.63, 0.31, 0.16, and 0.08 mg/ml proteinase K solutions.
2. Transfer a loopful of cells from a 18-h-old blood agar plate to a microcentrifuge tube containing 200 μl of sterile Tryptic soy broth.
3. Disperse cells by vortex agitation and adjust the A_{450} of the cell suspension to 1.0.
4. Dilute cells to 10^2–10^3 CFU/ml (ca. 1:100,000) in fresh TSB.
5. Transfer 180-μl aliquots of cells to the wells of a 96-well microtiter plate.
6. Supplement wells with 20 μl of each of the seven proteinase K solutions, or 20 μl of water as a control. Incubate the plate at 37 °C for 18 h.
7. Visualize biofilm colonies by light microscopy.

Figure 4 shows micrographs of *S. aureus* biofilm colonies cultured in the presence of increasing concentrations of proteinase K. Cells cultured in the absence of enzyme formed characteristic 1-mm-diam hemispherical colonies on the bottom of the well. The addition of 8 μg/ml proteinase K significantly inhibited the formation of biofilm colonies (*see* **Note 10**). Instead, the cells formed a thin film on the bottom surface of the well which was readily detached from the well by gentle rinsing. In contrast, cells cultured in the presence of higher concentrations of proteinase K (125–

500 μg/ml) produced biofilm colonies that were more compact than those produced by cells cultured in the absence of enzyme. Thus, the effect of proteinase K on biofilm development is highly dependent on the enzyme concentration.

4 Notes

1. Recombinant versions of bovine DNase I are also available from several suppliers including Life Technologies, Worthington Biochemical Corp. (Lakewood, NJ, USA), Fermentas (Thermo Fisher Scientific), and Roche Applied Science. Lyophilized forms of DNase I should be dissolved in water, aliquoted, and stored at −20 °C. Avoid repeated freezing and thawing.
2. In addition to the PNAG-producing bacteria that have already been identified [4], PNAG was recently shown to be produced by many bacteria that lack identifiable PNAG biosynthetic genes including *Streptococcus pyogenes*, *S. pneumoniae*, *Listeria monocytogenes*, *Neisseria meningitidis*, *N. gonorrhoeae*, non-typeable *Haemophilus influenzae* serogroup B, and by many *FUNGI* [26].
3. Dissolve dispersin B at 1–2 mg/ml in 50 mM sodium phosphate buffer (pH 5.8), 100 mM NaCl, 50 % glycerol. Store at −20 °C. Under these conditions the enzyme is stable for more than 1 year. Dispersin B is active in both PBS and broth. The working concentration should be 0.5–50 μg/ml. Although the optimal temperature for dispersin B is 30 °C, and the optimum pH is 5.0, the enzyme is active over a broader range of temperature and pH values. The half-life is >3 h at 37 °C.
4. Some commercially available native glycoside hydrolase enzymes are not highly purified and may be contaminated with DNases, proteases, or other enzymes. Check for DNase activity by treating bacterial genomic DNA with glycoside hydrolase enzyme in buffer containing 1 mM $CaCl_2$ and 10 mM $MgCl_2$. Monitor DNA degradation by agarose gel electrophoresis. To confirm that the antibiofilm activity of a commercial glycoside hydrolase preparation is not due to residual protease activity, assays can be carried out in the presence of protease inhibitors [11]. Use analytical grade, molecular biology grade, or recombinant enzymes whenever possible.
5. Proteinase K is stable for several months when dissolved in water at 1 mg/ml and stored at 4 °C. Biofilms produced by some bacteria are resistant to detachment by proteinase K even when tested at high concentrations [27].
6. Enzymes should be tested at a range of concentrations from 1 ng/ml to 1 mg/ml, if possible. Some biofilm matrix-degrading enzymes are capable of inhibiting or dispersing

biofilms in microtiter plate assays at low concentrations. For example, Pulmozyme® efficiently inhibits biofilm formation by *S. aureus* at 4 ng/ml and detaches preformed *S. aureus* biofilms in 4 min at 1 μg/ml [7]. Similarly, dispersin B efficiently detaches *S. epidermidis* biofilms in 6 min at 1 μg/ml [28] and detaches *Actinobacillus pleuropneumoniae* biofilms in 10 min at 20 μg/ml [29]. This property can be used to rapidly detach biofilms from microtiter plate wells prior to CFU enumeration [30].

7. New England Biolabs offers over 200 restriction endonucleases that are active in a single buffer (CutSmart™).
8. Other antimicrobial enzymes include lysozyme, bacteriophage lysins, oxidative enzymes, and quorum-quenching enzymes [25].
9. Recombinant human DNase I (Pulmozyme®) may be a much more potent antibiofilm agent than native bovine pancreatic DNase I. Pulmozyme® at 1 μg/ml detached *S. aureus* strain SH1000 biofilm in 4 min [7], whereas native bovine pancreatic DNase I at 100 μg/ml resulted in only partial detachment of SH1000 biofilms after 1 h [12].
10. Although biofilm matrix-degrading enzymes generally do not kill bacteria or inhibit their growth, the growth rate of bacteria in the presence and absence of enzyme should be measured to confirm that an observed reduction in biofilm formation in the presence of enzyme is due to biofilm inhibition and not to growth inhibition.

Acknowledgement

The author thanks Eun Sun Lee (New Jersey Dental School) for help with Fig. 3. This work was supported by NIH grant AI097182 and by American University.

References

1. Flemming H-C, Wingender J (2010) The biofilm matrix. Nat Rev Microbiol 8:623–633
2. Mah TF (2012) Biofilm-specific antibiotic resistance. Future Microbiol 7:1061–1072
3. Hymes SR, Randis TM, Sun TY, Ratner AJ (2013) DNase inhibits *Gardnerella vaginalis* biofilms *in vitro* and *in vivo*. J Infect Dis 207:1491–1497
4. Kaplan JB (2009) Therapeutic potential of biofilm-dispersing enzymes. Int J Artif Organs 32:545–554
5. Kaplan JB (2010) Biofilm dispersal: mechanisms, clinical implications and potential therapeutic uses. J Dent Res 89:205–218
6. Xavier JB, Picioreanu C, Abdul Rani S et al (2005) Biofilm-control strategies based on enzymic disruption of the extracellular polymeric substance matrix—a modelling study. Microbiology 151:3817–3832
7. Kaplan JB, LoVetri K, Cardona ST et al (2012) Recombinant human DNase I decreases biofilm and increases antimicrobial susceptibility in staphylococci. J Antibiot 65:73–77
8. Izano EA, Wang H, Ragunath C et al (2007) Detachment and killing of *Aggregatibacter actinomycetemcomitans* biofilms by dispersin B and SDS. J Dent Res 86:618–622

9. Lu TK, Collins JJ (2007) Dispersing biofilms with engineered enzymatic bacteriophage. Proc Natl Acad Sci U S A 104:11197–11202
10. Venketaraman V, Lin AK, Lea A et al (2008) Both leukotoxin and poly-*N*-acetylglucosamine surface polysaccharide protect *Aggregatibacter actinomycetemcomitans* cells from macrophage killing. Microb Pathog 45:173–180
11. Dashiff A, Kadouri DE (2011) Predation of oral pathogens by *Bdellovibrio bacteriovorus* 109J. Mol Oral Microbiol 26:19–34
12. Izano EA, Amarante MA, Kher WB, Kaplan JB (2008) Differential roles of poly-*N*-acetylglucosamine surface polysaccharide and extracellular DNA in *Staphylococcus aureus* and *Staphylococcus epidermidis* biofilms. Appl Environ Microbiol 74:470–476
13. Kaplan JB, Velliyagounder K, Chandran R et al (2004) Genes involved in the synthesis and degradation of matrix polysaccharide in *Actinobacillus actinomycetemcomitans* and *Actinobacillus pleuropneumoniae* biofilms. J Bacteriol 186:8213–8220
14. Izano EA, Sadovskaya I, Wang H et al (2008) Poly-*N*-acetylglucosamine mediates biofilm formation and detergent resistance in *Aggregatibacter actinomycetemcomitans*. Microb Pathog 44:52–60
15. Chen WJ, Liao TH (2006) Structure and function of bovine pancreatic deoxyribonuclease I. Protein Pept Lett 13:447–453
16. Nijland R, Hall MJ, Burgess JG (2010) Dispersal of biofilms by secreted, matrix degrading, bacterial DNase. PLoS One 12:e15668
17. Eckhart L, Fischer H, Barken KB et al (2007) DNase1L2 suppresses biofilm formation by *Pseudomonas aeruginosa* and *Staphylococcus aureus*. Br J Dermatol 156:1342–1345
18. Ramasubbu N, Thomas LM, Ragunath C, Kaplan JB (2005) Structural analysis of dispersin B, a biofilm-releasing glycoside hydrolase from the periodontopathogen *Actinobacillus actinomycetemcomitans*. J Mol Biol 349:475–486
19. Alkawash MA, Soothill JS, Schiller NL (2006) Alginate lyase enhances antibiotic killing of mucoid *Pseudomonas aeruginosa* in biofilms. APMIS 114:131–138
20. Friedman L, Kolter R (2004) Genes involved in matrix formation in *Pseudomonas aeruginosa* PA14 biofilms. Mol Microbiol 51:675–690
21. Friedman L, Kolter R (2004) Two genetic loci produce distinct carbohydrate-rich structural components of the *Pseudomonas aeruginosa* biofilm matrix. J Bacteriol 186:4457–4465
22. Wu J, Xi C (2009) Evaluation of different methods for extracting extracellular DNA from the biofilm matrix. Appl Environ Microbiol 75:5390–5395
23. Thallinger B, Prasetyo EN, Nyanhongo GS, Guebitz GM (2013) Antimicrobial enzymes: an emerging strategy to fight microbes and microbial biofilms. Biotechnol J 8:97–109
24. Brindle ER, Miller DA, Stewart PS (2011) Hydrodynamic deformation and removal of *Staphylococcus epidermidis* biofilms treated with urea, chlorhexidine, iron chloride, or dispersin B. Biotechnol Bioeng 108:2968–2977
25. Cordeiroa AL, Wernerb C (2012) Enzymes for antifouling strategies. J Adhes Sci Technol 25:2317–2344
26. Cywes-Bentley C, Skurnik D, Zaidi T et al (2013) Antibody to a conserved antigenic target is protective against diverse prokaryotic and eukaryotic pathogens. Proc Natl Acad Sci U S A 110:E2209–E2218
27. Ali Mohammed MM, Nerland AH, Al-Haroni M, Bakken V (2013) Characterization of extracellular polymeric matrix, and treatment of *Fusobacterium nucleatum* and *Porphyromonas gingivalis* biofilms with DNase I and proteinase K. J Oral Microbiol 5: 20015. doi:10.3402/jom.v5i0.20015
28. Kaplan JB, Ragunath C, Velliyagounder K et al (2004) Enzymatic detachment of *Staphylococcus epidermidis* biofilms. Antimicrob Agents Chemother 48:2633–2636
29. Izano EA, Sadovskaya I, Vinogradov E et al (2007) Poly-*N*-acetylglucosamine mediates biofilm formation and antibiotic resistance in *Actinobacillus pleuropneumoniae*. Microb Pathog 43:1–9
30. Kaplan JB, Izano EA, Gopal P et al (2012) Low levels of β-lactam antibiotics induce extracellular DNA release and biofilm formation in *Staphylococcus aureus*. MBio 3:e00198–12

Chapter 15

Efficacy Evaluation of Antimicrobial Drug-Releasing Polymer Matrices

Iolanda Francolini, Antonella Piozzi, and Gianfranco Donelli

Abstract

To assay in vitro antimicrobial activity of substances such as antibiotics or antiseptics, standard methods both in liquid and on solid media are available. These procedures cannot be adequate for testing antimicrobial-releasing or biocidal polymer systems.

This chapter is focused on the description of methods that the authors have developed to evaluate the antimicrobial activity of either antimicrobial agent-releasing polymers or biocidal polymers. These assays can be applied to different types of water-soluble or insoluble polymer matrices.

Key words Polymer antimicrobial activity, Antimicrobial-releasing polymers, Biocidal polymers, Microbial biofilm, Polymer surface-related infections

1 Introduction

Microbial contamination is a serious issue afflicting several applicative areas. In the food industry, bacterial adhesion to food products or product contact surfaces leads to important hygienic problems and economic losses due to food spoilage [1]. Biofilm formation on the surface of the pipe walls of drinking water distribution systems can be responsible for the decrease of water quality in terms of increasing bacterial levels, reduction of dissolved oxygen, and taste and odor change [2]. In the biomedical field, microbial biofilms are recognized to play a pivotal role in healthcare-associated infections (HAIs), especially those associated to implantable medical devices such as intravascular catheters, urinary catheters, and orthopedic implants [3].

The materials commonly used to package foods, to build drinking water pipes, or to manufacture medical devices are in majority polymers. Therefore, prevention or control bacterial colonization on these materials is an important challenge of current research.

Gianfranco Donelli (ed.), *Microbial Biofilms: Methods and Protocols*, Methods in Molecular Biology, vol. 1147,
DOI 10.1007/978-1-4939-0467-9_15, © Springer Science+Business Media New York 2014

Today, the development of antifouling or antimicrobial surfaces is considered as the best approach to prevent biofilm formation.

Antifouling surfaces are usually obtained by polymeric hydrophilic coatings (such as PEG) able to hamper microbial adhesion being the microbial surfaces hydrophobic. On the contrary, antimicrobial polymers can be developed either by impregnation with antibiotics or disinfectants or by providing polymers with functionalities exerting antimicrobial activity.

In the first class, the polymer acts as a carrier for one or more antimicrobial agent that once released from the polymer exerts its action (antimicrobial agent-releasing polymers).

Antibiotics, antiseptics, or heavy metals can be applied to the polymer by (a) physical adsorption [4–6], (b) entrapment into the polymer matrix [7–9], (c) complexation [10–12], or (d) conjugation [13]. In this latter case, the drug must be bound to the polymer via a hydrolytically labile linkage to allow its releasing in the surrounding environment.

Bactericidal functionalities such as quaternary amine compounds or phosphonium salts are instead possessed by intrinsically antimicrobial polymers (biocidal polymers) [14]. These polymers are "contact killing" since they are able to exert the killing action when microorganisms contact the surface. The main advantage of biocidal polymers is that, since they do not release antimicrobial substances, they do not exhaust their activity, at least in principle.

To assay in vitro antimicrobial activity of substances such as antibiotics or antiseptics, methods both in liquid and on solid media have been developed. The in-liquid medium assay allows determination of minimum inhibitory concentration (MIC). Instead, the test performed on solid media permits the evaluation of inhibition zone around a filter paper impregnated with the antimicrobial substance. The main requirement for the antimicrobial agent to be tested is its solubility in water and diffusion on the agar plate. These standardized procedures cannot be adequate for testing antimicrobial-releasing or biocidal polymer systems.

This chapter is focused on the description of methods to evaluate antimicrobial activity of either antimicrobial agent-releasing polymers or biocidal polymers. Particularly, standard microbiological assays have been properly adapted to test polymer systems. Besides polymer ability to release antimicrobial agents, also the solubility properties of the polymer must be taken into consideration before performing each test. In fact, some of the described assays are specific for water-soluble polymers and others for polymers insoluble in aqueous solution.

2 Materials

1. Ringer's solution: 112 mM NaCl, 6 mM KCl, 2 mM $CaCl_2$, 1 M $NaHCO_3$.
2. Phosphate buffer: 10 mM Na_2HPO_4, 1.8 mM KH_2PO_4, 2.7 mM KCl, 137 mM NaCl. If necessary, adjust the pH to 7.4 with NaOH.
3. 0.05 M sodium cacodylate buffer ($Na(CH_3)_2AsO_2 \times 3H_2O$, pH = 7): Prepare a 0.2 M stock solution of sodium cacodylate in water (4.28 g/100 mL). Add the 12.6 mL of 0.2 M HCl per 100 mL cacodylate stock solution, followed by the addition of water to a final volume of 400 mL.
4. 0.5 McFarland standard: to prepare a $BaSO_4$ suspension, add a 0.5 mL aliquot of 0.048 mol/L $BaCl_2$ (1.2 % w/v $BaCl_2 \times 2H_2O$) to 99.5 mL of 0.18 mol/L (0.36 N) H_2SO_4 (1 % v/v) with constant stirring to maintain a suspension. Measure the optical density of the turbidity standard using a spectrophotometer. Add a 5 mL aliquot into screw-capped glass tubes. The tubes should have the same size as those for preparing the bacterial suspension for inoculation. Store in the dark at room temperature.
5. Tetrahydrofuran, ACS reagent (high-quality chemical for laboratory use), ≥99 %.
6. Acetone, ACS reagent, ≥99.5 %.
7. Ethanol, ACS reagent, ≥99.5 %.
8. Glutaraldehyde solution 50 wt% in water.
9. Hexamethyldisilazane, reagent grade, ≥99 %.
10. D-(+)-Glucose, ACS reagent.
11. LIVE/DEAD® BacLight™ Bacterial Viability Kit.
12. Round-shaped Teflon plates, 2 cm in diameter.
13. 13 × 100 mm test tubes.
14. 24-wells polystyrene microtiter plates.
15. Petri plates, 87 mm diameter and 15 mm height.
16. Mueller-Hinton (MH) agar.
17. Biofilm-producing *Staphylococcus epidermidis* strain (ATCC 35984).
18. Biofilm-producing *Staphylococcus aureus* strain (10850).
19. Green fluorescent protein (GFP)-producing *P. aeruginosa* PAO1 strain carrying plasmid pMF230 Ca^r strain carrying a plasmid to produce 1945GFP*uvr*.
20. Incubator at 37 °C.
21. UV–VIS spectrophotometer.

22. Confocal Laser Scanning Microscope.
23. Rotary-pumped sputter coater (e.g., Quorum Technologies Ltd) suitable for non-oxidizing metals, such as gold (Au) and platinum (Pt).
24. Scanning electron microscope.

3 Methods

3.1 Polymers Insoluble in Waters

The assays described below are suitable for several types of polymers including polyurethanes, silicones, Dacron®, and Teflon®. The tests reported in Subheadings 3.1.2 and 3.1.3 can be performed only on antimicrobial agent-releasing polymer matrices. On the contrary, the test reported in Subheading 3.1.4 is suitable for both biocidal and antimicrobial agent-releasing polymers.

3.1.1 Preparation of Polymer Disks by Solvent Casting

1. Prepare a polymer solution at a 5 % wt/v. concentration in a low boiling point solvent, for example, tetrahydrofuran or acetone.
2. Pipet approximately 3 mL of solution in a round-shaped Teflon plate (2 cm in diameter).
3. Evaporate the solvent under vacuum at 30 °C.
4. To obtain a suitable polymer thickness (approximately 100 μm), repeat all the procedure for three times.
5. Detach the polymer disk by using metal tweezers and store it at 4 °C in plastic vessels (*see* **Note 1**). Sterilize the polymer by keeping it under the UV lamp in a laminar flow hood for 15 min.

3.1.2 Disk Diffusion Test (See **Note 2**)

1. Pour 20 mL of Mueller-Hinton (MH) agar into a Petri plate (87 mm diameter, 15 mm height) and allow the plate to set at room temperature.
2. To inoculate bacteria in the agar plate, use the streaking method [15]. Particularly, after overnight bacterial growth in broth, adjust the optical density (550 nm) at 0.125 (1×10^8 CFU/mL).
3. Dip a sterile cotton swab into the prepared bacterial suspension and remove excess fluid by pressing and rotating the swab against the wall of the tube above the fluid level.
4. Streak the inoculum evenly in three planes onto the surface of the agar. Roll the swab around the edge of the plate.
5. By using sterile tweezers, place the polymer disks (maximum four samples per plate) on the plate and gently press down to ensure contact.

6. Incubate the plate overnight at 37 °C.
7. The following day, record the inhibition zone (*see* **Note 3**).
8. To evaluate the durability of antimicrobial activity of the polymer matrix, repeat the test at 24 h intervals, by transferring the polymer disk to a freshly prepared plate, until the inhibition of bacterial growth is no longer noticed.

3.1.3 Assay of Polymer Antimicrobial Activity in Broth Culture

1. Fill three tubes (tubes 1, 2, and 3) with 2 mL of a bacterial culture, grown overnight in MH supplemented with 1 % (w/v) glucose, adjusted at 0.125 OD at 550 nm (A_0).
2. Use sterile 13 × 100 mm test tubes to conduct the test.
3. Use tube 1 as reference (control), while immerse the bare polymer disk into tube 2 and the antibiotic-impregnated polymer disk into tube 3 (*see* **Note 4**).
4. Incubate the control and the two test tubes overnight at 37 °C.
5. Following incubation, check the absorbance of each solution at 550 nm. Use clear broth as a blank. To assess durability of polymer antimicrobial activity, repeat the test daily by transferring the polymer disk to a freshly prepared broth until the absorbance value of the solution in the test tube ($A_{\text{test tube}}$) is comparable to the control (A_{control}). In fact, since the amount of adsorbed light depends on bacterial cell concentration, an absorbance value of the test tube lower than the control indicates the activity of the sample. When the two absorbances are similar, the polymer sample is no longer able to inhibit bacterial growth.
6. The percentage of killing effect (KE) at each day of experiment can be defined as follow;

$$KE = \left[1 - \left(\frac{A_{test\ tube} - A_0}{A_{control} - A_0}\right)\right] \times 100$$

7. By reporting KE versus time, different trends can be obtained mainly depending on the drug release rate (*see* **Note 5**). In Fig. 1 some of the possible trends are displayed.

3.1.4 Evaluation of Microbial Adhesion to the Polymer

1. To obtain massive exopolysaccharide production [16], use a bacterial culture, grown overnight in MH supplemented with 1 % (w/v) glucose, adjusted at 0.125 OD (1×10^8 CFU/mL).
2. Fill two test tubes (tubes 1 and 2) with 2 mL of this diluted culture and use tube 1 as control and tube 2 to immerse the polymer sample to be tested.
3. Following overnight incubation at 37 °C, collect the polymer and wash it twice with Ringer's solution to remove loosely adherent bacteria (*see* **Note 6**).

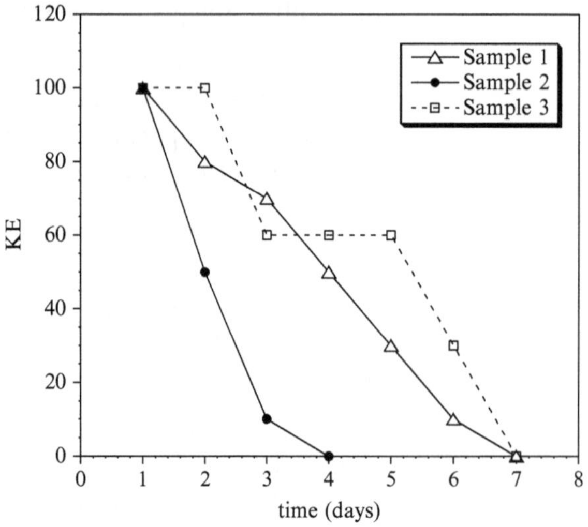

Fig. 1 Possible trends of the killing effect (KE) of polymer samples in broth

4. Insert the polymer sample into a tube containing 10 mL of Ringer's solution and sonicate for 5 min to detach the biofilm from the polymer surface.
5. Perform five serial dilutions by taking 0.5 mL aliquots from the solution in contact with the polymer (after vortexing for 10 s) and adding 4.5 mL of Ringer's solution.
6. Place three 10 μL aliquots of each dilution onto MH agar plates.
7. Allow the drops to dry at room temperature before inverting the plates for incubation.
8. Incubate the plates at 37 °C and after 18–22 h, count the colony in the first dilution in which they are well separated.
9. Considering the dilution factor and the surface area of the polymer disk, calculate the number of colony-forming units per polymer surface (CFUs/cm^2).

3.1.5 Assessment of Antibiofilm Activity

To assess the biofilm growth on polymer samples, different microscopy imaging techniques can be employed including scanning electron microscopy (SEM), fluorescence microscopy (FM), and confocal laser scanning microscopy (CLSM).

1. Incubate polymer disks in 24-well polystyrene microtiter plates containing 1 mL of bacterial culture grown in broth supplemented with 1 % glucose and adjusted to 0.125 OD.
2. After 24 h incubation at 37 °C, wash the polymers twice with phosphate buffer and collect them.
3. For SEM observations, perform bacterial fixation by putting in contact the polymer disk with 2.5 % glutaraldehyde in 0.05 M

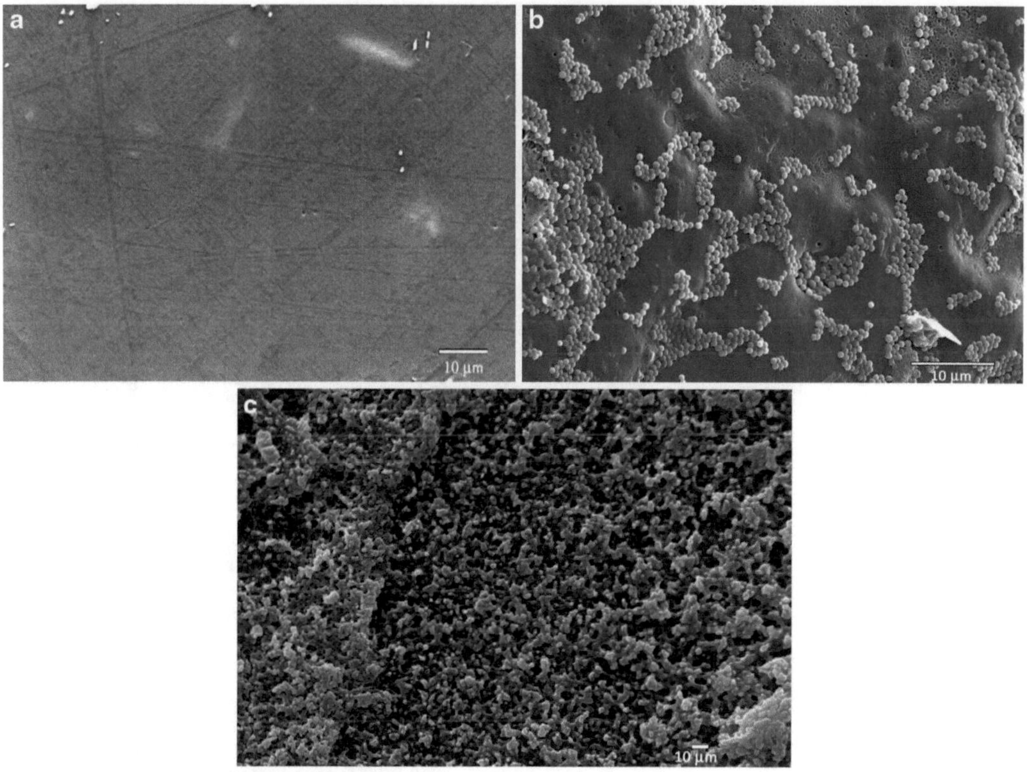

Fig. 2 SEM micrographs showing a polyurethane surface without colonization (**a**), with initial bacterial adhesion (**b**), and with a mature biofilm (**c**)

sodium cacodylate buffer (pH 7.4) at room temperature for 30 min. Wash the samples twice with cacodylate buffer, immerge them before in graded ethanol series (ethanol/water 30/70, 50/50, 70/30, 90/10, 100) for 15 min and then in hexamethyldisilazane for 30 s (*see* **Note 7**). In Fig. 2, SEM micrographs showing a polyurethane surface without colonization (a), with initial bacterial adhesion (b), and with a mature biofilm (c) are reported.

4. For FM or CLSM observations, use a GFP strain or stain the biofilm with fluorescent probes. By employing the GFP strains reported in Subheading 2, the 488 nm excitation laser of CLSM microscope must be used. An image of GFP-expressing *P. aeruginosa* biofilm grown on the surface of a polyurethane disk as an example is reported in Fig. 3. When a GFP strain is not used, the staining procedure can be performed with the LIVE/DEAD viability kit, by putting in contact a polymer disk with 1 mL of a staining solution containing both SYTO 9 (green) and propidium iodide (red) in a 1:1 M ratio. After 15 min incubation at room temperature in the dark, the residual stain can be rinsed off by the Ringer's solution (*see* **Note 8**).

Fig. 3 CLSM image of GFP-expressing *P. aeruginosa* biofilm grown on the surface of a polyurethane disk

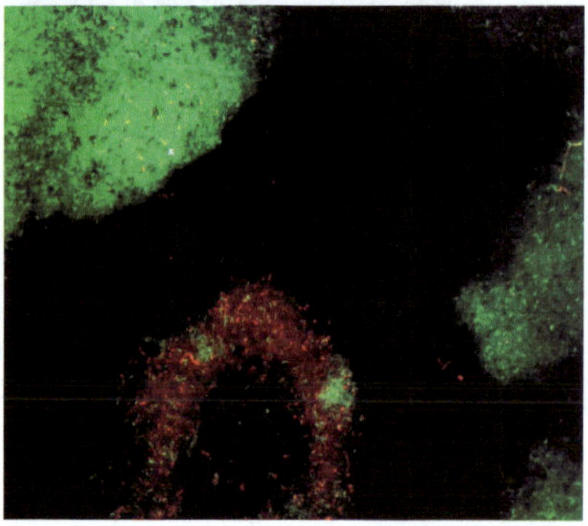

Fig. 4 CLSM image of *S. aureus* biofilm grown on the surface of a polyurethane disk after staining with LIVE/DEAD kit

In Fig. 4, CLSM image of *S. aureus* biofilm grown on the surface of a polyurethane disk after staining with LIVE/DEAD bacterial viability kit is reported. A biofilm-producing strain (10850) was used [17]. The image was taken by using the 488 nm excitation laser with a 488/568 nm/633 nm dichroic mirror. A band filter allowing wavelengths of 525–550 nm to pass to the first detector was used for the SYTO 9 stain. A long pass filter of 645 nm was used for imaging the propidium iodide stain.

3.2 Water-Soluble Polymers

Water-soluble antimicrobial polymers are gaining increasing interest since they could be considered as an alternative to existing biocides or even to antibiotics. Indeed, they could be used as food preservatives, agents for water sterilization, or fiber additives in healthcare and hygienic applications. Advantages of using antimicrobial polymers in place of low molecular weight compounds are the reduced toxicity, higher efficiency, and prolonged lifetime.

3.2.1 MIC Determination by Broth Dilution Method

1. The standard broth dilution method [18] can be adapted to polymers to determine the minimum inhibitory concentration. (*see* **Note 9**).
2. Prepare a bacterial inoculum at 1×10^6 CFU/mL with an optical density of 0.05 at 550 nm.
3. Add 1 mL of the inoculum to each test tube containing different concentrations of polymer solutions (1 mL). This will bring the final inoculum concentration to 5×10^5 CFU/mL. Use a control tube containing broth without antimicrobial agent.
4. Incubate the control and test tubes at 35 °C for 24 h.
5. Compare the amount of bacterial growth in the test tubes with that one in the control tube.
6. Identify the MIC as the lowest concentration of antimicrobial agent able to inhibit bacterial growth in the test tube by checking the absorbance at 550 nm of the control and test tubes.

4 Notes

1. The choice of the solvent to prepare polymer solution is not critical. However, avoid the use of solvent mixture that possessing different boiling points can impair homogeneity of the polymer disk. To store the polymer samples, plastic vessels are recommended in place of glass containers since some polymers can adhere strongly to glass.
2. The disk diffusion test can be performed only on polymer matrices releasing antimicrobial agents. In fact, as well known, this assay is routinely employed in microbiology laboratories for testing the antibiotic susceptibility of microorganisms. The standard procedure involves the inoculation of an agar plate with microorganisms, by the classic striking method, and the placement of an antibiotic-impregnated paper disk on the agar surface. Bacterial growth and antibiotic diffusion in the agar start simultaneously resulting in a circular zone of inhibition in which the drug amount exceeds the antibiotic minimum inhibitory concentration. The diameter of the inhibition zone measured after 24 h incubation depends on both drug

concentration in the disk and susceptibility of the microorganism. To evaluate polymer activity, the test was modified by replacing the antibiotic-impregnated paper disks with polymeric disks.

3. Caution should be used to interpret the data of inhibition zone. In fact, differently from the standard disk diffusion test (known as the Kirby Bauer test), in which for each microorganism the diameter of the inhibition zone is related to the antibiotic concentration in the paper disk, when testing polymers there cannot be a correlation between the width of inhibition zone and the amount of drug adsorbed to the polymer. In fact, if the drug interacts with the polymer, its release is hampered resulting in a small inhibition zone even when significant amount of drug is present in the polymer. In addition, usually the antimicrobial agent is not only adsorbed on the polymer surface but also entrapped in the polymer bulk. In fact, during the drug loading, the polymer can swell in water allowing the drug penetrating into the internal polymer layers. This further affects the drug release rate.
4. This test can be performed not only on polymer disks but also on nanoparticles. In this latter case, the sample must be weighed and homogeneous in size.
5. This assay is especially recommended when the antimicrobial agent is entrapped in the polymer bulk and is therefore released with difficulty by the polymer. The culture broth can penetrate into the polymer and favor drug release and action.
6. The choice of the buffer used to rinse polymer and detach bacteria is not critical. Ringer's solution or other isotonic media such as phosphate buffer can be used.
7. This dehydration procedure can be performed only if the polymer is not soluble in ethanol.
8. The staining procedure could not be suitable if the polymer adsorbs the stain. To reduce this effect, a low-thickness polymer film should be obtained by layering the polymer onto glass coverslips.
9. The MIC of biocidal polymers could be determined also on solid media by embedding cellulose disks with 10–20 μL of a polymer solution at the desired concentration and placing them on Petri plates previously seeded with 10^8 CFU/mL bacterial concentration. However, this method not always gives reproducible results. In fact, due to the high molecular weight, diffusion of polymers in the agar medium is usually hindered. Therefore, the diameters of inhibition zones could not be correlated to the concentration of the polymer solution.

References

1. Kumar C, Anand S (1998) Significance of microbial biofilms in food industry: a review. Int J Food Microbiol 42:9–27
2. Block J, Haudidier K, Pasquin J et al (1993) Biofilm accumulation in drinking water distribution systems. Biofouling 6:333–343
3. Francolini I, Donelli G (2010) Prevention and control of biofilm-based medical-device-related infections. FEMS Immunol Med Microbiol 59:227–238
4. Donelli G, Francolini I, Piozzi A et al (2002) New polymer-antibiotic systems to inhibit bacterial biofilm formation: a suitable approach to prevent central venous catheter-associated infections. J Chemother 14:501–507
5. Piozzi A, Francolini I, Occhiaperti L et al (2004) Antimicrobial activity of polyurethanes coated with antibiotics: a new approach to the realization of medical devices exempt from microbial colonization. Int J Pharm 280:173–183
6. Darouiche RO, Berger DH, Khardori N et al (2005) Comparison of antimicrobial impregnation with tunneling of long-term central venous catheters: a randomized controlled trial. Ann Surg 242:193–200
7. Francolini I, Norris P, Piozzi A et al (2004) Usnic acid, a natural antimicrobial agent able to inhibit bacterial biofilm formation on polymer surfaces. Antimicrob Agents Chemother 48:4360–4365
8. Crisante F, Francolini I, Bellusci M et al (2009) Antibiotic delivery polyurethanes containing albumin and polyallylamine nanoparticles. Eur J Pharm Sci 36:555–561
9. Schierholz JM, Rump A, Pulverer G (1997) New antiinfectious biomaterials. Ciprofloxacin containing polyurethanes as potential drug delivery systems to prevent foreign-body infections. Arzneimittelforschung 47:70–74
10. Francolini I, D'Ilario L, Guaglianone E et al (2010) Polyurethane anionomers containing metal ions with antimicrobial properties: thermal, mechanical and biological characterization. Acta Biomater 6:3482–3490
11. Francolini I, Taresco V, Crisante F et al (2013) Water soluble usnic acid-polyacrylamide complexes with enhanced antimicrobial activity against *Staphylococcus epidermidis*. Int J Mol Sci 14:7356–7369
12. Francolini I, Ruggeri V, Martinelli A et al (2006) Novel metal-polyurethane complexes with enhanced antimicrobial activity. Macromol Rapid Comm 27:233–237
13. Woo GL, Yang ML, Yin HQ et al (2002) Biological characterization of a novel biodegradable antimicrobial polymer synthesized with fluoroquinolones. J Biomed Mater Res 59:35–45
14. Kenawy E, Worley SD, Broughton R (2007) The chemistry and applications of antimicrobial polymers: a state-of-the-art review. Biomacromolecules 8:1359–1384
15. Acar J, Goldstein F (1996) Disk susceptibility test. In: Lorian V (ed) Antibiotics in laboratory medicine, 4th edn. Lippincott Williams & Wilkins, Baltimore, MD, pp 1–51
16. Baldassarri L, Donelli G, Gelosia A et al (1996) Purification and characterization of the staphylococcal slime-associated antigen and its occurrence among *Staphylococcus epidermidis* clinical isolates. Infect Immun 64:3410–3415
17. Donelli G, Francolini I, Romoli D et al (2007) Synergistic activity of dispersin B and cefamandole nafate in the inhibition of staphylococcal biofilm growth on polyurethanes. Antimicrob Agents Chemother 51:2733–2740
18. Cockerill F, Wikler M, Alder J et al (2012) Methods for dilution antimicrobial susceptibility tests for bacteria that grow aerobically, Approved standard 9th edn. Clinical and Laboratory Standards Institute M07-A9 (32):1–63

Chapter 16

Antibiotic Polymeric Nanoparticles for Biofilm-Associated Infection Therapy

Wean Sin Cheow and Kunn Hadinoto

Abstract

Polymeric nanoparticles are highly attractive as drug delivery vehicles due to their high structural integrity, stability during storage, ease of preparation and functionalization, and controlled release capability. Similarly, lipid–polymer hybrid nanoparticles, which retain the benefits of polymeric nanoparticles plus the enhanced biocompatibility and prolonged circulation time owed to the lipids, have recently emerged as a superior alternative to polymeric nanoparticles. Drug nanoparticle complex prepared by electrostatic interaction of oppositely charged drug and polyelectrolytes represents another type of polymeric nanoparticle. This chapter details the preparation, characterization, and antibiofilm efficacy testing of antibiotic-loaded polymeric and hybrid nanoparticles and antibiotic nanoparticle complex.

Key words Antibiofilm, Lipid–polymer hybrid nanoparticles, Antimicrobial testing, PLGA

1 Introduction

Biofilm cells are highly tolerant towards antimicrobial agents such as antibiotics due to (1) the majority of the biofilm cells being in a non-growing state causing them to be less susceptible to certain antibiotics (e.g., β-lactam); (2) the presence of the EPS matrix protecting the biofilm cells from antibiotics, therefore allowing timely adaptive stress responses; and (3) the existence of invulnerable subpopulation of persister cells regardless of the antibiotic concentration and exposure duration [1]. Encapsulating antibiotics into nanoscale carrier particles will allow the residence time of the antibiotic in the biofilm infection sites to be prolonged as a result of the controlled antibiotic release afforded by the carriers. In addition, the antibiotic can be released locally inside the biofilm matrix, hence increasing its residence time and concentration in the biofilm infection sites due to the efficiency with which nanoparticles can penetrate through the biofilm EPS matrix [2]. Furthermore, nanoparticles can effectively evade capture by the reticuloendothelial system (RES) resulting in

Gianfranco Donelli (ed.), *Microbial Biofilms: Methods and Protocols*, Methods in Molecular Biology, vol. 1147,
DOI 10.1007/978-1-4939-0467-9_16, © Springer Science+Business Media New York 2014

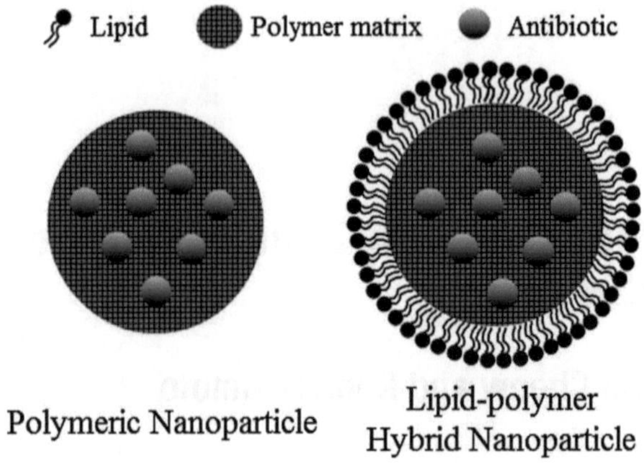

Fig. 1 Polymeric nanoparticle versus lipid–polymer hybrid nanoparticle

their high bioavailability [3, 4]. As a result, a higher antibacterial efficacy than that obtained by the non-encapsulated drug formulation is obtained.

Polymeric nanoparticles are highly attractive as drug delivery vehicles due to their high structural integrity, stability during storage, ease of preparation and functionalization, and controlled release capability [5]. In particular, polymeric nanoparticles prepared from biodegradable and biocompatible synthetic polymer such as poly-lactic-*co*-glycolic acid (PLGA) has been widely applied in drug encapsulation. The drug release can be precisely controlled by varying the polymer type, compositions, and molecular weight. More recently, lipid–polymer hybrid nanoparticle (LPN), which retains the benefits of polymeric nanoparticles in addition to improved biocompatibility and prolonged circulation time, has been developed [6]. Figure 1 shows the difference between polymeric and hybrid nanoparticles.

Another type of polymeric nanoparticles is the amorphous nanoparticle complex (nanoplex), in which charged polymers, i.e., polyelectrolytes, react with an oppositely charged drug molecule to form nanoplexes [7]. In contrast to the antibiotic-loaded nanoparticles in which the polymer matrix forms a large percentage of the nanoparticle, a majority of the nanoplex consists of the antibiotic. Furthermore, due to the amorphous form of the nanoplex, the release rate of the antibiotic is no longer controlled but is instead very rapid as a result of the higher energy level of the amorphous form of the antibiotic.

This chapter describes the preparation, characterization, and antibacterial efficacy testing of antibiotic-loaded polymeric and hybrid nanoparticles in sufficient detail to allow even first timers to carry out the studies. Levofloxacin (LEV), a fluoroquinolone, is used as the model antibiotic, while PLGA and phosphatidylcholine (PC) are used as the polymer and lipid, respectively. The natural polyanion, dextran sulphate, is used as the polyelectrolyte to interact with LEV to form the antibiotic nanoplex.

2 Materials

2.1 Nanoparticle Preparation

1. Poly(DL-lactide-*co*-glycolide) (PLGA, Purasorb 5004A) (PURAC Biomaterials, The Netherlands).
2. Dichloromethane (DCM), analytical grade.
3. Acetone, analytical grade.
4. Acetic acid.
5. Poly (vinyl alcohol) (PVA, MW = 23,000).
6. Pluronic F-68.
7. Sodium chloride, NaCl.
8. LEV.
9. Dextran sulphate (5,000 Da) (DXT).
10. PC, from egg yolk.
11. D-alpha tocopheryl polyethylene glycol 1000 succinate (TPGS).
12. Phosphate buffer saline (PBS) solution (pH = 7.4).
13. Dialysis membrane (molecular weight cutoff size of 12,400 g/mol).

2.2 Antimicrobial Testing

1. *E. coli.*
2. 96-well microplate.
3. 96-peg lid.
4. Mueller–Hinton broth (MHB).
5. Bacto agar.
6. Luria Bertani (LB) broth.

3 Methods

3.1 Preparation of Antibiotic-Loaded Polymeric Nanoparticles

The nanoprecipitation method to prepare nanoparticles is based on the work of Govender et al. [8] and results in LEV-loaded nanoparticles with a burst release profile. When the polymer and the highly water-soluble LEV (≈0.1 g/mL) are dissolved in water-miscible acetone and added into water, the acetone rapidly diffuses into the aqueous phase resulting in the formation of polymeric nanoparticles. LEV is predominantly adsorbed on the nanoparticle surface, contributing to the burst release profile.

The emulsification–solvent evaporation (ESE) method of Sung et al. [9] is employed to prepare LEV-loaded nanoparticles with the biphasic extended release profile. The polymer and LEV are dissolved in DCM, which is volatile and water immiscible. Upon addition of the polymer/LEV solution into water, an oil-in-water nano-emulsion is formed by ultrasonication. When DCM is evaporated, the nano-emulsion is transformed into a nanoparticulate

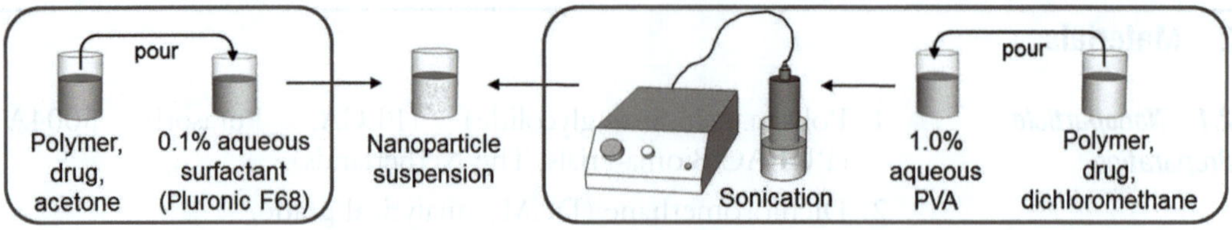

Fig. 2 Nanoprecipitation (*left*) versus emulsification–solvent evaporation (*right*) to prepare nanoparticles

suspension in which LEV is encapsulated inside the polymer matrix resulting in an extended release profile. Figure 2 demonstrates the difference between the nanoprecipitation and ESE methods. The double-ESE (DESE) method to prepare LEV-loaded nanoparticles is a modified version of the ESE, with the difference being an extra internal aqueous phase to result in a water-in-oil-in-water (w/o/w) emulsion.

1. To prepare antibiotic-loaded polymeric nanoparticles by the nanoprecipitation method, first dissolve (1) 80 mg of PLGA together with 8 mg of LEV in 5 mL of acetone to form the solvent phase (*see* **Note 1**) and (2) 10 mg of Pluronic F-68 in 10 mL of DI water to form the anti-solvent phase. Slowly add the solvent phase into the continuously stirred anti-solvent phase dropwise. A milky suspension should be observed to form immediately. Remove the acetone from the suspension by evaporation, either by stirring in an open beaker overnight or by using a rotary evaporator.
2. To prepare LEV-loaded nanoparticles by ESE, dissolve (1) 8 mg of LEV and 80 mg of PLGA in 2 mL of DCM to form the oil phase; (2) 60 mg of PVA in 6 mL of DI water to form a 1.0 % (w/v) aqueous PVA solution, which is the aqueous phase; and (3) 10 mg of PVA in 10 mL of DI water to form the diluting solution. Transfer the oil phase into the aqueous phase, and sonicate the mixture using a probe sonicator for 60 s (*see* **Note 2**). A milky suspension should form immediately. Pour the milky suspension into the constantly stirred diluting solution, and allow the DCM to evaporate off by overnight stirring in an open beaker.
3. To prepare nanoparticles by the DESE method, first dissolve (1) 8 mg of LEV in 300 μL of DI water to form the inner aqueous phase (w_1), (2) 90 mg of PLGA in 3 mL of DCM to form the oil phase (o), and (3) 120 mg of PVA in 12 mL of DI water to form a 1.0 % (w/v) aqueous PVA solution, which is the external aqueous phase (w_2). Transfer the w_1 phase into the water phase, and sonicate the mixture using a probe sonicator for 60 s. An emulsion should form at once, which is the w_1/o emulsion. Immediately transfer the emulsion to the w_2 phase and sonicate for a further 60 s to form the w_1/o/w_2 double emulsion.

Allow the DCM to evaporate off by overnight stirring in an open beaker. Note the final volume of the suspension after evaporation (V_{evap}).

4. For all the nanoparticles prepared, remove non-encapsulated antibiotic by three cycles of centrifugation (14,000 × *g*) and resuspension (*see* **Note 3**). Set aside the supernatant from the first centrifugation that contains the non-encapsulated antibiotic for encapsulation efficiency measurement (*see* Subheading 3.3). Use only 0.2 mL of DI water to resuspend the final nanoparticle pellet so that a concentrated stock nanoparticle suspension is obtained. Note the final volume of the stock suspension (V_{stock}) for yield calculation (*see* Subheading 3.4).
5. To prepare nanoparticles in larger quantities, multiple batches of nanoparticles should be prepared instead of merely scaling the volume as the sonication step is not easily scaled.

3.2 Preparation of Antibiotic-Loaded Lipid–Polymer Hybrid Nanoparticles

Lipid–polymer hybrid nanoparticles can be prepared by simple modification of the ESE and DESE methods, where the lipid PC is dissolved in the oil phase [10]. To ensure stability, an extra surfactant, TPGS, in included in a small amount together with PC (*see* **Note 4**).

1. To prepare DESE hybrid nanoparticles, dissolve 9 mg of the antibiotic in 300 μL of deionized water to form the internal aqueous phase. Co-dissolve 30 mg PC, 9 mg TPGS, and 90 mg PLGA in 3 mL of DCM to form the oil phase. Emulsify the aqueous antibiotic solution in the PLGA organic solution by sonicating for 60 s using a Vibra-Cell probe sonicator. Pour the resulting nano-emulsion into 12 mL of deionized water and sonicate again for 60 s to form the w/o/w double emulsion.
2. To prepare hybrid nanoparticles by the ESE method, dissolve 9 mg of LEV, 9 mg of TPGS, 30 mg of PC, and 90 mg of PLGA directly in the oil phase (DCM). Following that, emulsify for 60 s in 12 mL of deionized water.
3. For all the hybrid nanoparticles prepared, remove non-encapsulated antibiotic by three cycles of centrifugation (14,000 × *g*) and resuspension (*see* **Note 3**). Use only 0.2 mL of DI water to resuspend the final nanoparticle pellet so that a concentrated stock nanoparticle suspension is obtained.

3.3 Preparation of Antibiotic Nanoplex

1. Dissolve 6 mg of DXT and 5.8 mg of NaCl in 1 mL DI water to form the polyelectrolyte solution (*see* **Note 5**).
2. Dissolve 10 mg of LEV and 5.8 mg of NaCl in 1 mL DI water to form the drug solution. Add 2 μL of acetic acid to this drug solution to allow LEV to be ionized.

3. Add all the drug solution to the polyelectrolyte solution rapidly by pipetting using a 1,000 μL pipette. A white suspension should form immediately.
4. Allow the mixture to equilibrate for 30 min before washing by three cycles of centrifugation and resuspension in DI water.

3.4 Characterization of Antibiotic-Loaded Nanoparticles

1. Yield: The yield is the total amount of nanoparticles collected compared to the amount of polymer initially used in the preparation. Freeze-dry 50 μL aliquot of the stock to determine the dry mass ($M_{\mathrm{dry\ NP}}$). Using the dry mass and the final stock volume noted earlier, calculate the polymeric nanoparticle yield using Eq. 1:

$$\text{Yield} = \frac{M_{\mathrm{dry\ NP}}(\mathrm{mg}) \quad V_{\mathrm{stock}}(\mathrm{mL})/0.05}{\text{Mass of polymer initially added}} \quad 100\%. \tag{1}$$

In the calculation of yield for hybrid nanoparticles, take into account the mass of PC added.

2. Drug loading: The drug loading is the amount of drug per unit mass of the nanoparticles. Use the same dry nanoparticles in yield determination for the drug loading measurement. Dissolve the dry nanoparticles in 2 mL of DCM, and measure the absorbance of LEV in DCM at 254 nm. Compare the absorbance against the absorbance of a series of LEV standards in DCM to determine the concentration of DCM. Calculate the drug loading using Eq. 2:

$$\text{Drug loading} = \frac{\text{Concentration of LEV in 2mL} \quad 2\mathrm{mL}}{M_{\mathrm{dry\ NP}}(\mathrm{mg})} \quad 100\%. \tag{2}$$

3. Encapsulation efficiency: The encapsulation efficiency is defined as the amount of encapsulated drug compared to the amount of drug initially added. To determine the encapsulation efficiency, measure the drug concentration in the supernatant from the first cycle of centrifugation set aside in Subheading 3.1, **step 4** ($C_{\mathrm{supernatant}}$). Calculate the encapsulation efficiency using Eq. 3:

$$\text{Encapsulation efficiency} = \frac{\text{Mass of drug initially added - } \mathrm{C}_{\mathrm{supernatant}} \quad \mathrm{V}_{\mathrm{evap}}}{M_{\mathrm{dry\ NP}}(\mathrm{mg})} \quad 100\%. \tag{3}$$

4. Size and zeta-potential: Dilute an aliquot of the stock nanoparticle suspension by tenfold or until the suspension is nearly transparent to measure the size and zeta-potential of the nanoparticles prepared by using a Brookhaven 90Plus Nanoparticle Size Analyzer.

5. In vitro release: The in vitro release study determines the rate of drug release from the nanoparticles in PBS using the dialysis bag method under a sink condition (*see* **Note 6**). First, wet about 10 cm strip of the dry dialysis bag by immersing it in DI water for 30 min. Tie a short knot at one end of the dialysis bag, ensuring that there remains sufficient dialysis bag for tying a second knot at the opposite end. Open up the dialysis bag to form a tubing by gently rubbing the edges. Transfer 1.95 mL of PBS (pre-warmed in a 37 °C water bath) into the dialysis bag and 8 mL of the same PBS into a 15 mL bottle (external PBS) (*see* **Note 7**). Transfer 50 μL of stock nanoparticle suspension into the dialysis bag, being careful not to spill the content. Tie a knot to seal the remaining end of the dialysis bag, and place the dialysis bag into the 8 mL external PBS. Place the bottle containing the dialysis bag and nanoparticles into a shaking incubator set at 37 °C and 150 rpm. Immediately start the timer. At suitable time intervals (e.g., 12 h), collect 4 mL of the external PBS for drug concentration measurement, and replace the collected amount with fresh, pre-warmed PBS of the same volume. Determine the drug concentrations at different time intervals (i.e., LEV1, LEV2, and …) by using the UV–vis spectrophotometer and a corresponding standard curve. The in vitro drug release profile is conventionally presented as % drug released over time, where the total drug amount, i.e., 100 %, is calculated from the drug loading.

3.5 Antibiotic Susceptibility Testing

3.5.1 Determining Minimum Inhibitory Concentration of the Antibiotic Nanoparticles

The minimum inhibitory concentration (MIC), defined as the lowest antibiotic concentration that inhibits a visible planktonic bacterial growth after an overnight incubation at 37 °C, is used to determine the efficacy of the antibiotic-loaded nanoparticles prepared. An optical density measurement at 600 nm (OD_{600}) is typically used to examine the visible bacterial growth in which $OD_{600} < 0.1$ indicates a zero bacterial growth.

1. Adjust the concentration of an overnight *E. coli* suspension to 0.5 McFarland standard in MHB. Add 9.9 mL of MHB to 100 μL of the cell suspension to dilute the suspension by 100-fold to produce a bacterial cell suspension having approximately 1.0×10^6 colony-forming units (CFU)/mL. Add 100 μL of nanoparticle suspension in MHB at various concentrations to 100 μL of the bacterial cell suspension in each well of a 96-well microplate to yield a final cell concentration of 5.0×10^5 CFU/mL (*see* **Note 8**). For each nanoparticle concentration tested, prepare a well for reference OD by adding 100 μL of the nanoparticle suspension to 100 μL of MHB.
2. After a 24-h incubation at 37 °C, measure the OD_{600} of the mixed nanoparticles and cell suspension ($OD_{600,\ \mathrm{mixture}}$) as well

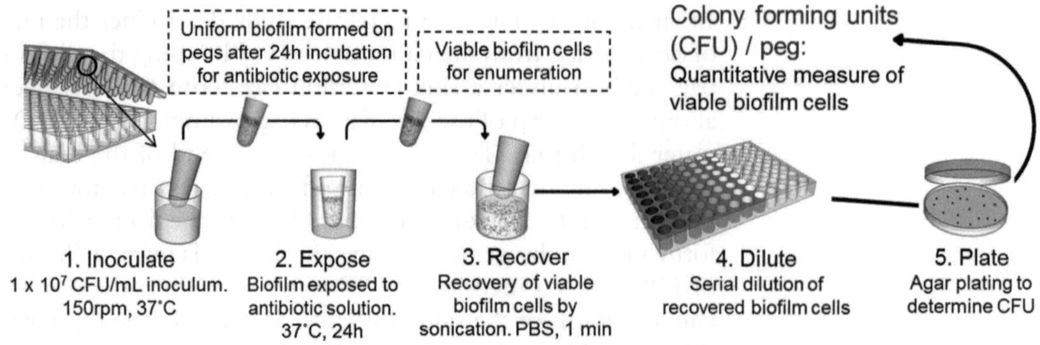

Fig. 3 Determination of MBIC

as the reference nanoparticle suspension ($OD_{600,\ reference}$) in each well by using a microplate reader (Synergy HT, Biotek, USA). Subtract off the OD_{600} contributed by the nanoparticles from the measured mixture OD_{600} to obtain the optical density due to the cell suspension alone ($OD_{600,\ cell}$), i.e., $OD_{600,\ cell} = OD_{600,\ mixture} - OD_{600,\ reference}$. The lowest antibiotic-loaded nanoparticle concentration that yields $OD_{600,\ cell} < 0.1$ is the MIC (*see* Fig. 3).

3.5.2 Determining Minimum Biofilm Inhibitory Concentration of the Antibiotic Nanoparticles

The minimum biofilm inhibitory concentration (MBIC) is defined as the lowest antibiotic concentration that inhibits a visible biofilm cell growth. The biofilm cultivation method and MBIC determination follow the methods of Harrison et al. [11] and Ceri et al. [12], respectively.

1. Adjust the concentration of an overnight cell suspension in MHB to 1.0 McFarland standard. Dilute the cell suspension by 30-fold in MHB to produce 1.0×10^7 CFU/mL. Transfer 150 μL of the inoculum to each well of a 96-well microplate, and fit a 96-peg lid on top of the microplate to provide a surface for the biofilm growth (*see* **Note 9**). After 24-h incubation (*see* **Note 10**) in a shaking incubator at 150 rpm and 37 °C, lift up the peg lid and rinse the biofilm formed on the pegs with PBS by fitting the peg lid onto another microplate filled with fresh PBS. The rinsing step is necessary to remove loosely attached planktonic cells. Break off two pegs to enumerate the amount of biofilm cells formed (*see* **Note 11**).
2. Expose the biofilm formed in the remaining 94 pegs to 200 μL of LEV solution in MHB in a new microplate (*see* **Note 12**). After a 24-h incubation at 37 °C, lift the peg lid and rinse the pegs with PBS to remove the dead biofilm cells. Fit the rinsed peg lid onto a new microplate (i.e., recovery plate) containing 200 μL fresh MHB, and detach the non-eradicated biofilm cells from the pegs into the wells by sonicating the peg-plate assembly for 5 min (*see* **Note 13**). Cover the microplate with a new lid (without pegs), and incubate the plate at 37 °C for 24 h.

3. Following that, measure the OD_{600} in each well of the recovery plate by using the microplate reader. The lowest LEV concentration that results $OD_{600} < 0.1$ is determined as the MBIC.
4. To investigate whether the nanoparticles exhibit any antibacterial activities, grow biofilm cells on a 96-peg lid following the same procedures used to determine the MBIC. Subsequently, expose the biofilm formed on the pegs to 40 μL of the blank nanoparticulate suspension in a microplate at the highest equivalent antibiotic-loaded nanoparticle concentration tested.
5. Serially dilute the recovered biofilm cells tenfold, and drop plate onto LB agar plates. Incubate overnight at 37 °C to determine the viable CFU. The biofilm-derived planktonic cells which are present in the wells of the microplate after the nanoparticle exposure can also be enumerated.
6. Prepare biofilm cells exposed to 40 μL of PBS and 160 μL of MHB as the experimental controls.
7. Compare the viable CFU counts of the biofilm-derived planktonic cells and the biofilm cells which have been exposed to the blank nanoparticles with those of the experimental controls to determine the percentage of surviving cells (*see* Fig. 3).

3.6 Time-Dependent Biofilm Susceptibility Testing

The time-dependent biofilm susceptibility testing allows biofilm grown to be exposed to a series of antibiotic solutions of different concentrations depending on the in vitro drug release profile. The separation of the antibiotic release and the susceptibility testing experiments is practical only when the nanoparticles do not exhibit any antimicrobial activity. In addition, the exclusion of the nanoparticles (using only the antibiotic solution released from the nanoparticles) allows the examination of the effect of release profiles on the biofilms to be performed easily without the complication of retaining nanoparticles while replacing the growth medium necessary to sustain the biofilm growth, particularly when antibiotics are released from the nanoparticles beyond 24 h.

1. Grow the biofilm on 96-peg lids following the same procedures used to determine the MBIC.
2. To test the antibiotic solution from the first time point, the biofilm cells from the pegs are exposed to an antibiotic solution in a microplate, which contains 20 μL of the LEV1 solution obtained from the in vitro drug release study and 180 μL of MHB. A solution containing 20 μL of PBS and 180 μL of MHB is used as the positive control. The fresh MHB must be supplied to sustain the biofilm existence over the time-kill period.
3. After a 12-h incubation at 37 °C (i.e., second time point of the time kill where time interval is 12 hourly), break off two

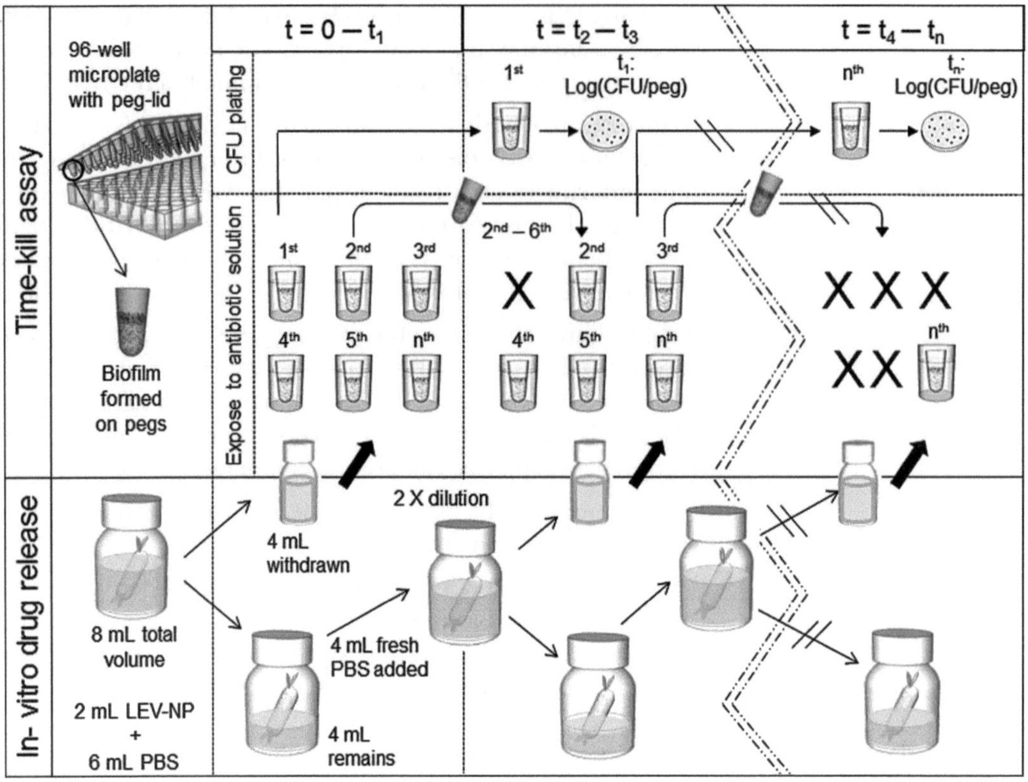

Fig. 4 An illustration of the time-kill study utilizing antibiotic solutions from the in vitro release study

pegs from the peg lid using a sterile plier. Immerse the pegs separately in two microcentrifuge tubes containing PBS to rinse the pegs and remove loosely attached biofilm cells. Transfer the pegs to new microcentrifuge tubes containing PBS, and sonicate the microcentrifuge tube (containing the pegs) in a sonicating bath for 5 min to transfer the biofilms from the pegs to the PBS. Determine the viable CFU by using drop plate counting.

4. Transfer the remaining biofilms on the peg lid to a new microplate containing antibiotic solutions from the next time interval, i.e., 20 μL of the LEV2 solution and 180 μL of MHB. Repeat these procedures until all the antibiotic solutions collected from the in vitro release study are used up (*see* Fig. 4).

4 Notes

1. Other water-miscible solvents that can dissolve the polymers can be used, such as acetonitrile. The solvent used will affect the nanoparticle characteristics. Any other antibiotics that can be dissolved in the solvent can also be used.

2. The DCM layer will be at the bottom; therefore, ensure that the sonicating probe is sufficiently deep to reach the DCM layer. Use a suitable container for sonication, such as a 15 mL bottle, to prevent splashing of the emulsion during sonication and also to avoid breaking the container due to strong sonication forces.
3. After centrifugation, all the supernatant should be removed, leaving behind an intact nanoparticle pellet. To resuspend the pellet, gently scrape the pellet against the centrifuge tube by using a pipette tip while pipetting up and down. A smooth suspension without any lumps should be formed after each resuspension step.
4. Other lipids such as DPPC can also be used. As an alternative to TPGS, PEGylated lipids can be used, e.g., DSPE-PEG.
5. Other antibiotics, such as ciprofloxacin, can be used.
6. Nanoplex dissolves immediately in PBS, so it should be excluded from in vitro release study.
7. Since fluoroquinolones are photodegradable, use opaque or amber bottles.
8. Use serial twofold dilutions to prepare a series of concentrations of antibiotic-loaded nanoparticles.
9. We find that the wells at the edges of the microplate, i.e., Rows A and G and Columns 1 and 12, are prone to evaporation, causing the biofilms formed at these locations to be slightly lower in density compared to the rest of the wells. To counter this problem, use the wells in these locations as an evaporation guard, where 200 μL DI water is filled into these wells. This way, any evaporation will not affect biofilm formation in the inner wells.
10. To form older (more mature) biofilms, incubation beyond 24 h can be done, but the medium needs to be changed daily. To expose the biofilm cells to fresh medium, simply lift up the peg lid and fit it onto a new microplate filled with MHB.
11. Breaking the pegs at the edges (instead of in the middle) is less likely to disrupt biofilms on other pegs. Breaking off the pegs might need some practice.
12. The volume of the inoculum is less than the total volume when tested against nanoparticle suspension so that the biofilm formed is completely submerged in the nanoparticle suspension.
13. The peg-plate assembly should be wrapped with parafilm to minimize contamination from the sonicating bath. In addition, only the lower half of the assembly, i.e., the plate base, should be submerged in the sonicating bath, while the upper peg lid

half should remain above the bath level. To achieve this, the volume of water in the sonicating bath should be reduced, or the "basket" of the sonicating bath should be elevated (e.g., by using retort clamps).

References

1. Stewart PS (2002) Mechanisms of antibiotic resistance in bacterial biofilms. Int J Med Microbiol 292:107–113
2. Meers P, Neville M, Malinin V et al (2008) Biofilm penetration, triggered release and in vivo activity of inhaled liposomal amikacin in chronic *Pseudomonas aeruginosa* lung infections. J Antimicrob Chemother 61:859–868
3. Albanese A, Tang PS, Chan WCW (2012) The effect of nanoparticle size, shape, and surface chemistry on biological systems. Annu Rev Biomed Eng 14:1–16
4. Alexis F, Pridgen E, Molnar LK, Farokhzad OC (2008) Factors affecting the clearance and biodistribution of polymeric nanoparticles. Mol Pharm 5:505–515
5. Peer D, Karp JM, Hong S et al (2007) Nanocarriers as an emerging platform for cancer therapy. Nat Nanotechnol 2:751–760
6. Zhang L, Chan JM, Gu FX et al (2008) Self-assembled lipid–polymer hybrid nanoparticles: a robust drug delivery platform. ACS Nano 2:1696–1702
7. Cheow WS, Hadinoto K (2012) Green amorphous nanoplex as a new supersaturating drug delivery system. Langmuir 28: 6265–6275
8. Govender T, Stolnik S, Garnett MC et al (1999) PLGA nanoparticles prepared by nanoprecipitation: drug loading and release studies of a water soluble drug. J Control Release 57:171–185
9. Sung J, Padilla D, Garcia-Contreras L et al (2009) Formulation and pharmacokinetics of self-assembled rifampicin nanoparticle systems for pulmonary delivery. Pharm Res 26: 1847–1855
10. Cheow WS, Hadinoto K (2011) Factors affecting drug encapsulation and stability of lipid-polymer hybrid nanoparticles. Colloids Surf B Biointerfaces 85:214–220
11. Harrison J, Ceri H, Yerly J et al (2006) The use of microscopy and three-dimensional visualization to evaluate the structure of microbial biofilms cultivated in the calgary biofilm device. Biol Proced Online 8: 194–215
12. Ceri H, Olson ME, Stremick C et al (1999) The calgary biofilm device: new technology for rapid determination of antibiotic susceptibilities of bacterial biofilms. J Clin Microbiol 37: 1771–1776

Chapter 17

Pharmacokinetics and Pharmacodynamics of Antibiotics in Biofilm Infections of *Pseudomonas aeruginosa* In Vitro and In Vivo

Wang Hengzhuang, Niels Høiby, and Oana Ciofu

Abstract

Although progress on biofilm research has been obtained during the past decades, the treatment of biofilm infections with antibiotics remains a riddle. The pharmacokinetic (PK) and pharmacodynamic (PD) profiles of an antimicrobial agent provide important information helping to establish an efficient dosing regimen and to minimize the development of antimicrobial tolerance and resistance in biofilm infections. Unfortunately, most previous PK/PD studies of antibiotics have been done on planktonic cells, and extrapolation of the results on biofilms is problematic as bacterial biofilms differ from planktonic grown cells in the growth rate, gene expression, and metabolism. Here, we set up several protocols for the studies of PK/PD of antibiotics in biofilm infections of *P. aeruginosa* in vitro and in vivo. It should be underlined that none of the protocols in biofilms have yet been certificated for clinical use or proved useful for guidance of antibiotic therapy.

Key words Biofilm, Antibiotics, PK/PD, *Pseudomonas aeruginosa*

1 Introduction

Dose regimes of antimicrobials have been established based on minimal inhibitory concentration (MIC) and minimal bactericidal concentration (MBC), which reflect the susceptibility of planktonically grown bacterial population [1, 2]. The traditional antibiotic regimes based on the MIC and MBC often failed to cure biofilm infections [3, 4] due to the intrinsic tolerance of biofilm-growing populations to antibiotics [5, 6]. Treatment strategies aimed at treating biofilm infections are urgently needed based on antibiotic susceptibility of biofilm-grown bacterial populations. Minimal biofilm inhibitory concentration (MBIC) and minimal biofilm eradication concentration (MBEC) are biofilm-related susceptibility parameters that should be used to guide the antibiotic treatment of biofilm infections.

Gianfranco Donelli (ed.), *Microbial Biofilms: Methods and Protocols*, Methods in Molecular Biology, vol. 1147,
DOI 10.1007/978-1-4939-0467-9_17,

MBIC and MBEC are static parameters established in vitro at a single time point. The dynamic of the interaction between antimicrobials and biofilms can be investigated by time-killing curves which integrate the antibiotic concentration and time of action [2, 7, 8].

Very few animal models are developed to study the PK/PD profiles of antimicrobials on biofilm infections. We established a biofilm lung infection model in order to provide an objective and quantitative evaluation of the PK/PD profile of antimicrobials [9]. To dissociate between the effect of antibiotics and of the immune system, the PK/PD studies of antimicrobials are traditionally conducted in neutropenic mice.

For an efficient treatment of biofilm infection, we propose to use the following PK/PD indices: area under the curve (AUC)/MBIC, AUC/MBEC, and time above MBIC or MBEC (T > MBIC and T > MBEC) as shown in Fig. 1. Comparing to planktonic cells, both higher concentrations and longer period treatment with antibiotics are required for biofilm cells [9–11]. Kinetics of colistin and imipenem in vitro (Fig. 4) and in vivo (Fig. 10) showed concentration-dependent and time-dependent killing, respectively, on biofilm-growing *P. aeruginosa*, and the elimination of the planktonic and biofilm bacteria in the lung was best correlated to AUC/MIC (AUC/MBIC) of colistin and T > MIC

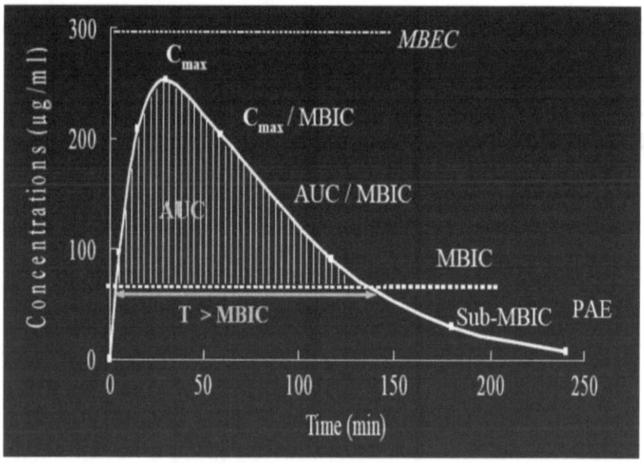

Fig. 1 PK–PD indice correlation with antibacterial efficacy on biofilms. The parameters are minimal biofilm inhibitory concentration (MBIC), minimal biofilm eradication concentration (MBEC), subminimum biofilm inhibitory concentration (sub-MBIC), post-antibiotic effect (PAE), maximum antibiotic concentration in serum (C_{max}), and total area under the time–concentration curve of antibiotics (AUC). The PK–PD indices are C_{max}/MBIC, AUC/MBIC, and the time that antibiotic concentration in serum remains above the MBIC (T > MBIC). Biofilms are notoriously difficult to be eradicated in vivo when the MBEC exceeded the C_{max}

(T > MBIC) of imipenem [9] in Fig. 11. Time-dependent killing of ceftazidime was observed in the *P. aeruginosa* PAO1 biofilms, but concentration-dependent killing activity was observed for β-lactamase-overproducing biofilms of PAΔDD h2Dh3 in vitro [11]. The PK/PD indices of AUC/MBIC and C_{max}/MBIC (C_{max}, maximum concentration) are probably the best parameters to describe the effect of ceftazidime in β-lactamase-overproducing biofilms of *P. aeruginosa* [9–11].

2 Materials

2.1 Bacterial Strains

P. aeruginosa PAO1 and other strains with or without fluorescent tag.

2.2 Medium

1. Minimum medium (ABTG medium): The medium consists of 1 mM $MgCl_2$, 0.1 mM $CaCl_2$, 15.1 mM $(NH_4)_2SO_4$, 33.7 mM $Na_2HPO_4 \cdot 2H_2O$, 22 mM KH_2PO_4, 51mMNaCl, 0.01 mM $FeCl_3$, 2.5 μg/ml thiamin, 0.5 % w/v glucose, and sterilized water.
2. Rich medium: Luria-Bertani (LB) media.

2.3 Preparation for Biofilm Susceptibility Assay and Kinetic Assay of Antibiotics on Biofilms

1. Microtiter plate: Flat-bottom 96-well microtiter plate and pegs of a modified polystyrene microtiter lid (Nunc A/S, Denmark).
2. 0.1 % crystal violet solution.
3. Ultrasonic cleaner: Bransonic 220.
4. Microtiter plate reader.

2.4 Preparation for PK/PD of Antibiotics on Biofilms In Vivo

1. Animal: 10-week-old female NMRI mice (Taconic, Denmark) weighing 32–34 g.
2. Cyclophosphamide.
3. NucleoCassette Device (NucleoCounter System, ChemoMetec A/S, Denmark).
4. Microisolation cage system.
5. Fentanyl and Fluanisone (Hypnorm, 10 mg/ml).
6. Midazolam (Dormicum, 5 mg/ml).
7. Trizma base.
8. Calcium chloride dihydrate.
9. Pronova.
10. SYTO 9.
11. *Streptococcus SP.* EB68.
12. *Bordetella bronchiseptica* ATCC4617.

3 Methods

3.1 PK/PD of Antibiotics on Biofilms In Vitro

3.1.1 Growth Conditions

The minimum medium (ABTG medium) is refreshed every 24 h in wells for biofilm cultivation. Antibiotics are of pharmaceutical rate.

3.1.2 Biofilm Susceptibility Assay by the Modified Calgary Biofilm Device Method [5, 6, 11] (Figs. 2 and 3)

1. Isolates are grown overnight in LB media.
2. After dilution of this culture to 10^7 CFU/ml with ABTG medium, 0.10 ml is transferred to all wells except the negative control of a flat-bottom 96-well microtiter plate.
3. Bacterial biofilms are formed by immersing the pegs of a modified polystyrene microtiter lid into the biofilm growth plate, followed by incubation at 37 °C for 1, 3, and 7 days without shaking, and relative humidity (RH) for biofilm growth plate is above 95 % to prevent the evaporation of medium (*see* **Note 1**).
4. Crystal violet (0.1 %) staining method is employed to check the biofilm formation for the quality control pegs (*see* **Note 1**).
5. Peg lids are rinsed three times in sterile water.
6. Place onto flat-bottom microtiter plates containing antibiotic, twofold dilutions in 0.12 ml of ABTG medium per well (antibiotic challenge plate) and cultivate for 20 h at 37 °C (*see* **Note 2**).
7. After antibiotic incubation, peg lids are again rinsed three times in sterile water (*see* **Note 3**).

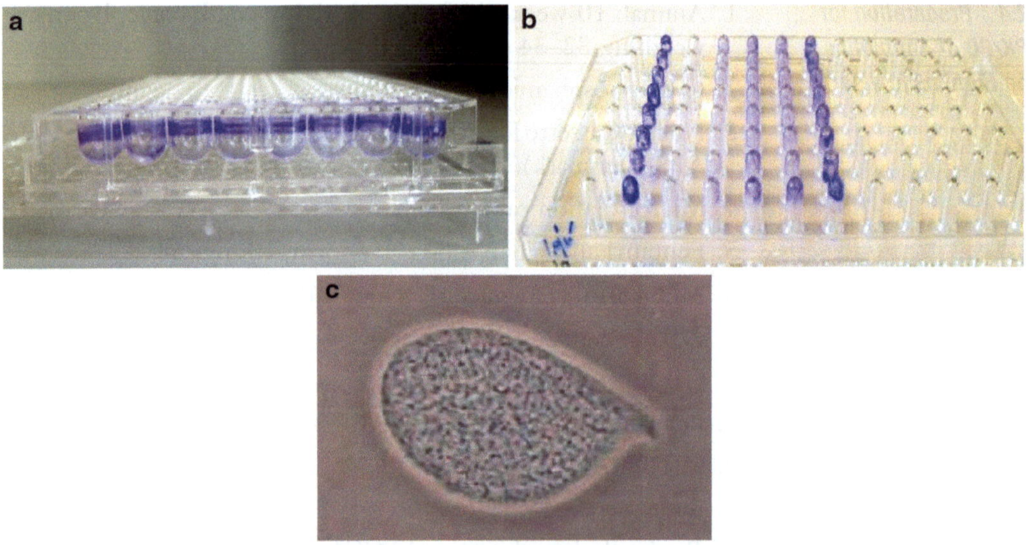

Fig. 2 The biofilm formation on wells (**a**) and peg lid of microplate (**b**) are stained by crystal violet. Artificial biofilm of alginate beads is shown (**c**) under microscopy (×60)

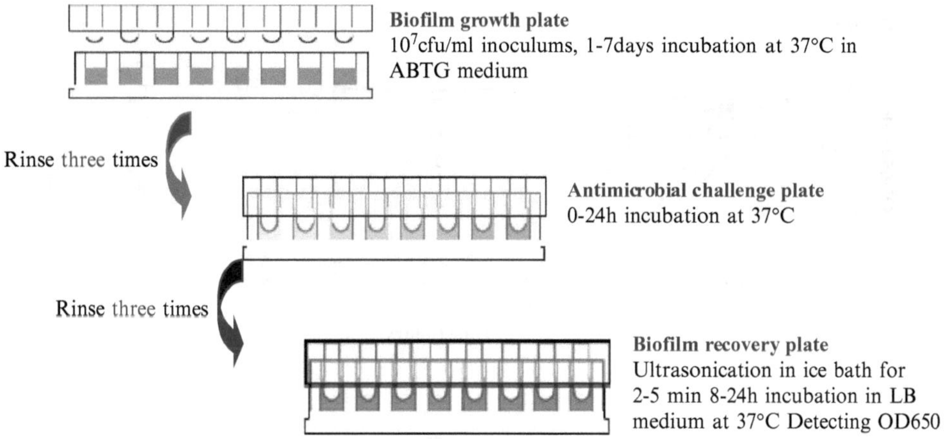

Fig. 3 Three steps of biofilm susceptibility assay including biofilm growth, single or combination antimicrobial challenge, and biofilm recovery. OD_{650} value is applied to evaluate the minimal biofilm inhibitory concentration (MBIC) and minimal biofilm eradication concentration (MBEC)

8. Peg lids are placed into antibiotic-free LB medium in a flat-bottom microtiter plate (biofilm recovery plate).
9. To transfer biofilms from pegs to wells, the sonication is performed at 4 °C for 2–5 min using Bransonic 220 (*see* **Note 4**). And then the biofilm cells in wells are cultivated for 8–24 h.
10. The optical density at 650 nm (OD_{650}) of well is measured by a microtiter plate reader, before and after incubation in LB medium at 37 °C for 8–24 h (*see* **Note 5**).
11. MBIC is defined as the concentrations of the drug that result in an OD_{650} difference at or below 10 % of the mean of three positive control well readings. The 10 % cutoff represents a 1-log 10 difference in growth after 8–12 h of incubation [6, 11]. MBEC is detected after 24-h recovery and defined as the concentrations of drug that resulted in no increase of OD_{650} or an increase of OD 650 < 0.05 [11].

3.1.3 Kinetics of Antibiotics on Biofilm Assay In Vitro (Figs. 3 and 4)

1. Biofilm cultivation and rinse are described as above for young biofilm and mature biofilm.
2. Peg lids are placed into flat-bottom microtiter plates containing different concentrations of antibiotics in 0.12 ml of ABTG medium per well (antibiotic challenge plate) and incubated for 0, 1, 2, 3, 4, 6, 8, 12, and 24 h at 37 °C with shaking (*see* **Notes 2** and **6**).
3. After antibiotic incubation, peg lids are rinsed three times and placed into antibiotic-free LB in a biofilm recovery plate with sonication (*see* **Notes 3** and 7).
4. OD_{650} is measured by a microtiter plate reader before and after incubation at 37 °C for 8–12 h to set up the dynamic time–kill curve of antibiotics (*see* **Note 5**).

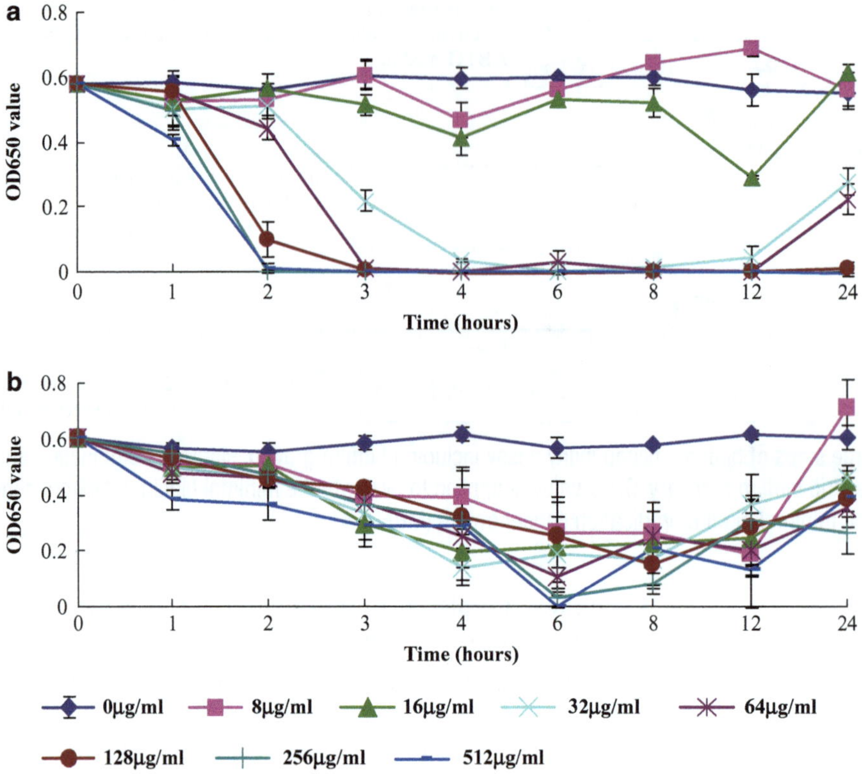

Fig. 4 Kill curves of biofilm-growing *P. aeruginosa* non-mucoid PAO1 with colistin (**a**) and imipenem (**b**) in vitro. Kinetics of colistin and imipenem in vitro showed concentration-dependent and time-dependent killing, respectively, on biofilm-growing *P. aeruginosa* (reprinted from [11] with permission)

3.1.4 Killing Curves of Antibiotics on Planktonic Cells

1. Planktonic *P. aeruginosa* 0.1 ml in approximately 10^6 CFU/ml is added into microtiter wells with LB medium.
2. Different concentrations of antibiotics (volume 0.1 ml) are added into different wells (total volume 0.2 ml), shaken, and cultivated at 37 °C for hours 0, 1, 2, 4, 8, 12, and 24.
3. Samples (0.1 ml) from wells are serially diluted and cultured overnight on plate, and then CFU is counted.
4. Killing curves of antibiotics on planktonic cells are plotted to compare with biofilms.

3.2 PK/PD of Antibiotics on Biofilms In Vivo

3.2.1 Animals

10-week-old female NMRI mice are used (*see* **Note 8**).

The mice are maintained on standard mouse chow and water ad libitum for 1 week before challenge. All animal experiments are performed under authorization from the National Animal Ethics Committee of Denmark.

3.2.2 Neutropenic Mouse Model of Biofilm Lung or Thigh Infection

1. Mice are rendered neutropenic by injecting three doses of cyclophosphamide intraperitoneally on day 1 (150 mg/kg of body weight), day 3 (100 mg/kg), and day 4 (100 mg/kg) prior to

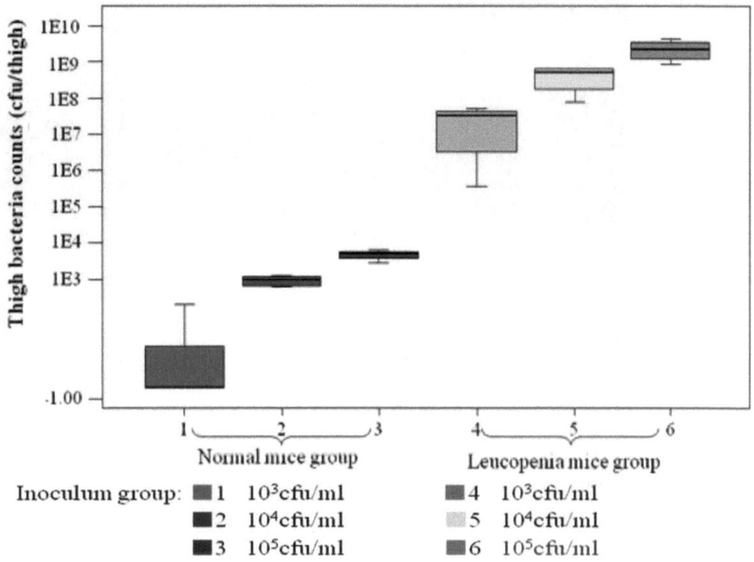

Fig. 5 The optimal challenge inoculums of planktonic *P. aeruginosa* in a neutropenic mouse thigh infection model at 24-h time points after bacterial challenge. The 10^5 CFU/ml is a candidate inoculum of *P. aeruginosa* for the neutropenic mouse thigh infection model

experimental infection. Optimization of the cyclophosphamide dose for mice is performed in pilot studies (*see* **Note 9**).

2. Blood is drawn from the tail (*see* **Note 10**), and leukocytes are counted with a NucleoCassette device. Blood smears are checked for the presence of granulocytes. Mice are severely granulocytopenic (absolute granulocyte count, <50/mm^3) by day 4 and remained so through day 5 and day 6 after the first injection of cyclophosphamide.
3. The bacterial infection is performed on day 5.
4. The neutropenic mice are raised in a microisolation cage system in a sterile environment.
5. Mouse operations are performed in a ventilated cabinet while the mice are under anesthesia (*see* **Note 11**).
6. Optimization of the inocula of planktonic and biofilm bacteria for lung or thigh infections is performed in pilot studies (*see* **Note 12**) shown in Fig. 5.
7. To prepare the biofilm bacteria, planktonic *P. aeruginosa* cells are immobilized in spherical alginate beads, as described [12–14] in Fig. 6.
8. The anesthetized mice are tracheostomized, a 0.04-ml inoculum of planktonic or biofilm bacteria adjusted to yield approximately 10^5 CFU/ml is instilled in the lower left lung using a curved bead-tipped needle (*see* **Note 13**), and then the incision is sutured as shown in Fig. 7. 0.1-ml inoculum of planktonic or biofilm bacteria is for thigh infection model.

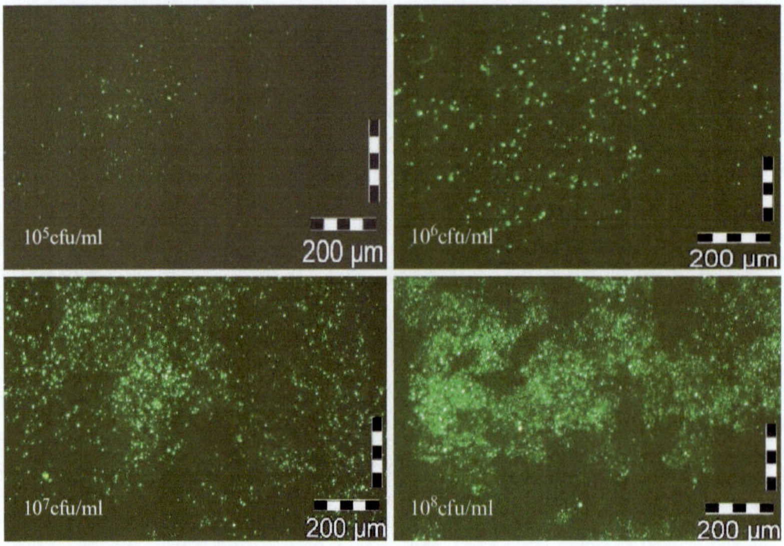

Fig. 6 Different concentrations of alginate bead staining by SYTO 9. The inoculum size is controlled precisely in the alginate beads from 10^5 to 10^8 CFU/ml

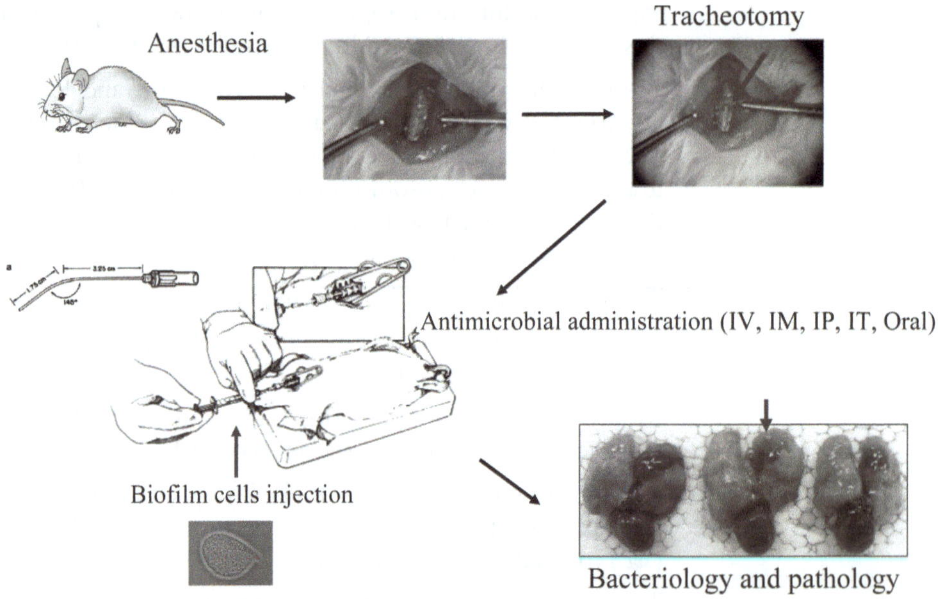

Fig. 7 Biofilm lung infection model. Biofilm cells are injected into lower left lung by trachea route when mouse is lightly anesthetized and receive a micro-operation with tracheotomy. Antibiotic administrations are included with intravenous (IV), intraperitoneal (IP), intramuscular (IM), intratracheal (IT), and oral administration. And then samples of lung are collected for the study of bacteriology, pathology, and immunology

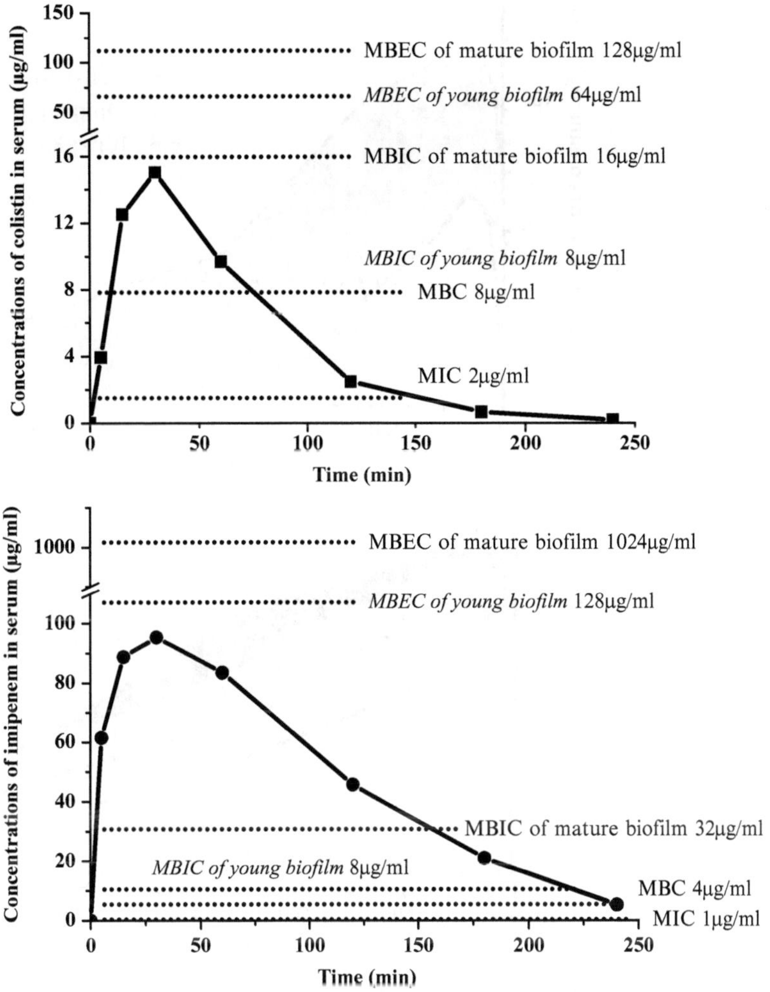

Fig. 8 Pharmacokinetics in mouse serum of colistin versus MIC, MBC, MBIC, and MBEC of *P. aeruginosa* PAO1 (*black square*) 16 mg/kg of colistin and (*black circle*) 64 mg/kg of imipenem with one-dose intraperitoneal administration. The concentrations of colistin and imipenem required to eradicate the biofilms in vivo are higher than the maximum concentration (C_{max}) of the drug in serum. It is difficult to reach the target of MBEC in serum when the drug is administrated systemically, but the target of MBIC is possible to be achieved in serum. Early treatment of antibiotics and its combinations in young biofilm are highly recommended in biofilm infections (reprinted from [11] with permission)

3.2.3 Pharmacokinetics of Antibiotics in a Mouse Model with Biofilm Infection for PK/PD Modeling (Figs. 8 and 9)

1. *P. aeruginosa* are immobilized in spherical alginate beads [12–14].
2. NMRI mice are lightly anesthetized with 0.15 ml mixture of fentanyl and fluanisone (Hypnorm, 10 mg/ml) and midazolam (Dormicum, 5 mg/ml) in 1:1.

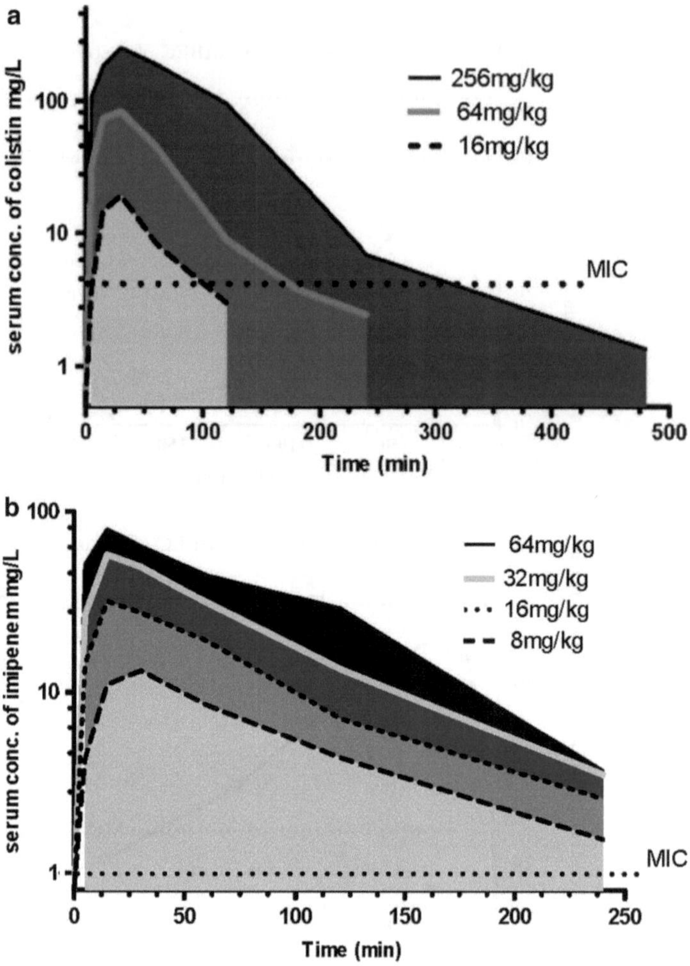

Fig. 9 Serum concentrations (*conc.*) of colistin (**a**) and imipenem (**b**) in mice after intraperitoneal treatment with doses from 8 to 256 mg/kg are shown. The *y*-axis in both panels is log 10. The MICs against *P. aeruginosa* PA01 are 4 mg/l for colistin and 1 mg/l for imipenem (reprinted from [9] with permission)

3. The mice are tracheostomized, and 0.04 ml inoculums adjusted to yield approximately 5×10^8 CFU/ml are installed in the lower left lung along the left bronchia using a curved bead-tipped needle.
4. The infected animals are treated with antibiotics intraperitoneally 2 h after infection (6 mice/group) (*see* **Note 14**).
5. An approximately 0.08-ml blood sample is collected from the tail (*see* **Note 10**) at 5, 15, 30, 60, 120, 180, and 240 min after antibiotic administration (*see* **Note 15**). At the end of the experiment, the mice are euthanized with pentobarbital/lidocaine.
6. Blood samples were centrifuged at $1{,}590 \times g$, and the serum is collected for measurement of antibiotic concentration. The concentrations of antibiotics in serum are measured by a biological

method with the indicator bacteria for example *Streptococcus sp.* EB68 for imipenem and *Bordetella bronchiseptica* ATCC4617 for colistin.

7. Time–concentration curves of antibiotics are established (shown in Fig. 9). The data running by PK programs is plotted against the value of MBIC and MBEC in Fig. 8 (*see* **Note 16**).

3.2.4 Time–Kill Study of Antibiotics in Planktonic and Biofilm Bacteria In Vivo (Fig. 10)

1. To establish time-killing curves of antibiotics, anesthetized neutropenic mice infected with planktonic bacteria (4 mice/point) or biofilm bacteria (4 mice/point) are treated at 2 h after infection in lung or thigh with a single intraperitoneal dose of antibiotics (*see* **Note 16**).
2. Control mice received the same volume of saline.
3. The mice are euthanized, and lungs or thighs are collected aseptically at −2, 0, 2, 4, 8, 12, and 24 h after bacterial challenge and homogenized in 5 ml of sterilized saline.
4. Humane endpoints are applied during the period.
5. 0.1 ml homogenate of tissue sample is serially diluted for plate cultivation.
6. The numbers of CFU are counted for plotting of the killing curves.

3.2.5 PK/PD Indices of Antibiotics in Planktonic and Biofilm Bacteria In Vivo (Fig. 11)

1. To establish PK/PD indices of antibiotics, anesthetized neutropenic mice infected with planktonic bacteria or biofilm bacteria are treated from the time point of 2 h after infection with multiple intraperitoneal doses of antibiotics.
2. The multiple dosages are administered at different time intervals ranging from 2 to 24 h after infection for periods of 12–24 h [15].
3. The mice are euthanized, and lungs or thigh are collected at the end of the experiment and homogenized in 5 ml of sterilized saline.
4. 0.1 ml homogenate of tissue sample is serially diluted for plate cultivation.
5. The numbers of CFU are counted for each lung or thigh and expressed as the log 10 number of CFU per lung or thigh.
6. The counts of viable bacteria for each regimen are plotted with the PK parameters.

4 Notes

1. Biofilm formation on pegs or wells is affected by several parameters of strains, medium, surface of pegs or wells, inoculum, cultivation period and temperature, and so on. Some clinical isolates are poor at forming biofilms in minimum medium, and

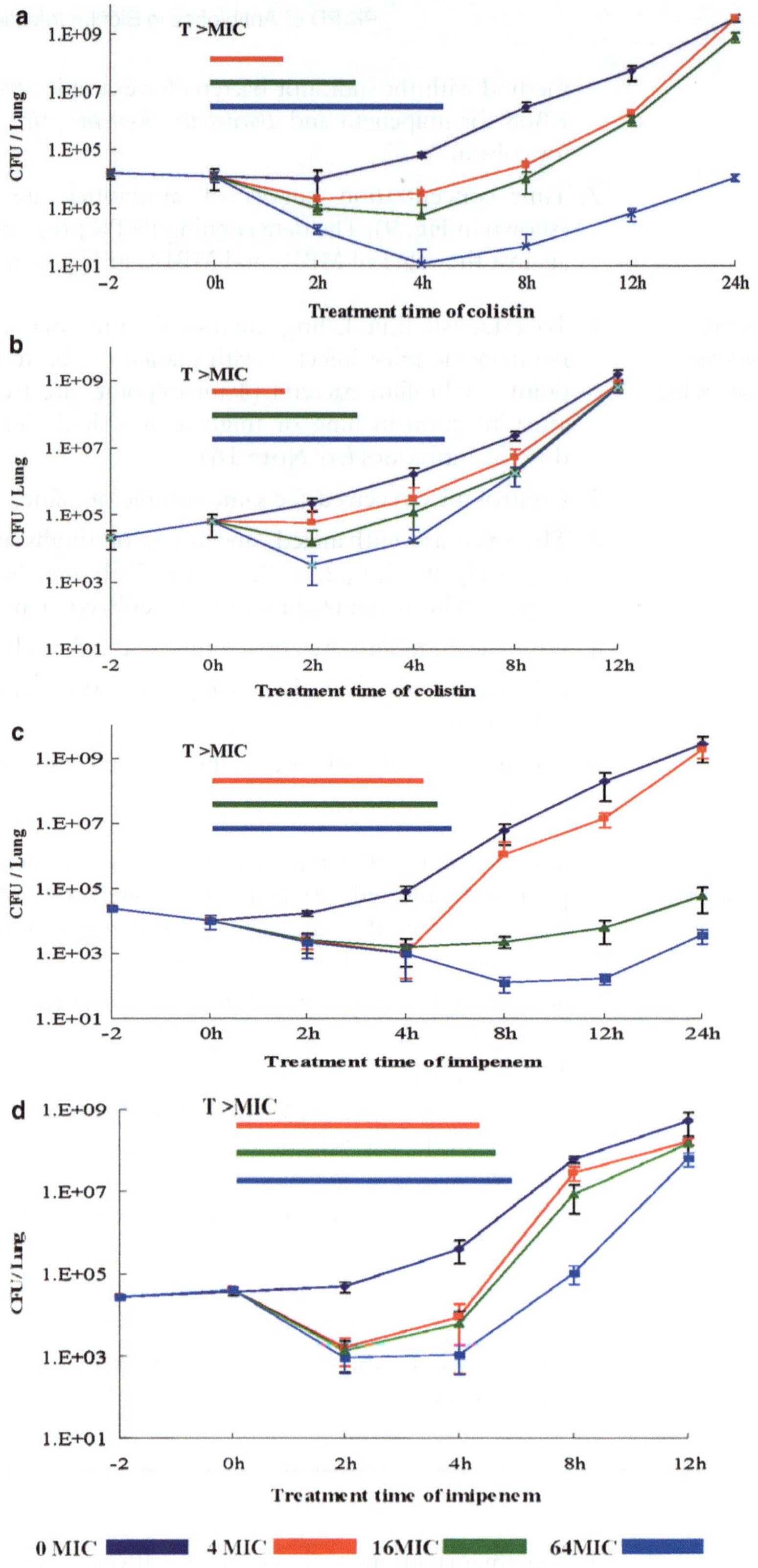

Fig. 10 Growth curves of control and antibiotic-exposed planktonic and biofilm *P. aeruginosa* PAO1 cells in the neutropenic mouse lung after a single intraperitoneal dose of colistin and imipenem. MIC of colistin, 4 mg/l; MIC of imipenem, 1 mg/l.

rich medium is used for some strains. The humidity incubator is used to prevent the fall of RH. Biofilms can be formed on pegs with or without shaking in cultivation.

2. Care should be taken to avoid contamination. The volume of antibiotic challenge (antibiotic challenge plate) should cover all biofilms, for example the volume of 0.1 ml for inoculum to form biofilms and 0.12 ml for antibiotic challenge at least.
3. Pegs should be rinsed softly three times at least before moving to recovery plate. The antibiotics on pegs can affect the recovery of biofilm cells if the challenge concentrations of antibiotics are very high.
4. 5-min sonication (80 W, 42 KHz) is generally enough for the biofilm detachment of *P. aeruginosa* on pegs.
5. Some clinical isolates of *P. aeruginosa* require a longer period than the 12 h for biofilm cell recovery after sonication. 6-h recovery of biofilm cells is not enough sometimes because of the post-antibiotic effect.
6. The variation of biofilm formation between well to well and plate to plate should be controlled.
7. The killing curve of antibiotics on biofilm is a recovery curve when the parameter of OD is applied. The option parameters are CFU or fluorescence scale. Samples of biofilm cells can also be serially diluted after biofilm detachment by sonication and then with the cultivation on plate for CFU counting or fluorescence detection.
8. Different animal species can be applied in the biofilm infection models.
9. NMRI mice required three doses of cyclophosphamide to set up the neutropenic mouse model, but two doses of cyclophosphamide are enough for most species of mouse.
10. Blood samples can also be collected by *heart puncture*.
11. You should be careful in the operation of cyclophosphamide solution, injection, and blood collection to prevent the toxic exposure from inhalation and skin contact.
12. The inocula for challenge are dependent on the species of bacteria in neutropenic mouse models.

Fig. 10 (continued) *Color bars* denote the interval that serum levels of antibiotic concentrations exceeded the MIC (T > MIC). (**a**) Colistin versus planktonic bacteria; (**b**) colistin versus biofilms; (**c**) imipenem versus planktonic bacteria; (**d**) imipenem versus biofilms. Kinetics of colistin and imipenem in vivo showed concentration-dependent and time-dependent killing, respectively, on biofilm-growing *P. aeruginosa*. Comparing to planktonic cells, both higher concentrations and longer period treatment with antibiotics are required to kill biofilm cells in vivo (reprinted from [9] with permission)

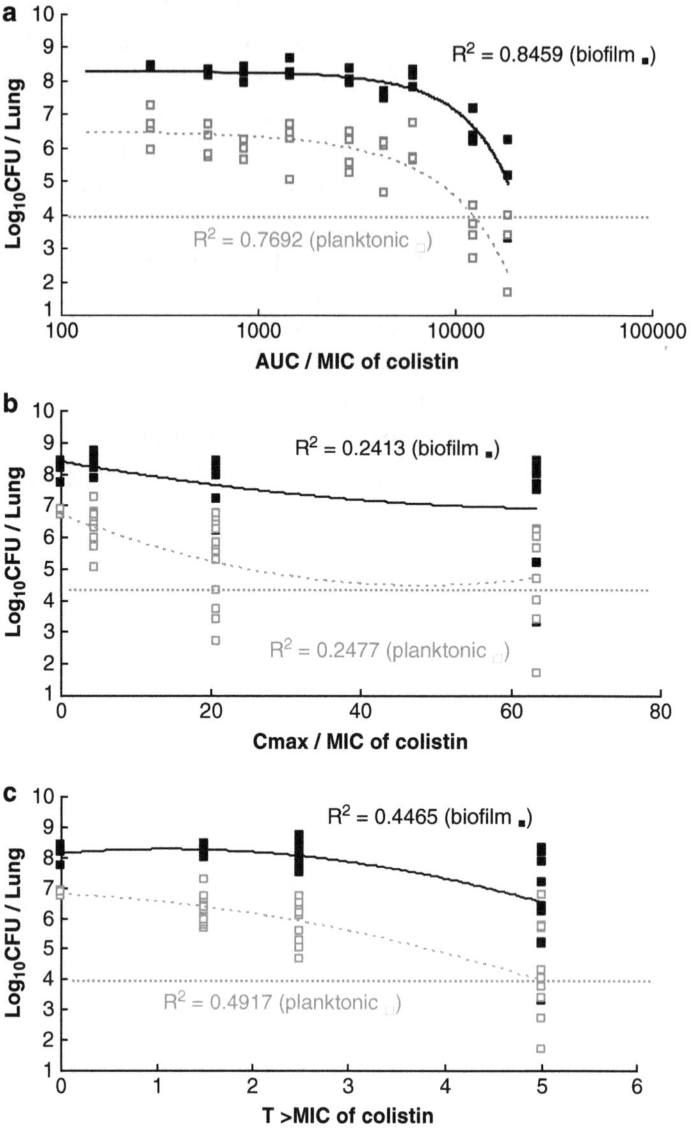

Fig. 11 Relationships for *P. aeruginosa* between the log 10 numbers of CFU per lung and PK/PD indices AUC/MIC, C_{max}/MIC, and T > MIC of colistin (**a**, **b**, **c**) and **Fig. 11** (continued) imipenem (**d**, **e**, **f**). Each *symbol* represents the data from a single lung. The *horizontal dotted lines* represent the mean bacterial burden in the lungs at the start of treatment. AUC/MIC (AUC/MBIC) is the best correlated parameter for colistin to describe the elimination of planktonic (biofilms) bacteria in the lung infection and T > MIC (T > MBIC) for imipenem (reprinted from [9] with permission)

13. The option of bacterial challenge is administrated by inhalation.
14. Some antibiotics show long half-life ($t_{1/2}$) in vivo, and the time points for blood collection should be arranged from 15 min to 24 h.

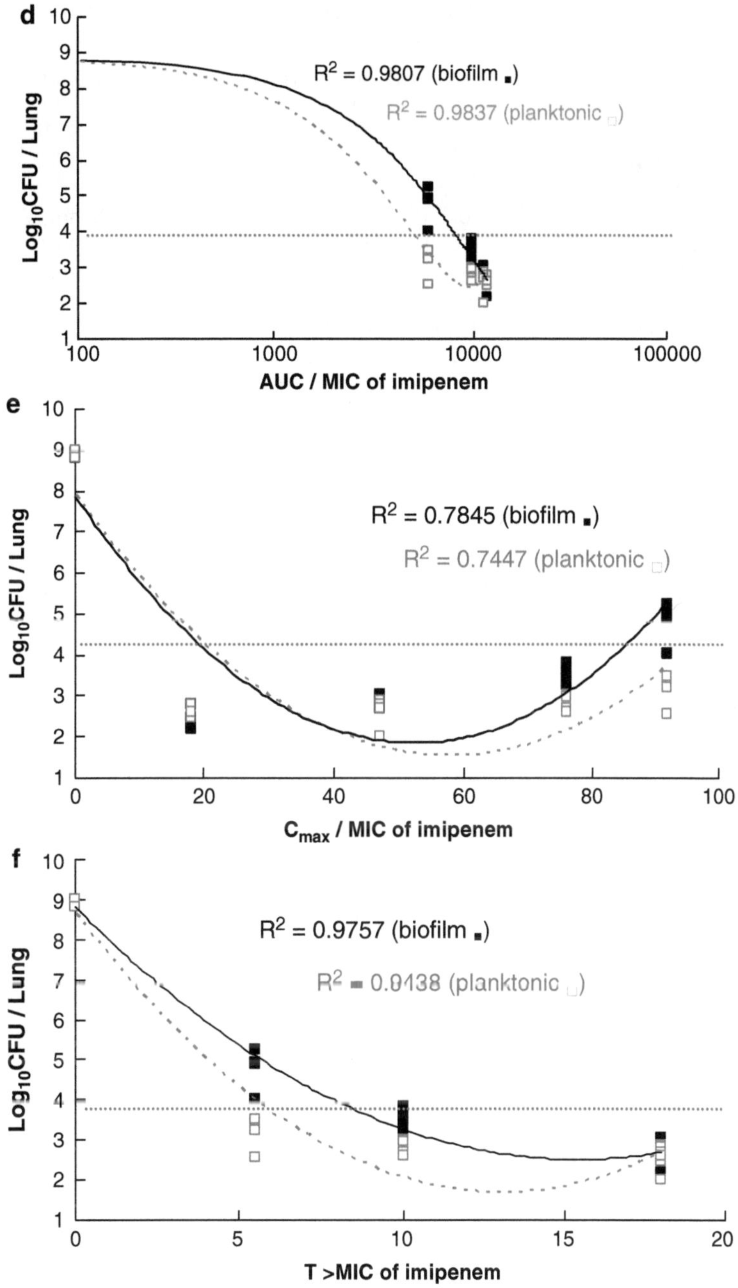

Fig. 11 (continued)

15. For the PK/PD modeling in vivo, the concentrations of antibiotics should be better with the similar or the equal dose application in patients. The application of the infected animal is advised to simulate the similar condition with patients under infection. PK study with normal animal is similar to that with healthy volunteers.

16. Antibiotic administrations are included with intravenous (IV), intraperitoneal (IP), intramuscular (IM), intratracheal (IT), and oral administration and other administration routes.

References

1. Moskowitz SM, Emerson JC, McNamara S et al (2011) Randomized trial of biofilm testing to select antibiotics for cystic fibrosis airway infection. Pediatr Pulmonol 46:184–192
2. Mueller M, de la Pena A, Derendorf H (2004) Issues in pharmacokinetics and pharmacodynamics of anti-infective agents: kill curves versus MIC. Antimicrob Agents Chemother 48:369–377
3. Høiby N, Krogh Johansen H, Moser C et al (2001) *Pseudomonas aeruginosa* and the in vitro and in vivo biofilm mode of growth. Microbes Infect 3:23–35
4. Werner E, Roe F, Bugnicourt A et al (2004) Stratified growth in *Pseudomonas aeruginosa* biofilms. Appl Environ Microbiol 70:6188–6196
5. Ceri H, Olson ME, Stremick C et al (1999) The Calgary Biofilm Device: new technology for rapid determination of antibiotic susceptibilities of bacterial biofilms. J Clin Microbiol 37:1771–1776
6. Moskowitz SM, Foster JM, Emerson J, Burns JL (2004) Clinically feasible biofilm susceptibility assay for isolates of *Pseudomonas aeruginosa* from patients with cystic fibrosis. J Clin Microbiol 42:1915–1922
7. Nielsen EI, Viberg A, Lowdin E et al (2007) Semimechanistic pharmacokinetic/pharmacodynamic model for assessment of activity of antibacterial agents from time-kill curve experiments. Antimicrob Agents Chemother 51: 128–136
8. Pankuch GA, Jacobs MR, Appelbaum PC (1994) Study of comparative antipneumococcal activities of penicillin-G, Rp-59500, erythromycin, sparfloxacin, ciprofloxacin, and vancomycin by using time-kill methodology. Antimicrob Agents Chemother 38:2065–2072
9. Hengzhuang W, Wu H, Ciofu O et al (2012) In vivo pharmacokinetics/pharmacodynamics of Colistin and Imipenem on biofilm *Pseudomonas aeruginosa*. Antimicrob Agents Chemother 56: 2683–2690
10. Hengzhuang W, Ciofu O, Yang L et al (2013) High β-Lactamase levels change the pharmacodynamics of β-lactam antibiotics in *Pseudomonas aeruginosa* biofilms. Antimicrob Agents Chemother 57:196–204
11. Hengzhuang W, Wu H, Ciofu O et al (2011) Pharmacokinetics/pharmacodynamics of Colistin and Imipenem on mucoid and nonmucoid *Pseudomonas aeruginosa* biofilms. Antimicrob Agents Chemother 55:4469–4474
12. Hoffmann N, Lee B, Hentzer M et al (2007) Azithromycin blocks quorum sensing and alginate polymer formation and increases the sensitivity to serum and stationary-growth-phase killing of *Pseudomonas aeruginosa* and attenuates chronic *P. aeruginosa* lung infection in Cftr(-/-) mice. Antimicrob Agents Chemother 51:3677–3687
13. Pedersen SS, Shand GH, Hansen BL, Hansen GN (1990) Induction of experimental chronic *Pseudomonas aeruginosa* lung infection with *Pseudomonas aeruginosa* entrapped in alginate microspheres. APMIS 98:203–211
14. Wu H, Song ZJ, Hentzer M et al (2000) Detection of N-acylhomoserine lactones in lung tissues of mice infected with Pseudomonas aeruginosa. Microbiology 146: 2481–2493
15. Gibaldi M, Perrier D (1982) Pharmacokinetics, 2nd edn. Marcel Dekker, New York, NY, pp 451–457

Chapter 18

Contribution of Confocal Laser Scanning Microscopy in Deciphering Biofilm Tridimensional Structure and Reactivity

Arnaud Bridier and Romain Briandet

Abstract

Confocal laser scanning microscopy (CLSM) became in last years an invaluable technique to study biofilms since it enables researchers to explore noninvasively the dynamic architecture and the reactivity of these biological edifices. The constant development of fluorescent markers and genetic tools along with the improvement of spatial, spectral, and temporal resolution of imaging facilities offers new opportunities to better decipher microbial biofilm properties. In this contribution, we proposed to describe the contribution of CLSM to the study of biofilm architecture and reactivity throughout two different illustrative approaches.

Key words Time-lapse confocal laser scanning microscopy, Biofilm architecture, Biofilm resistance mechanisms, Viability fluorescent labeling, Biocide activity

1 Introduction

In last decades, scientific research permits to become aware of the high complexity of biofilms and demonstrated that they should be considered as dynamic and organized communities of microorganisms rather than simple cell aggregates. It was demonstrated that the specific multicellular spatial organization of these biological edifices is crucial in the emergence of their functional properties [1, 2]. Therefore, a special interest should be given in the deciphering of biofilm tridimensional structures in order to obtain insights into the functioning of microbial communities. In this aim, the use of conventional fluorescence widefield microscopy, where biofilm is uniformly illuminated by a cone of light, proved to be not well adapted since fluorescence emitted from the whole thickness is collected and thus hinders the observation of a particular depth in the biofilm. Indeed, the poor depth resolution makes it possible to properly observe only very thin biofilms, whereas they are most of the time multicellular structures of many hundred microns of thickness.

Gianfranco Donelli (ed.), *Microbial Biofilms: Methods and Protocols*, Methods in Molecular Biology, vol. 1147, DOI 10.1007/978-1-4939-0467-9_18,

These last years, the emergence of confocal laser scanning microscopy (CLSM) in biofilm research has allowed overcoming these limitations and has opened new perspectives [3]. In confocal microscopes, laser excitation and the introduction of a pinhole enable to block out-of-focus light and thus to obtain images with a submicronic depth resolution (~0.3 μm) [4]. By taking series of such planed images throughout the depth of the sample, tridimensional biofilm structures can thus be scanned and then reconstructed and quantified after data processing using dedicated software [5–8]. Moreover, since the observation is not destructive, dynamic acquisitions can be made to follow biofilm growth for example. The application of CLSM led therefore to the discovering of important findings for the understanding of biofilm development. For example, Klausen et al. [9] used CLSM to depict the structural dynamic of *Pseudomonas aeruginosa* biofilms formed by wild-type, flagella, and type IV pili mutants and thus showed the different steps required for the settlement of the biofilm of this opportunistic pathogen. The constant development and the use of fluorescent markers able to target specific constituents of the biofilm as matrix components, nucleic acid, and protein residues or even to identify specific cellular physiological states give also the possibility to go further in the description of native architecture, composition, and cellular organization of biofilm. For instance, in a recent work, Zippel and Neu characterized the glycoconjugates of extracellular matrix of tufa-associated biofilm using fluorescence lectin binding analysis [10]. They showed that the spectrum of lectins binding to tufa-associated biofilms can be narrowed to a few with specificity for fucose, amino sugars, and sialic acid, and they identified different structural extracellular polymeric substance domains in biofilms which to some extent are in very close association with calcium carbonate crystals.

Such information is crucial to the understanding of the development and the functioning of microbial communities. In this perspective, the development of a high-throughput CLSM method, based on the use of a microtiter plate compatible with high-resolution imaging, gave the opportunity to amplify considerably the flow of such structural and informative data, participating in the deciphering of microbial community features [11]. The emergence of commercial systems including software dedicated to automated high-throughput screening now makes it possible to perform acquisition automatically, without an operator, further improving the flow of the technique.

Studies on biofilms have shown that one of their key features is their high heterogeneity because of the apparition of chemical and nutrient gradients due to the development of tridimensional structural patterns [1, 12]. Indeed, this chemical heterogeneity governs the emergence of multiple microenvironments in the structure within which cells experiment unique growth conditions and consequently express specific genes [13–15]. From this differential expression throughout the structure could emerge global biofilm functions. One of the most known specific properties of bacteria in biofilms is their amazing resistance to antimicrobial agents that lead to important concerns in terms of public health [16, 17]. While the precise mechanisms underlying this resistance are still poorly understood, it appears as a multifactorial process primarily related to the specific physiological state of included cells and structural characteristics of this heterogeneous edifice [18, 19]. The development of methods able to consider the local activity of biocides in native biofilm tridimensional structures could provide crucial information of cell resistance mechanisms in biofilms and thus constitute a prerequisite to the development of new and efficient treatments [3].

Conventional methods involving dilution and plate spreading after disinfection procedure as used in disinfectant regulatory standards proved to be not well adapted to biofilms as they are destructive and thus do not enable to consider local heterogeneity. Recently, methods have emerged by taking advantages of the combination of specific viability fluorescence staining and time-lapse CLSM and were developed to visualize, in real time, biocide activity within the biofilm of *Pseudomonas aeruginosa*, *Bacillus subtilis*, *Staphylococcus epidermidis*, or oral biofilm model for instance [20–24]. They are based on monitoring the loss of fluorescence in biofilm cluster that corresponds to the leakage of a fluorophore out of cells due to membrane permeabilization by the biocides. Such methods provided interesting data about spatial and temporal biocide activity patterns and thus on the spatial heterogeneity of bacterial susceptibilities enabling to better understand the mechanisms involved in the biofilm resistance.

In this chapter, in order to illustrate the contribution of confocal microscopy in the study of biofilm structure and reactivity, we chose to deal with two different CLSM approaches.

We first describe the protocol enabling the high-throughput analysis of biofilm tridimensional structure formed by 20 clinical strains of *Pseudomonas aeruginosa* using microtiter plate compatible with high-resolution imaging. We then depict the different steps enabling the direct visualization of biocide activity in biofilm tridimensional structures using CLSM. In this aim, we focus here on the reactivity of one clinical strain of *Pseudomonas aeruginosa* to benzalkonium chloride C14, a quaternary ammonium.

2 Materials

Some of the materials mentioned here such as growth medium or biocide are used with regard to our example and can be modified if required.

2.1 Reagents

1. Growth medium (here tryptone soy broth, TSB).
2. Sterilized distilled water.
3. Sterilized 150 mM NaCl.
4. Fluorescent nucleic acid markers as Syto9.
5. Chemchrome V6 esterasic fluorescent marker (*see* **Note 1**).
6. Chemsol B16 staining buffer (*see* **Note 2**).
7. Biocide powder or concentrated solution (here benzalkonium chloride C14, anhydrous >99 %) (*see* **Note 3**).

2.2 Equipment

1. Spectrophotometer.
2. Incubator at 30 °C.
3. Refrigerator at 4 °C.
4. 96-well µClear® polystyrene microtiter plates. Such microtiter plate is optically compatible with high-resolution CLSM observations due to its µClear® bottom (190 ± 5 µm).
5. Inverted confocal laser scanning microscope carrying 488 nm argon laser and objectives with numerical aperture from 0.8 to permit single-cell-scale observations with suitable resolution (*see* **Note 4**).
6. Computer with Excel and GinaFIT, a freeware add-in for Microsoft Excel developed by Geeraerd and colleagues [25] (*see* **Note 5**).
7. 3D reconstruction software: IMARIS® (Bitplane), AMIRA® (Visage imaging), ImageJ (http://rsb.info.nih.gov/ij/), DAIME [7].
8. Software for the quantification of biofilm structures: COMSTAT, PHLIP, ISA3D, DAIME [5–8].

3 Methods

Carry out all protocol steps at room temperature unless otherwise specified. Note that incubation temperature and time can vary depending on the strain studied and conditions required. Biofilm formation steps are common to both methods exposed after.

Note that for both methods, suggestions for confocal data analysis and presentation are given.

3.1 Biofilm Formation in Microtiter Plate

1. Inoculate 10 ml of TSB with 1 ml of a −20 °C bacterial stock and cultivate twice for 24 h at 30 °C under agitation.
2. Adjust the bacterial suspension to an OD600 nm = 0.01 in TSB using a spectrophotometer.

3. Fill the wells of the µClear® microtiter plate with 250 µl of adjusted suspension and keep for 1 h at 30 °C to enable bacterial adhesion. Three wells per each bacterial strain are a minimum.
4. After adhesion, rinse the wells twice with 250 µl of 150 mM NaCl to eliminate any non-adherent bacteria and refill wells with 250 µl of fresh TSB.
5. Incubate the microtiter plate at 30 °C for 24 h to enable biofilm development.

3.2 High-Throughput Structural Analysis of *P. aeruginosa* Biofilms

The structure of the biofilms formed by 20 strains of *P. aeruginosa* isolated from hospitalized patients were analyzed using microtiter plate compatible to high-resolution confocal imaging.

3.2.1 Fluorescent Labeling

1. After biofilm development, gently rinse microtiter wells with 250 µl of 150 mM NaCl.
2. Gently refill wells with 250 µl of TSB containing Syto 9 fluorescent nucleic acid marker (1:1,000 dilution from a commercial stock solution at 5 mM in DMSO) (*see* **Note 6**).
3. Incubate the microtiter in the dark at 30 °C for 20 min to enable fluorescence labeling of the bacteria.
4. Gently rinse the wells with 250 µl of fresh TSB medium.

3.2.2 Structural CLSM Analysis of Biofilms

1. Start the confocal microscope, and switch on the argon laser at least 1 h before the beginning of the experiment to reach power and stabilize.
2. In the CLSM dedicated software, select an objective with a numerical aperture of 0.8 or higher (*see* **Note 7**).
3. Set the image resolution on 512 × 512 pixels and the acquisition scanning rate at 400 Hz (*see* **Note 8**).
4. Set the argon laser on 25 % of its maximum intensity.
5. Set the software to collect emitted Syto9® fluorescence between 500 and 600 nm.
6. Carefully transfer the microtiter plate on the stage of the confocal microscope.
7. Set the photomultiplier (PMT) gain to adjust the level of the collected fluorescence.
8. Select in the CLSM software, the 3D acquisition mode enabling to make stacks of *XY* horizontal planed images throughout the depth of the biofilm.
9. Scan the biofilm on the *Z*-axis to find and fix the upper and lower limit of Z-scan (*see* **Note 9**).
10. Select a Z-step of 1 µm.
11. Select a line average of 2 (*see* **Note 10**).

12. Launch scan to localize a structure of interest in the biofilm and to readjust PMT gain of collected fluorescence if required.
13. Launch 3D scans at the selected area in the well.
14. Make at least two scans per well and for at least three different wells per each strain.

3.2.3 3D Reconstruction of Biofilms

Different software can be used to perform 3D reconstruction of biofilm directly from confocal image series and have their own procedure. We present here a typical sequential procedure used with the software IMARIS® (Bitplane), a commercial software widely used by biofilm researchers.

1. In IMARIS® software, open the image series of interest by clicking on "open" icon and selecting the corrected file series on the hard disk (*see* **Note 11**).
2. Click on "image processing," and select the item "background subtraction." Leave the settings as they are.
3. Click on the "Slice" tab, and select "Easy 3D" mode.
4. Go to the "Snapshot" mode by clicking on the corresponding item, and click on "snapshot" icon to take a picture.
5. Record the picture in the appropriate folder under the desired name.

Figure 1 shows 3D projections of biofilm structure obtained for the 20 clinical strains of *P. aeruginosa* from confocal Z-stacks using IMARIS software and "Easy 3D" mode.

3.2.4 Extraction of Structural Parameters from CLSM Images and Statistical Analysis

Different software enable the extraction of structural parameters directly from confocal image series. We present here the procedure performed with COMSTAT2 (http://www.comstat.dk/), a freeware which runs as a Java program under the software ImageJ (version 1.43) (*see* **Note 12**).

1. Launch ImageJ software.
2. In ImageJ, click on "Plugins" tab on menu bar and select Comstat2 item.
3. Open confocal file series by clicking on "Add" and selecting the folder containing the image series of interest.
4. In Comstat2 menu, select the options "Automatic thresholding (Otsu's method)" and "Connected Volume Filtering" and check the box corresponding to the parameter(s) that should be extracted from CLSM images series (for example "Biomass," "Maximum Thickness," or "Dimensionless Roughness Coefficient").
5. Click on the button "Go."
6. The results calculated are automatically recorded in a .txt files in the folder containing the corresponding image series.

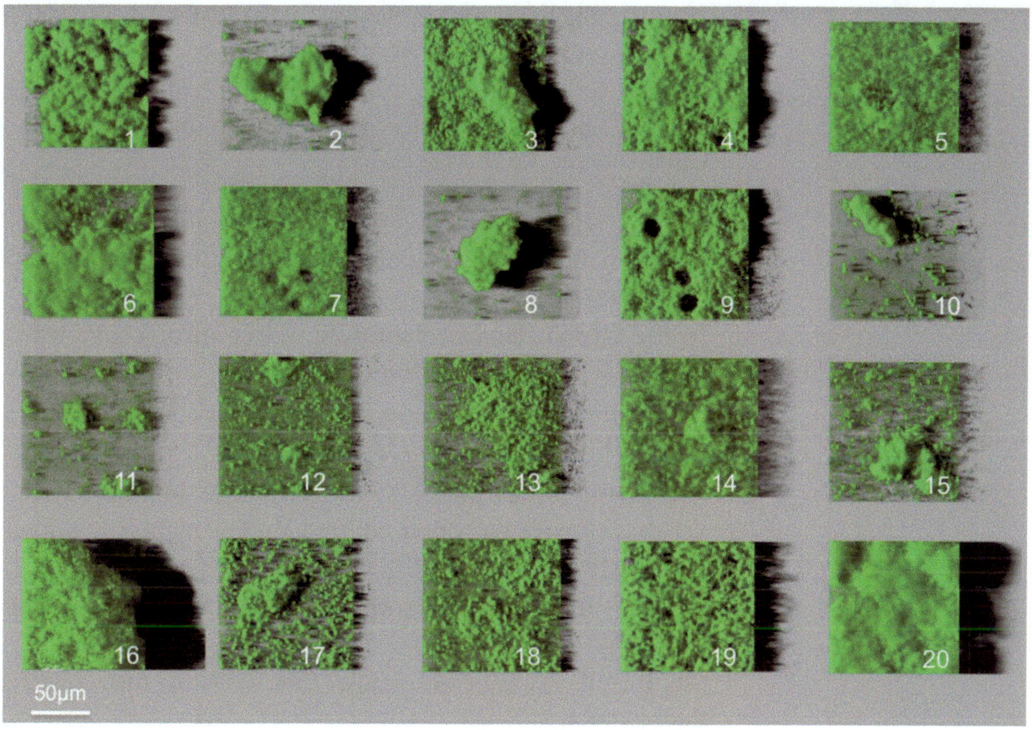

Fig. 1 3D reconstruction of biofilm formed by 20 clinical strains of *P. aeruginosa*

7. Statistical analysis as principal component analysis (PCA) for example can be performed on a large number of strains using the different parameter extract to cluster individual and to identify singular biofilm phenotypes.

3.3 Visualization of Spatiotemporal Activity Pattern of Biocides in Biofilms

In this part, we focus on the reactivity of one *P. aeruginosa* strain (*n*°3) to benzalkonium chloride C14 using a time-lapse CLSM method enabling the direct visualization of biocide activity within tridimensional biofilm structure.

3.3.1 Biocide Solution Preparation

1. Weigh the commercial biocide powder (here benzalkonium chloride C14, anhydrous, >99.0 %) and solubilize in distilled sterilized water to reach a concentration twofold higher than final desired concentration. In this example, 1 g of biocide powder was diluted in 100 ml of distilled water because the final challenge concentration desired was 5 g/l.
2. Keep the biocide solution at 4 °C until utilization (*see* **Note 13**).

3.3.2 Fluorescent Labeling

1. After biofilm development, gently rinse microtiter wells with 250 µl of 150 mM NaCl.
2. Gently refill wells with 100 µl of solution containing Chemchrome V6 (1:100 of commercial solution diluted in Chemsol B16 buffer) (*see* **Note 14**).

3. Incubate the microtiter in the dark at 30 °C for 1 h in order to reach fluorescence equilibrium (*see* **Note 15**).
4. After fluorescent labeling, gently rinse and refill the wells with 100 µl Chemsol B16 buffer in order to eliminate any excess of Chemchrome V6 marker.

3.3.3 Time-Lapse Monitoring of Biocide Activity Within Biofilm

1. Start the confocal microscope, and switch on the argon laser at least 1 h before the beginning of the experiment to reach power and stabilize.
2. In the CLSM dedicated software, select an objective with a numerical aperture of 0.8 or higher.
3. Set the image resolution on 512 × 512 pixels and the acquisition scanning rate at 800 Hz (*see* **Note 16**).
4. Set the argon laser on 10 % of its maximum intensity.
5. Set the software to collect emitted V6 Chemchrome fluorescence between 500 and 600 nm.
6. Carefully transfer the microtiter plate on the stage of the confocal microscope.
7. Set the PMT gain to adjust the level of the collected fluorescence.
8. Select in the CLSM software the acquisition mode enabling to scan biofilm in function of time (time-lapse scans).
9. Select a time of 15 s between two scans (*see* **Note 17**).
10. First, launch a time-lapse scan without adding biocide in order to check that confocal scanning alone did not lead to a decrease of fluorescence (*see* **Note 18**).
11. Once the CLSM parameters are checked launch scan to localize a structure of interest in the biofilm and to readjust PMT gain of collected fluorescence if required.
12. Launch time-lapse scans, and immediately after the completion of the first scan, very gently introduce 100 µl of biocide solution in the well (*see* **Note 19**).
13. Monitor the loss of fluorescence in the structure which is recorded.

3.3.4 Image Analysis and Data Processing

The processing of confocal images is highly dependent on the biological question which would like to be addressed. We suggest here a model of treatment to visualize the spatial patterns of biocide action in biofilms and to extract inactivation dynamic parameters from CLSM fluorescence data.

1. Visualize and quantify the intensity of Chemchrome V6 green fluorescence within the biofilm cluster directly from time-lapse images series using confocal microscope integrated software. In this aim, select different small square areas of 100 µm2 (10 × 10 µm) at different locations in the structure where fluorescence intensity data are captured in function of time.

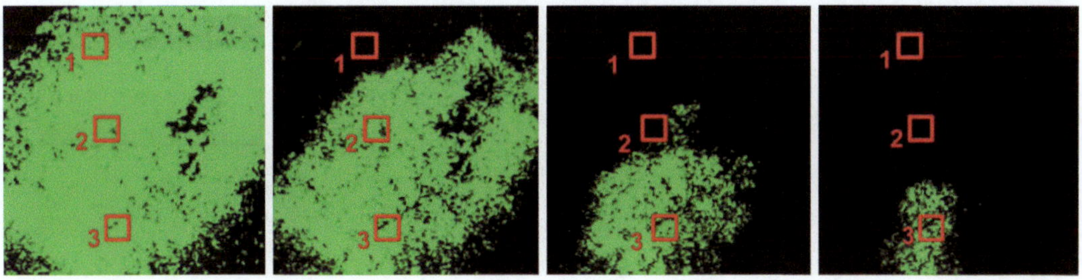

Fig. 2 Visualization of Chemchrome V6 fluorescence loss in *P. aeruginosa* biofilm during treatment with 0.5 % benzalkonium chloride C14 after 0, 60, 120, and 160 s. Images were recorded ~5 μm above the bottom of the well. Three *squares* are indicated in *red* and correspond to the three areas where fluorescent data were extracted

As an illustration, time-lapse image sequence obtained with a Leica SP2 confocal microscope and the Leica LCS software (Leica microsystems) and showing the action of benzalkonium chloride C14 (0.5 %) in *P. aeruginosa* biofilm is presented in Fig. 2.

2. Export the fluorescence intensity data in *.txt* file for the different areas selected.
3. Import the *.txt* file in Microsoft Excel® spreadsheet.
4. Use GinaFIT to model inactivation kinetics in the different areas selected. For this, choose the adequate survival model and fit it to the fluorescence intensity values. In this example, fluorescence intensity curves were fitted with the "shoulder + log-linear + tail" inactivation model.
5. Extract inactivation parameters from the fitting: here, Sl, the shoulder length which corresponds to the length of the lag phase, and k_{max} the inactivation rate.

   Figure 3 shows curves of normalized fluorescence in function of time in the three areas selected obtained with *P. aeruginosa* biofilms treated with benzalkonium chloride C14 (0.5 %). Fitted GinaFIT inactivation model and extracted inactivation parameters Sl and k_{max} are also indicated.
6. The quantitative kinetics parameters can then be used to make statistical analysis and compare different experiments, bacterial strains, or experimental conditions.

4 Notes

1. Chemchrome V6 is a nonfluorescent compound (carboxyfluorescein diacetate) that is taken up by metabolically active cells and cleaved by intracellular esterases to yield an intensely green fluorescent product (carboxyfluorescein). Conserve Chemchrome V6 commercial solution at 4 °C.
2. In Gram-negative strains, and particularly *Pseudomonas* sp., intense efflux pump activity can lead to the release of fluorescent residues from the cells, so that a stable and intense level

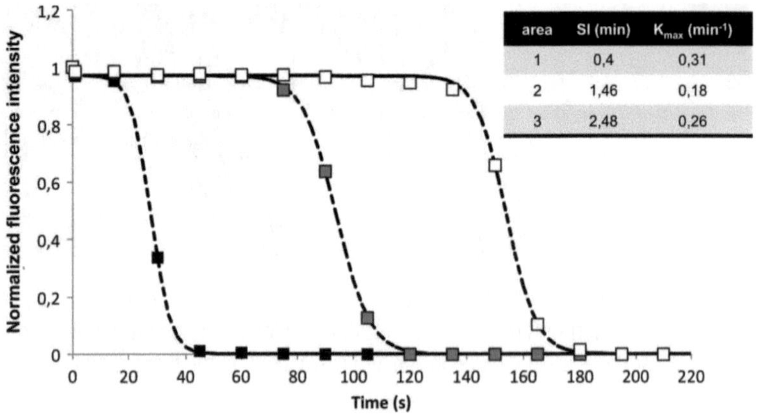

Fig. 3 Quantification of fluorescence intensity during benzalkonium chloride C14 treatment (0.5 %). The values shown represent the loss of fluorescence at three different areas in the biofilm: 1 (*black squares*), 2 (*grey squares*), and 3 (*white squares*). Two inactivation parameters, SI (shoulder length) and k_{max} (inactivation rate), were obtained after fitting GinaFIT "shoulder + log-linear + tail" inactivation model to fluorescence experimental data and are represented for each area (*dotted lines*)

of intracellular fluorescence cannot be achieved. Chemsol B16 buffer can block efflux pump activity and thus maintain fluorophores inside the cells.

3. This technique is only compatible with biocides that lead to the disruption of cell membrane integrity.
4. The microtiter plate biofilm device used here to grow biofilm required a stage to be mounted on the microscope which may not be provided with the confocal microscope. This stage can be provided by contacting the microscope manufacturers.
5. GinaFIT can be freely obtained at this link: http://cit.kuleuven.be/biotec/downloads.php. A manual detailing the installation of the freeware add-in in Microsoft Excel is also provided.
6. Syto9 is a nucleic acid fluorescent marker which intercalates DNA leading to a green fluorescence. A wide range of Syto® fluorescent markers are available and correspond to various excitation and emission wavelength. Commercial solution should be conserved at −20 °C.
7. Note that a numerical aperture of 0.8 is a minimum, and higher numerical apertures of 1.2 or 1.4 are preferred to obtain images with a good resolution.
8. Such pixel resolution and scan rate provide satisfying image quality standard for 3D reconstruction. However, image quality can be improved if needed by selecting a higher pixel resolution (1,024 × 1,024 or even 2,048 × 2,048) and also by decreasing scan rate to 200 Hz. Acquisition time will thus be increased.

9. When fixing the upper and lower Z-limit of the scan, it is recommended to take a ten of micrometers more than the apparent limits.
10. Line average was used to decrease noise and thus increase image quality, especially when the laser gain is high due to the low fluorescent signal. A line average of 2 means that the confocal microscopy takes two scans and then makes the average of the two scan to constitute on image. This value proved to be well adapted (and most of the time line average of 1 is sufficient) in the case of Syto9 because the fluorescent signal is generally good.
11. IMARIS supported many file formats as those obtained with different microscopes commonly used in biofilm research (Leica, Zeiss, Olympus, Nikon, etc.).
12. COMSTAT2 program and installation files are available at this address: http://www.comstat.dk/.
13. Prepare a fresh biocide solution for each new experiment due to the potential loss of activity for certain compounds in diluted solutions, and keep it at 4 °C between preparation and utilization.
14. Prepare a fresh staining solution for each experiment, and keep it at 4 °C between preparation and utilization.
15. This incubation time and temperature were determined to reach a stable and high level of fluorescence for *Pseudomonas aeruginosa*. By studying other bacteria, these parameters may be modified, and we recommend performing several tests before experiments to determine specifically the optimal time and temperature for staining with respect to the bacteria strain used.
16. We found that such pixel resolution image and scanning rate enable to obtain a good resolution along with sufficient speed of acquisition comparing to the dynamic of activity of benzalkonium chloride C14 in *P. aeruginosa* biofilm. However, the use of other biocides and/or bacteria is required to slightly adapt image resolution or scanning rate to find optimal settings if the inactivation dynamic is faster or slower.
17. This *t*-step time should be modified in function of the dynamic of action of the biocide used. It is thus required to perform preliminary experiments to determine the speed and the duration of the inactivation dynamic and thus to fix this value.
18. A loss of fluorescence inferior to 5 % for duration of experiment of 30 min is appropriate. Beyond this threshold, laser intensity should be decreased in order to avoid photobleaching due to the acquisition scan.
19. The introduction of the biocide solution in the well is a critical step which can disturb biofilm structure. Therefore, it is recommended to add very gently the biocide into the well with the help of the edge of the well.

References

1. Stewart PS, Franklin MJ (2008) Physiological heterogeneity in biofilms. Nat Rev Microbiol 6:199–210
2. Tolker-Nielsen T, Molin S (2000) Spatial organization of microbial biofilm communities. Microb Ecol 40:75–84
3. Bridier A, Tischenko E, Dubois-Brissonnet F et al (2011) Deciphering biofilm structure and reactivity by multiscale time-resolved fluorescence analysis. Adv Exp Med Biol 715: 333–349
4. Palmer RJ, Sternberg C (1999) Modern microscopy in biofilm research: confocal microscopy and other approaches. Curr Opin Biotechnol 10:263–268
5. Heydorn A, Nielsen AT, Hentzer M et al (2000) Quantification of biofilm structures by the novel computer program COMSTAT. Microbiology 146:2395–2407
6. Xavier JB, White DC, Almeida JS (2003) Automated biofilm morphology quantification from confocal laser scanning microscopy imaging. Water Sci Technol 47:31–37
7. Daims H, Lucker S, Wagner M (2006) Daime, a novel image analysis program for microbial ecology and biofilm research. Environ Microbiol 8:200–213
8. Beyenal H, Donovan C, Lewandowski Z, Harkin G (2004) Three-dimensional biofilm structure quantification. J Microbiol Methods 59:395–413
9. Klausen M, Heydorn A, Ragas P et al (2003) Biofilm formation by *Pseudomonas aeruginosa* wild type, flagella and type IV pili mutants. Mol Microbiol 48:1511–1524
10. Zippel B, Neu TR (2011) Characterization of glycoconjugates of extracellular polymeric substances in tufa-associated biofilms by using fluorescence lectin-binding analysis. Appl Environ Microbiol 77:505–516
11. Bridier A, Dubois-Brissonnet F, Boubetra A et al (2010) The biofilm architecture of sixty opportunistic pathogens deciphered using a high throughput CLSM method. J Microbiol Methods 82:64–70
12. Wimpenny J, Manz W, Szewzyk U (2000) Heterogeneity in biofilms. FEMS Microbiol Rev 24:661–671
13. Rani SA, Pitts B, Beyenal H et al (2007) Spatial patterns of DNA replication, protein synthesis, and oxygen concentration within bacterial biofilms reveal diverse physiological states. J Bacteriol 189:4223–4233
14. Beloin C, Ghigo JM (2005) Finding gene-expression patterns in bacterial biofilms. Trends Microbiol 13:16–19
15. Resch A, Rosenstein R, Nerz C, Götz F (2005) Differential gene expression profiling of *Staphylococcus aureus* cultivated under biofilm and planktonic conditions. Appl Environ Microbiol 71:2663–2676
16. Hall-Stoodley L, Stoodley P (2009) Evolving concepts in biofilm infections. Cell Microbiol 11:1034–1043
17. Lebeaux D, Ghigo JM (2012) Management of biofilm-associated infections: what can we expect from recent research on biofilm lifestyles? Med Sci (Paris) 28:727–739
18. Bridier A, Briandet R, Thomas V, Dubois-Brissonnet F (2011) Resistance of bacterial biofilms to disinfectants: a review. Biofouling 27:1017–1032
19. Hoiby N, Bjarnsholt T, Givskov M, Molin S, Ciofu O (2010) Antibiotic resistance of bacterial biofilms. Int J Antimicrob Agents 35: 322–332
20. Bridier A, Dubois-Brissonnet F, Greub G et al (2011) Dynamics of the action of biocides in *Pseudomonas aeruginosa* biofilms. Antimicrob Agents Chemother 55:2648–2654
21. Bridier A, Sanchez-Vizuete Mdel P, Le Coq D et al (2012) Biofilms of a *Bacillus subtilis* hospital isolate protect *Staphylococcus aureus* from biocide action. PLoS One 7:e44506
22. Corbin A, Pitts B, Parker A, Stewart PS (2011) Antimicrobial penetration and efficacy in an in vitro oral biofilm model. Antimicrob Agents Chemother 55:3338–3344
23. Davison WM, Pitts B, Stewart PS (2010) Spatial and temporal patterns of biocide action against *Staphylococcus epidermidis* biofilms. Antimicrob Agents Chemother 54: 2920–2927
24. Takenaka S, Trivedi HM, Corbin A et al (2008) Direct visualization of spatial and temporal patterns of antimicrobial action within model oral biofilms. Appl Environ Microbiol 74:1869–1875
25. Geeraerd AH, Valdramidis VP, Van Impe JF (2005) GInaFiT, a freeware tool to assess non-log-linear microbial survivor curves. Int J Food Microbiol 102:95–105

Chapter 19

Chip Calorimetry for Evaluation of Biofilm Treatment with Biocides, Antibiotics, and Biological Agents

Frida Mariana Morais, Friederike Buchholz, and Thomas Maskow

Abstract

Any growth or bioconversion in biofilms is accompanied by the release of heat. The heat (in J) is tightly related to the stoichiometry of the respective process via law of Hess, and the heat production rate (in W or J/s) is additionally related to the process kinetics. This heat and the heat production rate can nowadays be measured by modern calorimetry with extremely high sensitivity. Flow-through calorimetry allows the measurement of bioprocesses in biofilms in real time, without the need of invasive sample preparation and disturbing of biofilm processes. Furthermore, it can be applied for long-term measurements and is even applicable to turbid media. Chip or miniaturized calorimeters have the additional advantages of extremely short thermal equilibration times and the requirement of very small amounts of media and chemicals. The precision of flow-through chip calorimeters (about 3 mW/L) allows the detection of early stages of biofilm development (about 10^5 bacteria cm^{-2}).

Key words Calorimetry, Antibiotic, Biocides, Antimicrobial agents, Activity measurement, Susceptibility testing, Noninvasive, Real-time monitoring

1 Introduction

Different monitoring technologies that were developed over the last years are explained in detail in this book. These technologies deliver information on different levels (L1, deposition of material, but no differentiation between organic and inorganic material and microorganisms; L2, distinction between biotic and abiotic components; L3, detailed information about the chemical composition of the deposit or direct access to the microorganisms; L4, discrimination between living and dead microorganisms). A detailed overview about these information levels and the related monitoring technologies is given by Flemming [1]. Calorimetry is a unique method because it provides direct access to the activity of the microorganisms and thus complements these other methods perfectly. For instance, viability (determined by plate counts), cell numbers (quantified from confocal laser scanning microscopy images), or biomass

Gianfranco Donelli (ed.), *Microbial Biofilms: Methods and Protocols*, Methods in Molecular Biology, vol. 1147,
DOI 10.1007/978-1-4939-0467-9_19,

(determined with crystal violet) combined with calorimetry provides valuable information about biofilm resistance against treatments with antibiotics or biological agents. Furthermore, widely used methods to investigate the antibiotic effect on biofilms (e.g., plate counts and ATP measurements) are adapted from the methods for planktonic cells. This adaptation requires the disintegration of the biofilm structure which can lead to inaccurate results. ATP measurements are related to the biomass deposit and to the energy charge of the cells. Plate counts give information about viability. Both are not necessarily related to the metabolic activity. Calorimetry uses heat as measuring parameter, and heat is in case of aerobic growth—via the oxycaloric equivalent (about −460 kJ/mol O_2 [2])—directly related with the catabolic side of the metabolism (i.e., electron transport phosphorylation). The reaction heat (in J) of biofilm processes is, via law of Hess, related to the stoichiometry of the total rates of substrate consumption and formation of any products (biomass, EPS, signalling molecules etc.). The heat production rate (in W) equals the product of reaction heat and respective reaction rate. Thus, alterations in metabolic activities in the biofilm are immediately reflected by changing heat production rates [3].

Despite the clear advantages of calorimetry and the proof of the principle since more as 150 years ago [4, 5], the first applications to biofilms were reported in the early 1990s [6–9]. Buchholz et al. [10] evaluated in detail the role and level of information of calorimetry in comparison to other biofilm investigation methods.

Undesired biofilms are very frequently treated with chemical agents (biocides or antibiotics) but also with biological agents (e.g., grazing microorganisms, phages, enzyme solutions). Antibiotic treatments, for instance, influence the heat production rate of biofilms in two different directions. Firstly, the expected effect of antibiotic treatment is, in case of bactericides, the killing of cells, which leads to the decrease of heat production rate simply due to the reduction of cell number. Also in case of bacteriostatics, heat production rate is expected to decline because further growth of biofilm is inhibited and thus the number of dying cells is not compensated by the formation of new cells. Basically similar effects are expected from the treatment of biofilms with biological agents (e.g., predatory bacteria or phages).

Secondly, the metabolically active microorganisms in biofilms resist antibiotic treatment by means of energy-demanding efflux pumps, by target modification, or by chemical modification of the antibiotic. Most of these countermeasures are energy dependent and need ATP or reduction equivalents. These requirements drive the electron transport phosphorylation and thereby enhance the cell-specific heat production rate. Both contradicting heat effects overlap in the final heat signal. Thus, careful interpretation of the heat signal and comparison of the heat signal with viability, number of cells, or ATP content provide some information on the respective antibiotic resistance mechanism. This is more deeply analyzed and explained by Buchholz et al. [11] and Mariana et al. [12]. In case of biofilm treatment with predatory bacteria, the cell-specific heat production rate is reduced due to the disintegration of the cell membrane by the predatory bacteria as reported by Buchholz and co-workers [13]. In the following, chip calorimetry is introduced as a method to evaluate biofilm treatments.

A chip calorimeter is a miniaturized calorimeter, designed as a flow-through system which serves the purpose of biofilm establishment ([14]; Fig. 1). The centerpiece of the device is a silicon chip

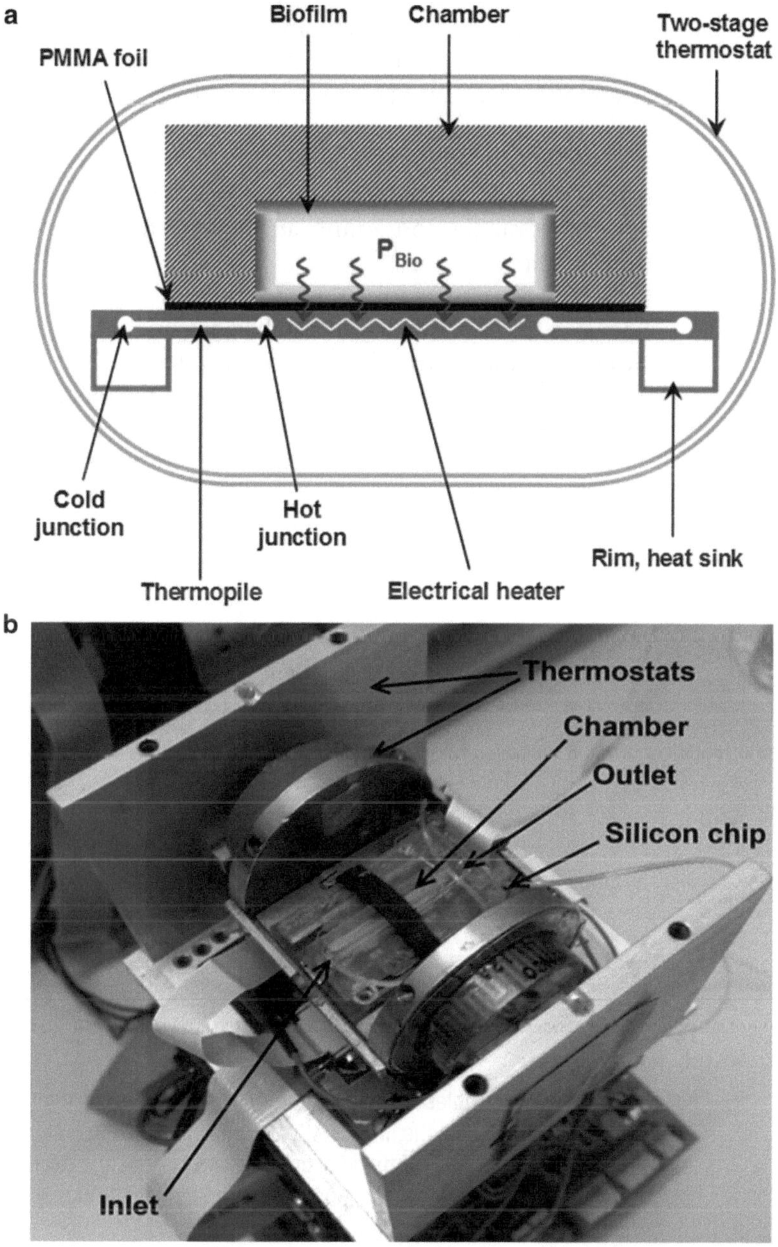

Fig. 1 The schematic front view of the chip calorimetric core (**a**); the calorimetric unit with opened lids (**b**)

with membrane-integrated thermophiles, which are responsible for the conversion of heat dissipated from the measuring chamber into voltage signals. The chamber will be placed on the chip during the measurement. The static air condition inside the calorimeter provides good insulation, so the heat transfer focuses in the contact area. The lids of the calorimeter serve as thermostat and maintain the working temperature with an accuracy of $\pm 10^{-5}$ K. The device is connected to a high-precision syringe pump to allow a regulated flow-through system. Working temperature, the flow pattern of the fluidic, and the data acquisition are computer controlled and predefined in the MATLAB program prior to the experiments. A more detailed overview about the working principles of chip calorimetry, the strengths and weaknesses, and the potential applications is given by Maskow et al. [15].

For the desired applications, the precultivated chamber is placed on the chip, connected to the fluidic system to allow the flow through of the growth media (for the biofilm maturation step) or media containing antimicrobial agents (for the treatment step).

The first injection is made after the thermal equilibrium is achieved, which is visible as a stabile baseline. The injection of media causes a short signal disturbance (seen in Fig. 2 as an endothermic peak). The metabolism of the nutrients in the media is accompanied by a heat release; thus, the exothermic metabolism governs the signal after the disturbance. Oxygen limitation in the chamber ends the metabolism and brings the signal back to the baseline until the next injection (Fig. 2). The signal shift is determined by the

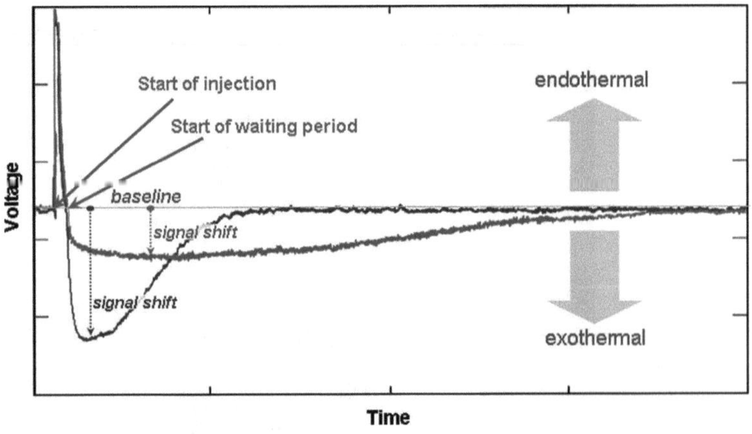

Fig. 2 Two examples of signal shifts induced by the injection of the medium. During the injection, the imperfect working heat exchanger causes an endothermal signal. After stop of the injection (waiting period), the exothermal metabolic heat production defines the signal. The magnitude of the signal shifts reflects the degree of biofilm activity

microbial activity in the chamber. Active biofilms have higher metabolic rate; thus, the signal shift is higher and the oxygen depletes faster, accordingly. Less active biofilms (biofilms with less active cells or biofilms with a decreased specific activity per cell) show shallower signals and take longer to return to the baseline.

2 Materials

Here, we exemplary illustrate the application of chip calorimetry to evaluate the treatment of *Pseudomonas putida* biofilm with antibiotics. Other biofilms can be similarly produced. Biofilm treatment with other agents such as biocides, grazing microorganisms, or phages follows the identical procedure (and can be modified if desired).

1. Test strain (*Pseudomonas putida* PaW340).
2. Suitable complex media (Luria-Bertani/LB media) and defined media (NH_4Cl (764), $CaCl_2.6H_2O$ (5.5), KH_2PO_4 (340), K_2HPO_4 (435), $MgCl_2.6H_2O$ (59), $ZnCl_2$ (0.21), $MnCl_2$ (0.46), $CuCl_2$ (0.42), $Na_2MoO_4.2H_2O$ (0.25), $FeCl_3.6H_2O$ (4.0), sodium benzoate (600), and tryptophan (20). All numbers are in mg/L).
3. Syringe pump (*see* **Note 1**).
4. Chip calorimetric chambers + PMMA foil (50-μm thickness) (*see* **Note 2**). Only one chamber is needed for calorimetric measurement. If reference experiments are planned, prepare several chambers in parallel to attain identical conditions.
5. Tygon tube (diameter 0.25 mm) to connect the chambers (e.g., to the syringes or in the calorimeter).
6. Disposable syringes of 5 and 10 mL.
7. Cannula.
8. Disinfectant: mixture of ethanol/water/sulfuric acid (1:0.41:0.01).
9. Phosphate buffer.
10. Antibiotic stock solutions.
11. Liquid paraffin.
12. Chip calorimetric device.

3 Methods

3.1 Precultivation of Biofilms

1. Cultivate the test strain in 20 mL suitable medium using 100-mL Erlenmeyer flask overnight on the rotation shaker (150 rpm at suitable temperature, e.g., 30 °C).

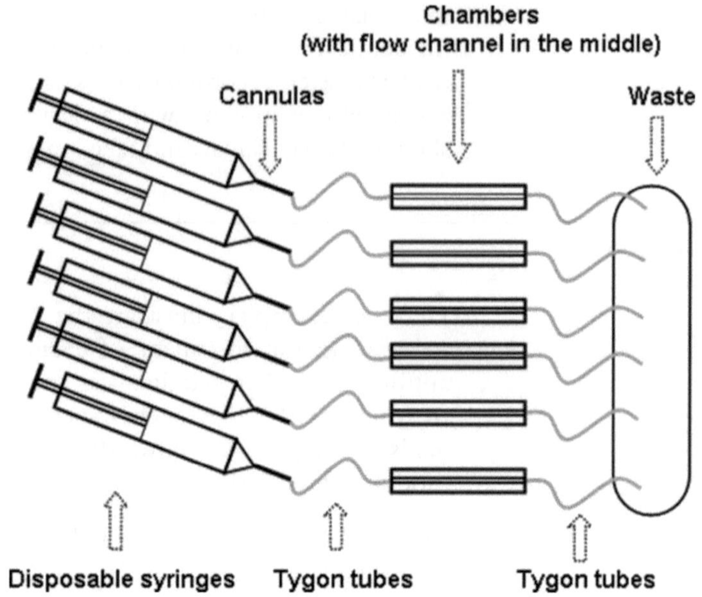

Fig. 3 Schematic arrangement of the biofilm precultivation. Several chambers are prepared at the same time and treated identically. One of them will later be measured chip calorimetrically, and the others serve for reference analysis

2. Prepare the precultivation arrangement as seen in the Fig. 3.
3. Disinfection process: Fill the 5-mL syringes with 3 mL disinfectant solution. Pump the disinfectant solution through the arrangement for 20–40 min (*see* **Note 3**) with a continuous flow rate of 3 mL/h. Afterwards, replace the syringes aseptically with syringes filled with 5 mL phosphate buffer. Pump the phosphate buffer at the same flow rate through the setup for 30 min. Check with pH paper if the neutralization was sufficient.
4. Meanwhile, prepare cell solution with cell concentration of about 7×10^7 cells/mL (OD 0.2 in case of *P. putida*) in defined media. Fill 5-mL syringes with 3 mL cell suspension. Additionally, pull 1 mL aseptic air into the syringes to preclude limitation of oxygen.
5. Initial attachment process: Pump cell suspension through the chamber continuously at flow rate of 3 mL/h for 90 min.
6. Biofilm maturation process: Replace the syringes by 10-mL syringes containing cell-free media. Pump the media in stop-flow modus consisting of waiting periods and 30-s injection periods at flow rate of 6 mL/h (=50 μL suspension) every 30 min. After 24 h, disconnect one chamber to put it into the calorimeter (*see* Subheading 3.2); the biofilm maturation is continued for the next 24 h. Chambers for reference analysis are treated identically as the one for calorimetric analysis.

3.2 Preparation for Chip Calorimetric Measurement

1. Prior to any measurements, flush the fluidic system of the calorimeter with ethanol/water/sulfuric acid (1:0.41:0.01) for 30 min for disinfection followed by neutralization with phosphate buffer for 30 min. Confirm with pH paper (*see* **Note 4**).
2. Before connecting the precultivated chamber, flush the fluidic system with cell-free medium once to ensure that the complete fluidic system is only filled with the medium, providing comparable conditions to the arrangement outside the calorimeter.
3. Connect the precultivated chamber, grease the foil side with liquid paraffin, and place it on the silicon chip with the foil side facing down and in direct contact to the thermopile membrane. The thin film of liquid paraffin improves the contact between the cover foil and the thermopile membrane, thereby maximizing heat transfer.
4. Close the calorimeter and allow 20–30 min for thermal equilibration.
5. After thermal equilibration is achieved, start the injection of cell-free media to continue the biofilm maturation process. This should be identical to the maturation process outside the calorimeter: injection of 50 μL medium at a flow rate of 6 mL/h every 30 min (*see* **Note 5**).

3.3 Chip Calorimetric Analysis of Antibiotic Treatments (See Note 6)

After 48 h of biofilm maturation, replace the medium with antibiotic solution (*see* **Note 7**). The same injection/waiting periods are used. The experiment can be terminated when the biofilm eradication of biofilms is successful, indicated by the decrease of the heat signal (*see* **Note 8**).

3.4 Reference Analysis of Antibiotic Treatments

The chambers outside the calorimeter serve as reference. Depending on the reference methods chosen (e.g., CLSM, ATP content, or plate counts), detach one or more reference chambers and process them as needed.

3.5 Analysis and Interpretation of the Calorimetric Data

Use the MATLAB program to collect signal shifts and generate a profile of heat production rate throughout the experiment and compare it to the result of reference analysis to get more insight into metabolic changes (Fig. 4).

4 Notes

1. A peristaltic pump can be used instead of syringe pump. However, the dosage is more precise with a syringe pump.
2. The chambers should be prepared in advance. Apply epoxy-based resin (Epo-Tek® 301, Epoxy Technology, Inc., MA. It consists of two components, which are mixed in 4:1 ratio

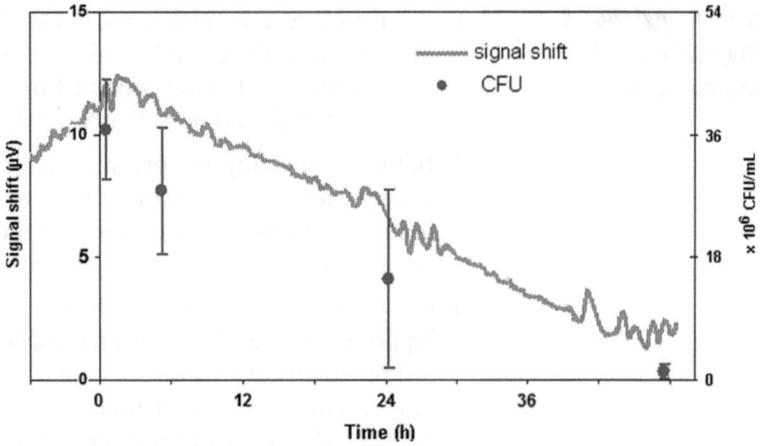

Fig. 4 Signal shifts gained from every injection over the measurement duration are collected and plotted against the elapsed time to generate an activity profile. For further calculation, the signal shift (in μV) can be converted into heat production rate (in μW) using a factor accounting for the sensitivity of the instrument (0.1205 V/W for our instrument). The sensitivity can be determined by chemical calibration

by weight prior to the application, as stated in the product guideline) to fix the cover foil (25 × 4.5 mm) on the chamber. Let the chambers dry for 1–2 h at 60 °C.

3. Flushing the chamber with disinfectant solution too long might disperse the glue between the chamber and the cover foil, thus causing leaks.
4. For the disinfection process of the fluidic system in the calorimeter, connect the ends of the open tubes, where the chamber will be connected later.
5. The injection/waiting period can be adjusted if necessary. However, it is important to keep it identical with the reference outside the calorimeter. The waiting period chosen should allow the injection signal to return to the baseline.
6. Besides the study of biofilms treated with antibiotics, this method can be used to study a variety of inhibiting or promoting influences (e.g., biological agents like phages or predatory bacteria, other chemicals, etc.). To do so, use the desired agent instead of the antibiotic solution and proceed as described.
7. The antibiotic solutions can be prepared in desired concentrations by mixing antibiotics stock with medium. The solution should not contain alcoholic solvent because their addition to solvent-free medium causes considerable mixing heat. If it is necessary, prepare medium that contains identical fraction of

solvent and perform a control experiment in the calorimeter to ensure that no mixing heat is generated.

8. To study the regrowth of biofilm after antibiotic treatment, prolong the experiment duration as necessary. If required, change the antibiotic solution to antibiotic-free medium.

References

1. Flemming H-C (2003) Role and levels of real-time monitoring for successful anti-fouling strategies—an overview. Water Sci Technol 47:1–8
2. Gnaiger E, Kemp RB (1990) Anaerobic metabolism in aerobic mammalian cells: information from the ratio of calorimetric heat flux and respirometric oxygen flux. Biochim Biophys Acta 1016:328–332
3. Maskow T, Harms H (2006) Real time insights into bioprocesses using calorimetry: state of the art and potential. Eng Life Sci 6:66–277
4. Dubrunfaut M (1856) Note sur al chaleur et le travail mecanique produits lar la fermentation vineuse. C R Acad Sci 42:945–948
5. Hill AV (1911) A new form of differential microcalorimeter, for the estimation of heat production in physiological, bacteriological or ferment reactions. J Physiol 43:260–285
6. Weppen P, Ebens J, Müller BG et al (1991) On-line estimation of biological oxygen demand using direct calorimetry on surface attached microbial cultures. Thermochim Acta 193:135–143
7. Wentzien S, Sand W, Albertsen A et al (1994) Thiosulfate and tetrathionate degradation as well as biofilm generation by *Thiobacillus intermedius* and *Thiobacillus versutus* studied by microcalorimetry, HPLC, and ion-pair chromatography. Arch Microbiol 161:116–125
8. von Rège H, Sand W (1998) Evaluation of biocide efficacy by microcalorimetric determination of microbial activity in biofilms. J Microbiol Methods 33:227–235
9. Peitzsch M, Kiesel B, Harms H et al (2008) Real time analysis of *Escherichia coli* biofilms using calorimetry. Chem Eng Process 47:1000–1006
10. Buchholz F, Harms H, Maskow T (2010) Biofilm research using calorimetry—a marriage made in heaven? Biotechnol J 5:1339–1350
11. Buchholz F, Wolf A, Lerchner J et al (2010) Chip calorimetry for fast and reliable evaluation of bactericidal and bacteriostatic treatments of biofilms. Antimicrob Agents Chemother 54:312–319
12. Mariana F, Buchholz F, Lerchner J et al (2013) Chip-calorimetric monitoring of biofilm eradication with antibiotics provides mechanistic information. Int J Med Microbiol 303:158–165
13. Buchholz F, Lerchner J, Mariana F et al (2012) Chip-calorimetry provides real time insights into the inactivation of biofilms by predatory bacteria. Biofouling 23:351–362
14. Lerchner J, Wolf A, Wolf G et al (2006) A new micro-fluidic chip calorimeter for biochemical applications. Thermochim Acta 445.144–150
15. Maskow T, Schubert T, Wolf A et al (2011) Potentials and limitations of miniaturized calorimeters for bioprocess monitoring. Appl Microbiol Biotechnol 92:55–66

Chapter 20

Bacteriophage Attack as an Anti-biofilm Strategy

Sanna Sillankorva and Joana Azeredo

Abstract

Bacteriophages are bacterial viruses. Bacteriophages replicate inside their target host whether this is in planktonic or biofilm forms. Here, we describe the methods used to control readily formed biofilms using bacteriophages.

Key words Bacteriophage, Biofilm, Control

1 Introduction

Bacteriophages are bacterial viruses that infect bacteria and can be used as biocontrol agents targeting bacterial cells either in suspension or in biofilm. The vast majority of known bacteriophages (96 %) belong to the *Caudovirales* order which encompasses dsDNA-tailed bacteriophages [1]. These phages start their infection cycles upon specifically binding to bacterial receptors displayed at the cell surface and inject their nucleic acid into the cell cytoplasm. There are two main bacteriophage life cycles—lytic and lysogenic. Lytic or virulent bacteriophages reproduce within the host and induce lysis, resulting in the release of progeny bacteriophages that start another round of infection. Temperate bacteriophages integrate their prophage into the chromosome or other replicon of the host bacteria which, through cell division, pass the bacteriophage genome (prophage) to daughter cells [2]. Only lytic bacteriophages are of interest for bacterial biocontrol [3]. The bacteriophage plaques can vary in turbidity and size depending on the bacteriophage characteristics, the host physiological state, and the growth conditions. Furthermore, plaques can sometimes be surrounded by a halo, an indicative of genes encoding for extracellular polymeric substances (EPS) depolymerase within the bacteriophage genome [4]. Bacteriophages that exhibit this type of characteristic are particularly useful for biofilm control because the

Gianfranco Donelli (ed.), *Microbial Biofilms: Methods and Protocols*, Methods in Molecular Biology, vol. 1147,
DOI 10.1007/978-1-4939-0467-9_20, © Springer Science+Business Media New York 2014

depolymerases released after cell lysis, by disrupting the biofilm matrix, help bacteriophage to gain access to the inner layers of the biofilm [5]. Besides the diffusional barrier to bacteriophage penetration imposed by the biofilm matrix, the presence of slow-growing cells constitutes an extra challenge to bacteriophage efficacy in biofilms since bacteriophages need the enzymatic machinery of actively growing cells, replicating faster in exponentially growing cells [6]. Nevertheless, there are bacteriophages that display an active life cycle in stationary growing cells, such as T7-like phages, which is undoubtedly a useful characteristic for biofilm control [7].

Biofilm control can be done by adding a bacteriophage suspension to already formed biofilm. Bacteriophages will diffuse through the biofilm and cause cell lysis. Along with cell lysis, the number of bacteriophages increases, and frequently, small portions of biofilms detach to the planktonic phase. The infection progresses both in the remaining biofilm and detached cells [8]. Biofilm control can also be made prior to biofilm formation, and in this case, the substratum surface needs to be coated with bacteriophages. The virus particles on the surfaces prevent the colonization of bacterial cells and thus prevent biofilm formation [9].

In this chapter, we describe all necessary steps for biofilm control and prevention assays using bacteriophages.

2 Materials

In order to use bacteriophages for biofilm control, one must produce them to have high-titer bacteriophage stocks. Therefore, the procedure to obtain these stocks has been included in this chapter.

2.1 Bacteriophage Propagation

1. Bacteriophage stock.
2. Tissue culture flask(s) (100 cm^2) containing a thin agar layer of LB: Weigh 8.0 g of LB and 4.8 g of agar and pour in a 500 mL bottle. Adjust to 400 mL with distilled water and autoclave at 121 °C for 15 min. Once the LB agar has cooled to 42 °C, pour on the tissue culture flasks (*see* **Note 1**).
3. Overnight-grown bacteria: Transfer a loopful of the host bacterium to 100 mL Erlenmeyers containing 25 mL of sterile LB and incubate 16 h at the proper host growth temperature.
4. Sterile molten top agar (MTA): Weigh 8.0 g of LB and 2.4 g of agar and pour in a 500 mL bottle. Adjust to 400 mL with distilled water and autoclave at 121 °C for 15 min and store accordingly (*see* **Note 2**).
5. Sterile saline magnesium buffer (SM buffer): Prepare 1 M Tris–HCl buffer (pH 7.5) in a 100 mL bottle. Weigh 6.06 g of Tris base, add 50 mL of water, and adjust the pH with HCl

to 7.5. To a 1 L bottle, add 5.8 g of NaCl, 2.0 g of $MgSO_4{\cdot}7H_2O$, and 50 mL of the prepared 1 M Tris–HCl (pH 7.5) and make up to 1 L with water (*see* **Note 3**). Autoclave at 121 °C for 15 min.

6. Sterile 50 mL Falcon tubes.
7. Sterile 200 mL bottle.
8. Filters 0.2 μm.

2.2 Bacteriophage Concentration and Purification

1. Sterile 15 and 50 mL Falcon tubes.
2. Sterile 500 mL Erlenmeyers.
3. Sterile 100 and 200 mL bottles.
4. Filters 0.2 μm.
5. DNase I.
6. RNase A.
7. Solid NaCl.
8. Solid polyethylene glycol 8000 (PEG 8000).
9. Chloroform (≥99.8 %).
10. Sterile saline magnesium buffer (SM buffer).

2.3 Biofilm Formation

1. 24-well microplates (*see* **Note 4**).
2. LB medium: Weigh 8.0 g of LB and adjust to 400 mL with distilled water. Autoclave at 121 °C for 15 min.
3. Overnight-grown bacteria: Transfer a loopful of the host bacterium to 100 mL Erlenmeyers containing 25 mL of sterile LB and incubate 16 h at the proper host growth temperature.
4. Sterile cell scrapers.
5. Sonication bath.

2.4 Prevention of Biofilm Formation Using Bacteriophages

1. Bacteriophage-conditioned 24-well microplates: Add a high concentration of bacteriophage ($>1\times10^9$ PFU per mL) to 24-well microplates and leave at least 1 h at room temperature. Remove all bacteriophage prior to use and wash the 24-well microplate twice with SM buffer.
2. LB media.
3. Overnight-grown bacteria.
4. Sterile cell scrapers.
5. Sonication bath.

2.5 Biofilm Control with Bacteriophages

1. 24-well microplates containing biofilms.
2. Bacteriophage.
3. LB medium.

4. Sterile cell scrapers.
5. Sonication bath.

2.6 Bacteriophage Titration

1. Sterile SM buffer.
2. 96-well microplates.
3. LB agar plates.
4. Overnight-grown bacteria.

2.7 Biofilm Cell Enumeration

1. LB agar plates (20 plates).
2. 96-well microplates.
3. Sterile saline solution: To a 1 L bottle, add 9.0 g of NaCl and make up to 1 L with water. Autoclave at 121 °C for 15 min.

3 Methods

Carry out all procedures at room temperature unless otherwise specified.

3.1 Bacteriophage Propagation

1. Add 1 mL of an overnight-grown culture and 1 mL of diluted bacteriophage (approximate titer 1×10^5 PFU per mL) in 5 tissue culture flasks (100 cm^2) containing an agar layer of LB and mix gently.
2. Incubate for 15 min at the proper temperature to allow bacteriophages to adsorb to the host bacterium.
3. Add 30 mL of MTA (47 °C) and let it harden.
4. Incubate overnight, without inverting, at the proper temperature (*see* **Note 5**).
5. Add 30 mL of SM buffer to the tissue culture flask(s) (100 cm^2) and incubate a minimum of 5 h at 4 °C under slow agitation (*see* **Note 6**).
6. Remove the SM buffer with the eluted bacteriophages to 50 mL Falcon tubes.
7. Centrifuge (9,000 × *g*, 4 °C, and 10 min).
8. Carefully collect the supernatant, and filter (0.2 μm) to sterile or 200 mL bottles.
9. Store at 4 °C until further use (*see* **Note 7**).

3.2 Concentration and Purification of Bacteriophage Lysates

This protocol was adapted from Sambrook and Russell [10]; however, it includes some minor modifications and does not include an ultracentrifugation step.

1. To a 200 mL bottle, add 100 mL of bacteriophage lysate (*see* Subheading 3.1 **step 9**) and DNase I and RNase A (1 μg/mL each).

2. Incubate for 30 min at room temperature.
3. Add 5.84 g of NaCl and mix gently until the NaCl has dissolved.
4. Incubate at 4 °C or on ice under agitation (50–90 rpm) for 1 h.
5. Pour onto 50 mL Falcon tubes and centrifuge (9,000 × *g*, 4 °C, 10 min).
6. Collect the supernatant to a 500 mL Erlenmeyer.
7. Add PEG to the supernatant (10.0 g PEG per 100 mL solution).
8. Incubate 5 h to overnight at 4 °C under gentle agitation (50–90 rpm).
9. Pour onto 50 mL Falcon tubes and centrifuge (9,000 × *g*, 4 °C, 10 min).
10. Discard the supernatant and resuspend the pellet, containing the precipitated bacteriophage particles, in SM buffer containing 2 % gelatin. Use 6 mL of SM buffer for each 50 mL of centrifuged sample.
11. Transfer the bacteriophage solution to 15 mL Falcon tubes and add chloroform (*see* **Note 8**) in a proportion of 1:4 (v/v). Vortex for 30 s.
12. Centrifuge (3,500 × *g*, 4 °C, 5 min).
13. Recover and filter (0.2 μm) the aqueous phase (upper phase) which contains the bacteriophage to a 100 mL Falcon tube and store at 4 °C.
14. Determine the bacteriophage titer as described in Subheading 3.6.

3.3 Biofilm Formation on Microplates

1. Add to a 24-well microplate 1 mL of sterile LB media (*see* **Note 4**).
2. Add 10 μL of a bacterial culture adjusted to an OD_{600} of 1.0.
3. Incubate at appropriate temperature conditions and under agitation (120 rpm or at static conditions (0 rpm)) during the desired period of time (e.g., 24 h—7 days), replacing the media every 12 h to remove planktonic bacteria and enhance biofilm formation.
4. At the end of the desired biofilm formation period, remove all media.
5. Wash twice the wells with 1 mL of sterile LB media.
6. Add 1 mL of LB media.
7. Use a cell scraper to scrape the biofilm from the well surface.
8. Put the 24-well microplate on a sonication bath for 30 min.
9. Quantify the viable cells present in the biofilms as described below (*see* Subheading 3.7) (*see* **Note 9**).

3.4 Prevention of Biofilm Formation with Bacteriophages

1. Add 1 mL of LB to a 24-well microplate conditioned with bacteriophages.
2. Follow all steps described above for biofilm formation (*see* Subheading 3.3).
3. Quantify the number of bacteriophages and viable cells as described in Subheadings 3.6 and 3.7.

3.5 Biofilm Control with Bacteriophages

After quantifying the numbers of viable cells in the biofilms, in at least three independent assays performed in triplicate, the efficacy of bacteriophages for biofilm control can be evaluated. In order to maintain the infection parameters identical along the experiments, a constant initial multiplicity of infection (MOI) must be used. MOI is calculated according to the number of bacteriophages per number of viable host cells, and, for instance, an MOI of 1 represents that there is 1 bacteriophage to each host cell (*see* **Note 10**).

1. After biofilm formation and washing (Subheading 3.3 step 4), add 950 μL of LB media and 50 μL of bacteriophage at a proper concentration in order to have the desired constant multiplicity of infection (MOI).
2. Incubate at the proper temperature during at least 4 h (*see* **Note 11**).
3. Take samples (Fig. 1a) to quantify the numbers of bacteriophages and viable cells (*see* Subheadings 3.6 and 3.7), respectively, present in the planktonic stage (*see* **Note 12**).
4. Remove the spent media and wash twice with sterile LB media.
5. Add 1 mL of sterile LB media.
6. Use a cell scraper to scrape the biofilm from the surface (Fig. 1b).
7. Put the 24-well microplate on a sonication bath for 30 min.
8. Take samples to quantify the numbers of bacteriophages and viable cells (*see* Subheadings 3.6 and 3.7), respectively, present in the biofilm stage (Fig. 1c).

3.6 Bacteriophage Titration

After infection, bacteriophages can be found in both planktonic and biofilms phases (Fig. 1) and should be quantified.

For bacteriophages producing small plaques (<1–2 mm in diameter), use the double agar overlay method (Subheading 3.6.1) [11], and for bacteriophages producing large plaques (>2 mm in diameter), use the small drop plaque assay (Subheading 3.6.2) [12].

3.6.1 Bacteriophage Enumeration by Double Agar Overlay

1. Prepare successive serial dilutions (1:10) in SM buffer of the bacteriophage solutions (add 20 μL of bacteriophage solution and 180 μL of SM buffer to a 96-well microplate).
2. Add to a test tube 100 μL of bacteriophage solution, 100 μL of overnight-grown bacteria, and 3–5 mL of MTA (47 °C) and tap gently.

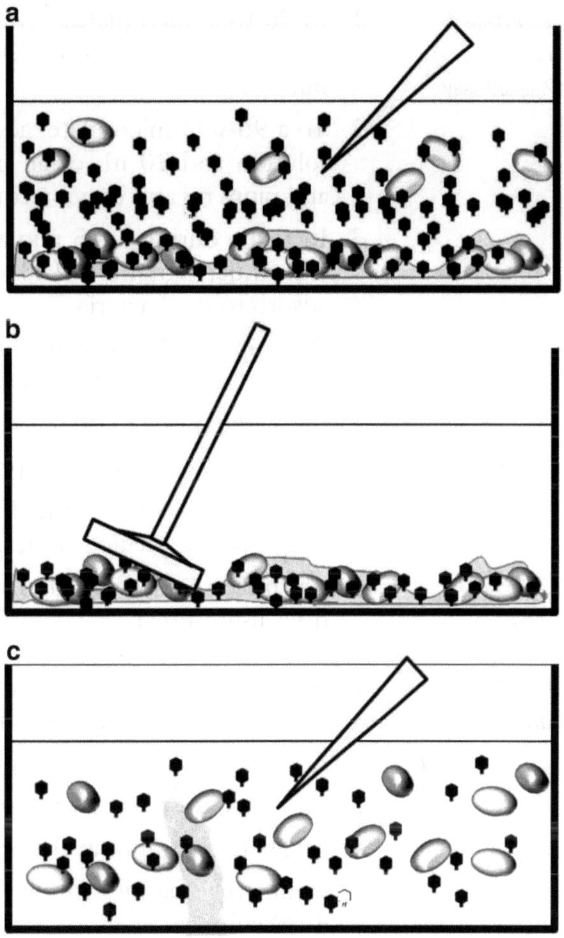

Fig. 1 Final stage of a biofilm control with bacteriophages. (**a**) Removal of samples to quantify the bacteriophages and bacteria in the planktonic phase; (**b**) using a cell scraper to scrap the biofilm; (**c**) removal of samples to quantify the bacteriophages and bacteria that were present only in the biofilm phase

3. Pour on an agar plate with LB and swirl carefully.
4. Let the plates dry for 1–2 min.
5. Incubate inverted overnight under optimal temperature conditions.
6. Count the bacteriophage plaques in the dilution which resulted in 20–200 plaques.
7. Determine the titer of triplicate preparations according to Eq. 1:

$$\text{Bacteriophage titre}\left(\text{PFU per mL}\right)=\frac{\text{No. of plaques}\times\text{Dilution factor}}{\text{Volume of phage sample}\left(\text{mL}\right)} \tag{1}$$

3.6.2 Bacteriophage Enumeration Using the Small Drop Plaque Assay

1. In 96-well microplates, serially dilute the bacteriophage samples in sterile SM buffer (20 μL of sample in 180 μL of SM buffer).
2. To a 96-well microplate, add 20 μL of diluted bacteriophage solution and 20 μL of an overnight-grown bacterial culture and pipet up and down a couple of times.
3. Incubate during 5–15 min at the proper growth temperature and under agitation (120 rpm) to allow bacteriophages to adsorb to the bacteria.
4. Pipet up and down a few times and add a drop of 20 μL of the dilution mixture onto an agar plate with LB.
5. Allow the drops to dry completely.
6. Incubate overnight at the proper growth temperature.
7. Count the plaques formed in the drop of the dilution with 3–30 bacteriophage plaques.
8. Calculate the titer of the bacteriophage of triplicate preparations using Eq. 1.

3.7 Biofilm Cell Enumeration

1. In 96-well plates, serially dilute the bacterial samples in sterile saline solution (20 μL of sample in 180 μL of SM buffer).
2. Add a drop of 20 μL of sample on an agar plates containing LB.
3. Allow the drop to dry completely.
4. Incubate overnight at the proper growth temperature.
5. Count the colonies formed in the drop of the dilution with 3–30 colonies.
6. Calculate the number of viable cells using Eq. 2:

$$\text{Nr. of viable cells}\left(\text{CFU per mL}\right) = \frac{\text{No. of colonies} \times \text{Dilution factor}}{\text{Volume of sample}\left(\text{mL}\right)} \qquad (2)$$

4 Notes

1. Bacteriophage propagation can alternatively be done using Petri dishes. Additionally, the propagation can also be done by infecting mid-exponential bacterial cells with bacteriophages. LB is the suggested medium, however alternative media can be used.
2. Leave the bottle with MTA at 65 °C if the media is to be used within 24 h. If planned to be used after 24 h, it is preferable to let the MTA solidify. Use a water bath (100 °C) or microwave oven to melt the MTA.
3. Optional: 2 % of gelatin (w/v) can be added to SM buffer. Gelatin is known to preserve bacteriophages and thus can be used in the later steps of bacteriophage purification.

4. Biofilms can also be formed on a variety of different substrata. In this case, coupons of the different materials (stainless steel, rubber, silicone, acrylic, etc.) can be cut and placed on the wells. Alternatively, other microplates can be used (e.g., 6-well, 48-well, and even 96-well microplates).
5. In some circumstances, for example, with T7-like phages, 7 h is enough, and after this, SM buffer can be added.
6. The bottle can alternatively be left overnight at 4 °C.
7. Frequently lysates, such as those obtained after this step, are used without any further purification.
8. Chloroform is used to extract PEG; however, it should be used with caution as some bacteriophages may be sensitive to it.
9. It is necessary to quantify the number of viable cells in order to allow the researcher to use a constant MOI throughout all the infection assays.
10. The MOIs commonly used vary between 0.1 and 1,000, but other bacteriophage–host ratios can be tested.
11. According to the results from different authors, the maximum cell lysis is obtained after 4–5 h, and after this point there can be observed an increase of cell growth due to the present of phage-resistant phenotypes.
12. The infection of biofilms with bacteriophages results often in a release of cell clusters to the planktonic phase, and therefore, this should be assessed.

References

1. Ackermann HW (2009) Phage classification and characterization. Methods Mol Biol 501: 127–140
2. Summers WC (2012) The strange history of phage therapy. Bacteriophage 2:130–133
3. Pirnay JP, De Vos D, Verbeken G et al (2011) The phage therapy paradigm: prêt-à-porter or sur-mesure? Pharm Res 28:934–937
4. Cornelissen A, Ceyssens PJ, T'Syen J et al (2011) The T7-related *Pseudomonas putida* phage phi15 displays virion-associated biofilm degradation properties. PLoS One 6: e18597
5. Sillankorva S, Oliveira R, Vieira MJ et al (2004) Bacteriophage Phi S1 infection of *Pseudomonas fluorescens* planktonic cells versus biofilms. Biofouling 20:133–138
6. Hadas H, Einav M, Fishov I, Zaritsky A (1997) Bacteriophage T4 development depends on the physiology of its host *Escherichia coli*. Microbiology 143:179–185
7. Sillankorva S (2008) The use of bacteriophages to control biofilms. PhD thesis in chemical and biological engineering, University of Minho
8. Sillankorva S, Neubauer P, Azeredo J (2010) Phage control of dual species biofilms of *Pseudomonas fluorescens* and *Staphylococcus lentus*. Biofouling 26:567–575
9. Curtin JJ, Donlan RM (2006) Using bacteriophages to reduce formation of catheter-associated biofilms by *Staphylococcus epidermidis*. Antimicrob Agents Chemother 50:1268–1275
10. Sambrook J, Russell DW (2001) Molecular cloning—a laboratory manual, 3rd edn. Cold Spring Harbor Laboratory Press, Cold Spring Harbor, NY
11. Adams MH (1959) Bacteriophages. Interscience Publishers, INC, New York
12. Mazzocco A, Waddell TE, Lingohr E, Johnson RP (2009) Enumeration of bacteriophages using the small drop plaque assay system. Methods Mol Biol 501:81–85

Chapter 21

Photodynamic Therapy as a Novel Antimicrobial Strategy Against Biofilm-Based Nosocomial Infections: Study Protocols

Francesco Giuliani

Abstract

Hospital-acquired infections (HAIs), also known as nosocomial infections, are one of the most serious health-care issues currently influencing health-care costs. Among them, those sustained by microbial biofilm represent a major public health concern. Here, we describe the experimental protocols for microbial biofilm inactivation relying on antimicrobial photodynamic therapy (APDT) as a new strategy for the control of these kinds of infections.

Key words Antimicrobial photodynamic therapy (APDT), Photosensitizer (PS), RLP068/Cl, *Staphylococcus aureus*, *Pseudomonas aeruginosa*

1 Introduction

Health-care-associated infections are the fourth leading cause of disease in industrialized country and represent one of the major health issues [1]. Nosocomial (such as hospital-acquired) infections are the most common complications affecting hospitalized patients and can lead to longer hospitalization with reduced bed number availability and of course increasing health-care costs [1]. Many HAIs are associated to microbial biofilms especially those related to chronic wound and to the use of medical devices such as urinary catheters, intravascular catheters, and orthopedic implants [2]. The common hallmark of biofilm-related infections, including those of indwelling medical devices, is their intrinsically resistance to host immunity, conventional antimicrobial agents, and biocides [3, 4]. Bacterial biofilms in fact are known to tolerate levels of antibiotics 10 1,000 times higher than the MICs of the corresponding planktonic form. In addition, biofilms facilitate the spread of antibiotic resistance by promoting horizontal gene transfer [4]. For these reasons, standard antimicrobial treatments often

Gianfranco Donelli (ed.), *Microbial Biofilms: Methods and Protocols*, Methods in Molecular Biology, vol. 1147,
DOI 10.1007/978-1-4939-0467-9_21, © Springer Science+Business Media New York 2014

fail in biofilm eradication and chronic infections can arise. Today, the development of alternative and effective drugs and/or therapies, at a time when the pipeline for new antimicrobials is drying up, against biofilm-related infections represents one of the major public health concerns. One example of these relatively novel strategies (therapies) is antimicrobial photodynamic therapy (APDT) which is expected to be useful in the treatment of acute and chronic localized infections [5, 6]. APDT is a treatment that utilizes a combination of visible light, a chemical known as a photosensitizer (PS) that is capable of being activated by light, and oxygen, to achieve a cytotoxic effect. The process involves delivering harmless visible light of the appropriate wavelength to the PS molecule to bring to its excited singlet state, which subsequently crosses to a more stable, lower-energy triplet state. The interaction between the PS excited states and the endogenous oxygen in the proximity of the target cells provides the cytotoxic effects through the production in situ of reactive oxygen species (ROS) following two pathways, type I and type II, leading to the formation of free radicals and singlet oxygen, respectively ([7], Fig. 1). The main advantages of APDT are high target specificity provided by the fact that the PS is localized in the microorganisms without major involvement of the surrounding tissues or cells and by the selection of a suitable illumination protocol [8], with few undesired side effects (the drug is inactive in the dark and becomes active only when exposed to light) [9].

A standard APDT protocol, in fact, includes incubation of the target cells or diseased tissue with the PS (which is intrinsically non-

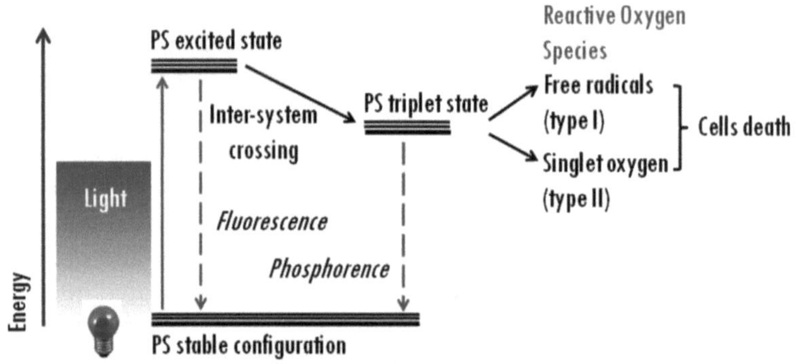

Fig. 1 Mechanism of action of APDT. Production of ROS can follow two different pathways after light activation. Upon absorption of a photon, the PS goes through its ground state to an excited state. The excited state is transitory and can undergo intersystem crossing to a more stable lower-energy triplet state or, alternatively, can return to the ground state by fluorescence emission or heat or both. This PS triplet state may undergo electron transfer (type I) or energy transfer (type II) to oxygen to form ROS as hydrogen peroxide (H_2O_2), hydroxyl radical ($HO^\bullet$), and superoxide anion (O_2^-) or singlet oxygen, ($^1O_2$), respectively, which can kill both Gram-positive and Gram-negative bacteria

toxic), followed by a short illumination step to generate ROS responsible for irreversible oxidative damage to essential cellular constituents of the microorganisms to which the PS is bound. Different to antibiotics, the short time of incubation of the PS with the microorganisms and the typically multitarget nature of photosensitized inactivation processes, usually involving a number of membrane proteins and lipid domains, prevent the possible expression of protective factors (e.g., the biosynthesis of stress proteins), thus minimizing the risk of the emergence of resistant strains [10].

2 Materials

The organisms described in the Subheading 2.1 are all classified as biosafety level two (BL2) with no health risk to laboratory staff if managed taking standard universal precautions (gloves, laboratory coat, glasses, and mask) and using a class 2 biosafety cabinet.

All the solutions and media are prepared using Milli-Q water (prepared by purifying deionized water to obtain a sensitivity of 18 MΩ cm at 25 °C) and analytical grade reagents. Store all reagents and media at room temperature unless otherwise specified. Waste material should be disposed according to all waste disposal regulations.

2.1 Microorganisms

1. Gram-positive bacterium *Staphylococcus aureus* ATCC 6538.
2. Gram-positive bacterium *Staphylococcus aureus* ATCC 43300.
3. Gram-negative bacterium *Pseudomonas aeruginosa* PAO1.

In this chapter, we will describe inactivation protocols of biofilm sustained by *S. aureus* ATCC 6538 and *P. aeruginosa* PAO1 for in vitro experiments and by *S. aureus* ATCC 43300 for in vivo procedures (*see* **Note 1**).

2.2 Equipment

1. Shaking incubator.
2. Stationary incubator.
3. Class 2 biological safety cabinet.
4. Autoclave.
5. Centrifuge.
6. Light source.
7. Vortex mixer.
8. Spectrophotometer.

2.3 Buffers, Reagents, Solutions

1. Tryptic soy broth (TSB). Prepared according to the manufacturer's instructions.
2. Tryptic soy agar (TSA). Prepared according to the manufacturer's instructions.

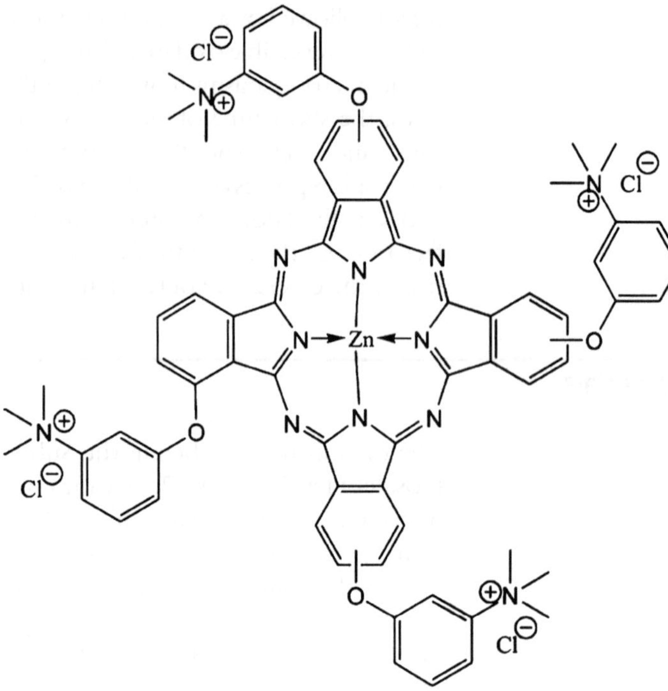

Fig. 2 Structural formula of RLP068 chloride. Molecular formula, $C_{68}H_{64}Cl_4N_{12}O_4Zn$; molecular weight, 1,320.52

3. Saline solution. For 1 L of 1× saline: 9 g of sodium chloride (NaCl) dissolved in Milli-Q water, to a total volume of 1,000 ml.
4. Phosphate-buffered saline (PBS). For 1 L of 1× PBS, prepare as follows: Start with 800 ml of distilled water. Add 8 g of NaCl. Add 0.2 g of KCl. Add 1.44 g of Na_2HPO_4. Add 0.24 g of KH_2PO_4. Adjust the pH to 7.4 with HCl. Add distilled water to a total volume of 1 L. Dispense the solution into aliquots and sterilize by autoclaving (20 min, 121 °C, liquid cycle). Store at room temperature.
5. RLP068/Cl. For in vitro procedures, RLP068/Cl (Fig. 2), a tetracationic Zn(II) phthalocyanine chloride (Molteni Therapeutics, Scandicci, Firenze), was used as the photosensitizer (PS) [11]. RLP068/Cl was dissolved in sterile Milli-Q water to give a 2.0 mM stock solution and stored at −20 °C in the dark until used (*see* **Note 2**). For in vivo experiments, RLP068/Cl was used as a gel formulation for topical application (0.3 % w/w). The gel used to deliver compound RLP068 was administered at various percentages of active product (0.01 %, 0.1 %, 0.3 %, 0.5 % w/w) containing pharmaceutically acceptable excipients. The proprietary composition is based on the use of a mixture of alcohols gelified with an appropriate amount of a carboxymethyl cellulose polymer. Mice were treated locally with 25 μL of the gel, corresponding to a dose of 75 μg

of RLP068/Cl. RLP068/Cl is light sensitive and exposure to ambient light should be avoided (*see* **Note 3**).

2.4 Light Source

1. Noncoherent halogen lamp (PDT 1200, Waldmann, Villingen-Schwenningen, Germany) with a band-pass filter isolating the 600- to 700-nm-wavelength range for in vitro experiments.
2. Diode laser at 689 nm (Ceralas PDT 689 nm/3 W, CeramOptec GmbH, Bonn, Germany) for in vivo experiments.

2.5 Disposable Plasticware

1. Sterile 50-ml tubes for carrying out the starting inoculum.
2. Microcentrifuge 1.5-ml tubes for carrying out serial dilutions.
3. Sterile plastic Petri dishes 9 cm in diameter are used to grow the serial dilutions onto agar medium in order to count bacterial colonies.

2.6 Animals

CD1 mice (weight 28–30 g; 4 weeks old) anesthetized by an intraperitoneal (i.p.) injection of ketamine 87 mg/kg, acepromazine 2 mg/kg, and atropine 0.066 mg/kg had the hair of the back shaved and the skin cleansed with 10 % povidone-iodine solution. Using a 0.8-cm (diameter) template, one full-thickness wound was established through the panniculus carnosus on the back subcutaneous tissue of each animal. Only one wound per animal was performed [12].

3 Methods

3.1 In Vitro Biofilm APDT Treatment

1. Biofilm formation. Prepare liquid media (tryptic soy broth [TSB]) and autoclave. Prepare solid media (tryptic soy agar [TSA]), autoclave, and pour into Petri dishes (9 cm diameter). Use a sterile loop to pick a single colony from the agar plate and put into a 50-ml tube containing 5 ml of TSB. Bacterial cells were grown at 37 °C with orbital shaking overnight to allow adequate aeration. Cells were harvested by centrifugation (3,220×*g* for 10 min) and washed twice in phosphate-buffered saline (PBS) pH 7.4. The cells were then resuspended in TSB containing 0.5 % glucose to a final concentration of 10^6 (*see* **Note 4**). For each biofilm experiment, 24 wells, 3 for PS concentration tested, of a flat-bottomed commercially available presterilized polystyrene 24-well microtiter plate were inoculated with 1 ml of these cells suspension in duplicate (illuminated and nonilluminated samples) and 3 control wells of another plate were filled with sterile medium. Following 6 h of adhesion, the supernatant was removed and wells were washed twice with 2 ml of PBS to remove nonadherent cells. Subsequently, 1 ml of fresh medium was added to each well and the plates were further incubated for 24 h. After incubation, the medium was removed and the well rinsed as described

above. 1 ml of fresh medium was added to each well and the plates were incubated for 24 h more. After 24 h, again remove the medium, wash the well (*see* **Note 5**), and leave the microtiter plates to air-dry for 60 min at room temperature in an inverted position into a biological sterile cabinet for biofilm fixation (*see* **Note 6**).

2. Incubation with PSs. Several concentrations of RLP068/Cl in PBS ranging from 0.5 to 50 μM (0, 0.5, 1.0, 2.5, 5, 10, 25, and 50 μM) were used for the experiments. The incubation time was 60 min at 37 °C (*see* **Note** 7). It is recommended to protect the microbial biofilm treated with the PS from external light all along the incubation phase.
3. Light delivery. One of the most important issues is to wash the PS biofilm before illumination, to eliminate the PS excess, which due to the high extinction coefficient may act as a light shield thus preventing from the activation of the specifically bound PS to the biofilm. In our experience, we carry out illumination in PBS after three washes of the samples with the same buffer to remove the excess of PS. The delivered light energy was 60 J/cm^2 (resulting from 10-min illumination at 100 mW/cm^2) (*see* **Note 8**).
4. Sonication. Place the 24-well microtiter plate containing the microbial biofilms into an ultrasonic bath (the sonicator). Sonicate on the setting "high" for 5–20 min (the time required depends on the microorganism being assayed). The vibrations created by the sonicator transfer first to the water and finally to the microtiter plate to disrupt biofilm from the bottom of the wells into the PBS.
5. Serial dilution. Once sonication is complete, to obtain a dose–response curve, illuminated and nonilluminated samples (to quantify dark toxicity of the PS) were assayed by five serially tenfold dilution in PBS. This will provide tubes with dilutions of 1×, 10×, 100×, 1,000×, 10,000×, and 100,000×. Twenty-five microliter of each dilution is plated on solid growth medium. After incubation of the plates at 37 °C for 24 h, the number of CFU was counted and the results multiplied by the appropriate power of ten to obtain the number of CFU/ml (Fig. 3).

3.2 In Vivo APDT Treatment of Infection Sustained by Biofilm in Mouse Model

All experiments were performed according to the Principles of Laboratory Animal Care. The experimental protocols were also approved by the Institutional Animal Care Committee of the Ministry of Health, Italy. All animals were housed in individual cages under constant temperature (20 ± 4 °C) and humidity (55 ± 10 %) with a 12-h light/dark cycle and had access to food and water ad libitum throughout the study. The environment was temperature and humidity controlled, with lights on and off at 06:30 and 18:30 h.

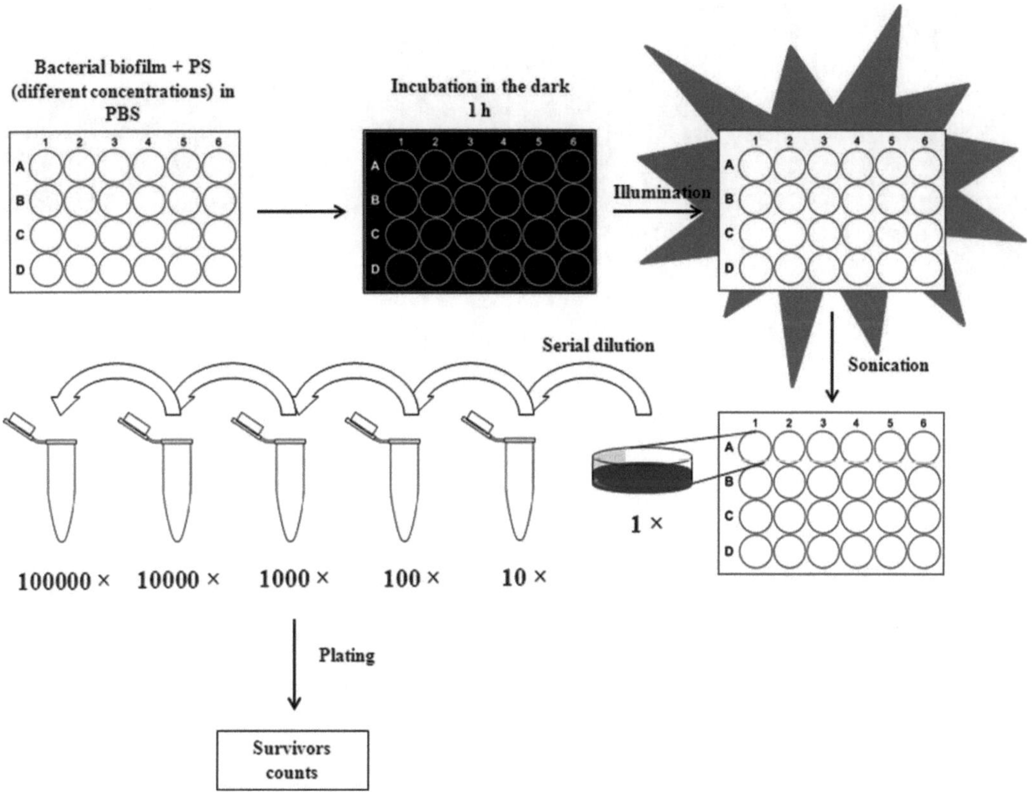

Fig. 3 Schematic cartoon illustrating an in vitro APDT experiment with RLP068/Cl versus bacterial biofilm

1. Preparation of mice. Male CD1 mice were anesthetized by an intraperitoneal (i.p.) injection of ketamine/acepromazine/atropine cocktail and shaved on the back. One full-thickness wound was performed through the panniculus carnosus on the back subcutaneous tissue of each animal using a template.
2. Establishment of a mouse skin infection (Fig. 4). The MRSA ATCC 43300 was grown in brain–heart infusion broth. When bacteria were in the log phase of growth, the suspension was centrifuged at 1,000×*g* for 15 min, the supernatant was discarded, and the bacteria were resuspended and diluted into sterile saline to achieve a concentration of approximately 10^8 CFU/ml. A gauze (approximately 1 cm^2) placed over each wound was dropped with 100 μL of bacterial suspension (5×10^7 CFU/ml). The lesion overlaid with the infected gauze was closed by means of skin clips, resulting in a local abscess after 2 days. This procedure, as previously described by Kugelberg et al. [13], was performed to establish a superficial skin infection and to enable better definition and quantification of the infection. The animals were returned to individual cages and thoroughly examined daily.
3. Experimental design. RLP068/Cl gel formulation was initially tested at concentration of 0.01 %, 0.1 %, 0.3 %, and 0.5 % in order to determine the dose–response curve in terms of local

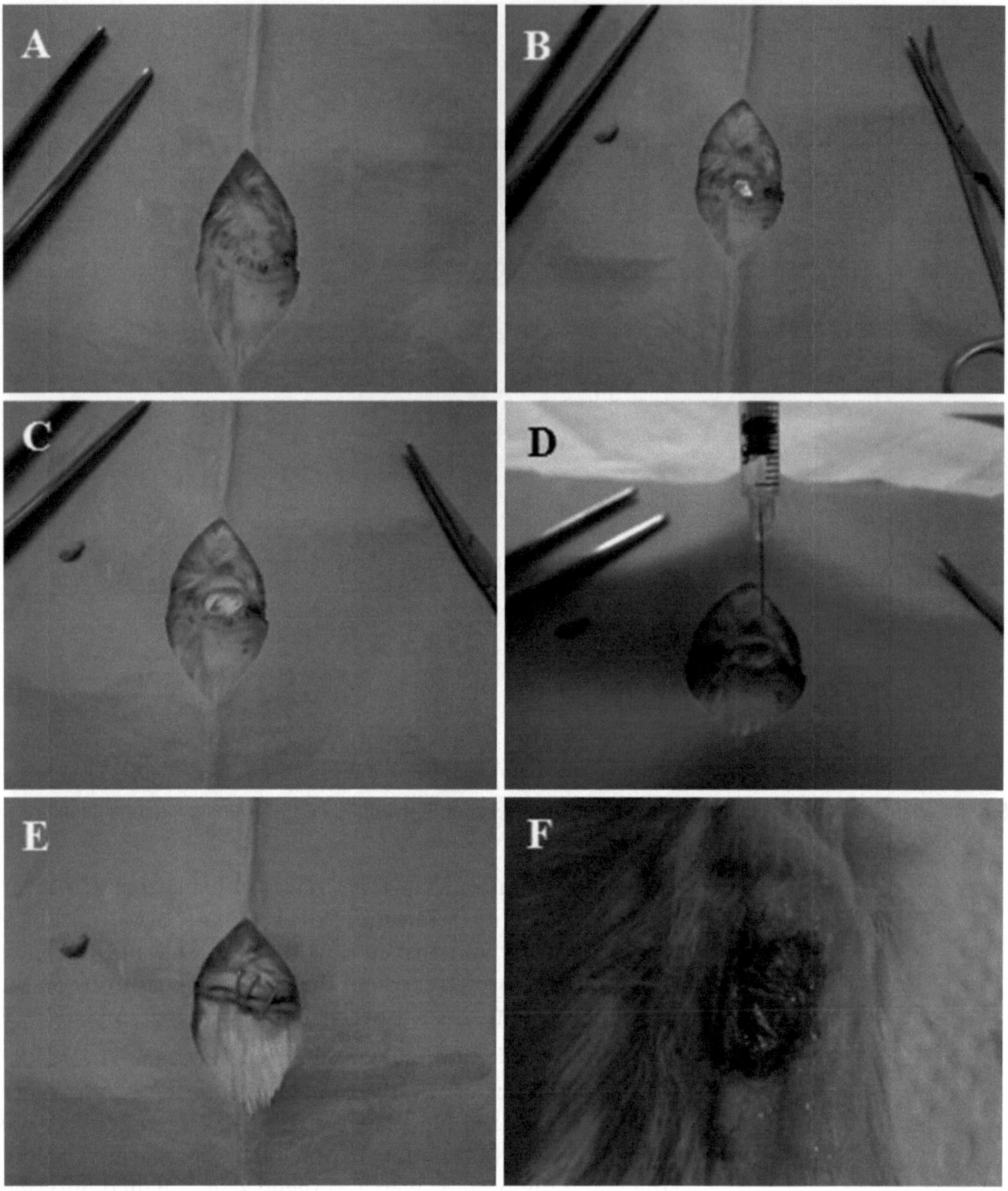

Fig. 4 Procedure for in vivo mouse skin infection. (**a**) Anesthetized mouse is shaved on the back. (**b**) Following a template, a full-thickness wound is performed. (**c**) A gauze is placed over the full-thickness wound and (**d**) is dropped with 100 μL of a 5.0×10^7 CFU/ml bacterial suspension. (**e**) The infected lesion is surgically closed using sterile skin clips. (**f**) After 2 days from infection, a local abscess is observable

infection inhibition, immediately after APDT of the infected wound, at day 2. Subsequently in the main study, APDT was carried out with 0.3 % RLP068/Cl gel formulation. The study included a group of infected mice not receiving any treatment and two groups of infected mice whose wound were treated, respectively, with (1) APDT with RLP068/Cl gel corresponding

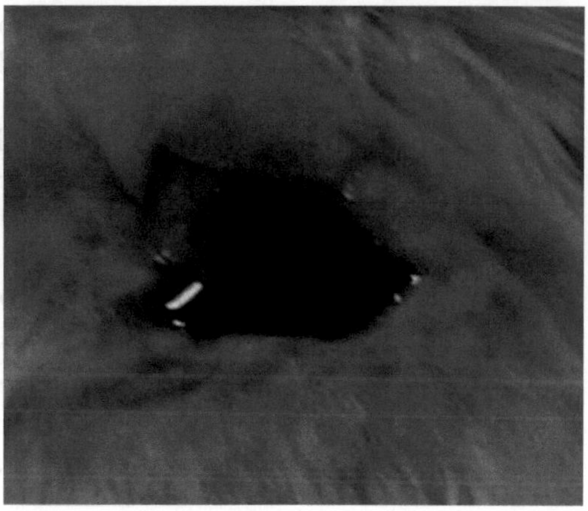

Fig. 5 25 μL of RLP068/Cl 0.3 % gel is applied to the wound with a pipette tip and spread uniformly all over the lesion

to 75 μg and (2) light with placebo gel (the placebo gel contains all the same excipients of the gel used to deliver RLP068 without the PS).

4. Addition of RLP068/Cl gel formulation (Fig. 5). On day 2 from infection, mice were anesthetized as described above and, in all animals, the wounds were opened, the gauzes removed, and the treatment started. The gel containing RLP068/Cl 0.3 % or the placebo gel was applied to the wound of each animal. The gel formulation was set at 25 μL as this amount was determined to be sufficient to fully cover the lesion without excess of product, thus avoiding potential attenuation of the light intensity during illumination by the gel itself, and thereby attenuating this parameter. The gel was spread uniformly on the surface of the lesion and the illumination was performed after 1 h from the application. This time was considered sufficient to allow the diffusion of RLP068 from the gel and to target the bacteria present in the lesion. All topical treatments were given to a group of ten animals, while a further group consisting of ten animals remained untreated as the infection control.
5. Light delivery. Mice are illuminated with a diode laser at 698 nm at a suitable distance from the skin in order to have a light spot corresponding to a diameter of 4.6 cm. The fluence rate at the spot level was determined to be 120 mW/cm^2. The light dose of 60 J/cm^2 was administered according to previous experiments in which this value was determined to be safe for APDT. The maximum wavelength of the photosensitizer is 690 nm, overlapping with the laser emission of 689 ± 3 nm (*see* **Note 8**).
6. At the end of experiment, all the animals were killed with excess CO_2. A 1 × 2 cm area of skin, including the wound, was

excised aseptically and homogenized in 2 ml saline solution (NaCl 0.9 %) using an ULTRA-TURRAX (IKA-Werke GmbH, Staufen, Germany). Quantification of viable bacteria was performed by culturing serial dilutions (0.1 ml) of the bacterial suspension on blood agar plates. For this purpose, all plates were incubated at 37 °C for 48 h and evaluated for the presence of the staphylococcal strain. The bacteria were quantified by counting the number of CFU per plate. The limit of detection for this method was approximately 10 CFU/ml.

7. Statistical analysis. All results are presented as mean ± SD. Statistical analysis was performed using student's t test. Significance was accepted at $p < 0.05$.

4 Notes

1. Microbiological strains. Reference strains of most species of microorganisms can be obtained from several nonprofit bioresource centers and research organizations. The American Type Culture Collection (ATCC) (Manassas, VA, USA) is one of the well-recognized vendors worldwide.
2. Antimicrobial PS. There are a large amount of compounds that have been reported to work as PS. Most of them are cationic molecule bearing one or more quaternary nitrogen atoms. These molecules are based on tetrapyrrole nucleus like porphyrins [14, 15], phthalocyanines [16, 17], or chlorine [18] backbones. Other important groups of compounds include cationic synthetic dyes such as phenothiazinium dyes or triarylmethane dyes. In this chapter, we describe RLP068/Cl, a tetracationic Zn(II) phthalocyanine chloride (Molteni Therapeutics, Scandicci, Firenze) [11].
3. Photosensitizer efficacy. It is well known that the PSs in aqueous solution, even stored in the dark, are not stable and can aggregate losing effectiveness as antimicrobial agents. For this reason, it is recommended to prepare the solutions of PS shortly before their use.
4. Microbiological culture. To reduce the possibility of contamination, this step should be performed using good aseptic technique or into a biological safety cabinet. Since biofilm formation is highly susceptible to various in vitro conditions, in order to minimize errors and provide reliable analysis, it is very important to test each PS concentration in triplicate in a single experiment. In addition, the experiment should be carried out three times.
5. Washing biofilm is another critical step since it is believed to remove all nonadherent cells while simultaneously providing preservation of biofilm integrity. There are two important

points to consider: number of washing and the technique used for washing. It is clear that insufficient washing may lead to false-positive results as well as excessive washing may lead to false-negative results.

6. After washing, the mature bacterial biofilm should be fixed. There are many methods for fixation: Exposing biofilm to hot air at 60 °C for 60 min appears to be the method of choice due to its safety. It is known that Bouin's reagent is the most reliable biofilm fixative followed by air-drying (1 h at 60 °C) [19], but it contains explosive chemicals. Fixation by methanol can also be used since it has proven to be as effective as Bouin's reagent [19]. We found that exposing microtiter plate to air-dry for 60 min at room temperature into a biological sterile cabinet leads to a well-fixed biofilm and provides greater protection for contamination.
7. Choice of incubation time. Incubation time of bacterial cells with PS can be a very free parameter depending on chemical nature of PS and on the microorganism producer of biofilm being treated. It has been found that, using RLP068/Cl as the PS, 1 h is a reasonable time for both *S. aureus* and *P. aeruginosa* biofilms.
8. Light delivery. The delivered light energy required in the APDT of microbial biofilm again depends on the bacterial strain producer of the biofilm and on the technical specifications of the light source used for the illumination phase.

References

1. Guggenbichler JP, Assadian O, Boeswald M, Kramer A (2011) Incidence and clinical implication of nosocomial infections associated with implantable biomaterials—catheters, ventilator-associated pneumonia, urinary tract infections. GMS Krankenhhyg Interdiszip 6:Doc18
2. Francolini I, Donelli G (2010) Prevention and control of biofilm-based medical-device-related infections. FEMS Immunol Med Microbiol 59:227–238
3. Hall-Stoodley L, Stoodley P (2009) Evolving concepts in biofilm infections. Cell Microbiol 11:1034–1043
4. Fux CA, Costerton JW, Stewart PS, Stoodley P (2005) Survival strategies of infectious biofilms. Trends Microbiol 13:34–40
5. Hamblin MR, Hasan T (2004) Photodynamic therapy: a new antimicrobial approach to infectious disease? Photochem Photobiol Sci 3: 436–450
6. O'Riordan K, Akilov OE, Hasan T (2005) The potential for photodynamic therapy in the treatment of localized infections. Photodiagnosis Photodyn Ther 2:247–262
7. Demidova TN, Hamblin MR (2004) Photodynamic therapy targeted to pathogens. Int J Immunopathol Pharmacol 17:245–254
8. Soncin M, Fabbris C, Busetti A et al (2002) Approaches to selectivity in the Zn(II)-phthalocyanine-photosensitized inactivation of wild-type and antibiotic-resistant *Staphylococcus aureus*. Photochem Photobiol Sci 1:815–819
9. Winckler KD (2007) Special section: focus on anti-microbial photodynamic therapy (PDT). J Photochem Photobiol B 86:43–44
10. Giuliani F, Martinelli M, Cocchi A et al (2010) In vitro resistance selection studies of RLP068/Cl, a new Zn(II) phthalocyanine suitable for antimicrobial photodynamic therapy. Antimicrob Agents Chemother 54:637–642
11. Fabris C, Soncin M, Mazzon E et al (2005) A novel tetracationic phthalocyanine as a potential skin phototherapeutic agent. Exp Dermatol 14:675–683

12. Simonetti O, Cirioni O, Orlando F et al (2011) Effectiveness of antimicrobial photodynamic therapy with a single treatment of RLP068/Cl in an experimental model of *Staphylococcus aureus* wound infection. Br J Dermatol 164:987–995
13. Kugelberg E, Norstrom T, Petersen TK et al (2005) Establishment of a superficial skin infection model in mice by using *Staphylococcus aureus* and *Streptococcus pyogenes*. Antimicrob Agents Chemother 49:3435–3441
14. Oliveira A, Almeida A, Carvalho CM et al (2009) Porphyrin derivatives as photosensitizers for the inactivation of *Bacillus cereus* endospores. J Appl Microbiol 105:1986–1995
15. Caminos DA, Spesia MB, Pons P, Durantini EN (2008) Mechanisms of Escherichia coli photodynamic inactivation by an amphiphilic tricationic porphyrin and 5,10,15,20-tetra(4-N, N, N trimethylammoniumphenyl) porphyrin. Photochem Photobiol Sci 7:1071–1078
16. Mantareva V, Kussovski V, Angelov I et al (2007) Photodynamic activity of water-soluble phthalocyanine zinc(II) complexes against pathogenic microorganisms. Bioorg Med Chem 15:4829–4835
17. Scalise I, Durantini EN (2005) Synthesis, properties, and photodynamic inactivation of *Escherichia coli* using a cationic and a non-charged Zn(II) pyridyloxyphthalocyanine derivatives. Bioorg Med Chem 13: 3037–3045
18. Schastak S, Gitter B, Handzel R et al (2008) Improved photoinactivation of gram negative and gram-positive methicillin resistant bacterial strains using a new near infrared absorbing meso-tetrahydroporphyrin: a comparative study with a chlorine e6 photosensitizer photolon. Methods Find Exp Clin Pharmacol 30: 129–133
19. Baldassarri L, Simpson WA, Donelli G, Christensen GD (1993) Variable fixation of staphylococcal slime by different histochemical fixatives. Eur J Clin Microbiol Infect Dis 12:866–868

Part III

Investigations on Biofilms in the Environment and Manufacturing Plants

Chapter 22

Capturing Air–Water Interface Biofilms for Microscopy and Molecular Analysis

Margaret C. Henk

Abstract

Described here is a simple and versatile technique for capturing small samples of the thin biofilm that is located at the meniscus of almost every air–water interface (AWI), an electrostatically distinct aquatic domain/habitat. The method uses a microscope slide (and other supporting surfaces) coated in the lab with a collodion membrane, which has a strong affinity for the upper surface of the AWI biofilm. The structural integrity of the biofilm is maintained during the capture process, and components of the biofilm are effectively separated from the subtending liquid. The captured thin biofilm can be analyzed in many ways including almost any form of light, electron, and atomic force optics; and spatially significant molecular analyses may be performed on the captured biofilm or its components.

Key words Air–water interface, Biofilm, Microscopy, Microbiology, Ecomicrobiology

1 Introduction

One of the more frequently encountered types of biofilm that exist in nature is also one of the more often overlooked. Underappreciated partly because the AWI biofilm is thin and difficult to see, a large percentage of the microbes that reside here may be underestimated or completely escape detection, either because the biofilms adhere to the sides of the collection container and thus are lost from the aquatic sample analysis or because they are purposely avoided and discarded as fouling "scum."

The distinctive electrostatic properties of a water meniscus are responsible for concentrating, physically supporting, and activating molecules as well as larger living and nonliving particles of many different kinds, conceivably even providing conditions suitable for evolution of new life forms here at this thin, mobile interface between two fluid environments—the atmosphere and the hydrosphere. Subjected to extreme variations in physical and chemical conditions, those organisms living at the AWI have developed

Gianfranco Donelli (ed.), *Microbial Biofilms: Methods and Protocols*, Methods in Molecular Biology, vol. 1147, DOI 10.1007/978-1-4939-0467-9_22, © Springer Science+Business Media New York 2014

specializations for this ecological niche, and their community makeup differs from the bulk aquatic community directly subtending the meniscus.

This capture method capitalizes on the special affinities of the hydrophobic upper, aerial surface of any floating natural or lab-grown AWI biofilm community [1].

Though primarily designed for applications using glass microscopy slides, the many possible applications for biofilms captured in this way include:

(a) Collecting and observing organisms in introductory microbiology or ecomicrobiology labs to demonstrate the ubiquity and diverse morphologies of microbes in nature as well as providing the ideal sample for introduction to proper use of the student microscope [2]; (b) characterizing microbes with routine microbiology staining protocols, including Gram staining (slightly modified); (c) using fluorescent dyes to identify specific organisms, their components, or biological processes; and (d) observing living microbes in situ, usually with interference contrast (Nomarski) microscopy, to detect cell division, close microbial associations, and predation by affiliated protozoa, e.g., (e) using scanning and transmission electron microscopy, as well as atomic force microscopy for physical and ultrastructural analysis. Portions of the collodion/biofilm may be removed for (f) analyzing the community as a whole using molecular analyses or (g) analyzing specific components of the community selected with laser dissecting microscopy. In addition, the fact that the captured biofilm is effectively separated from the rest of the aquatic sample allows for (h) characterizing and comparing the two separate communities using any of the above techniques [2].

2 Materials

2.1 Reusable "Strainer" Apparatus Components for Making Capture Slides

Petri dishes, polystyrene, size 100 mm × 15 mm (Sigma Aldrich).

Simple soldering iron or heat pen with tip about 5 mm in diameter.

2.2 Components for the Capture Slide for Light Optical Applications

Super Up-Rite glass microscope slides (Fisher Scientific); Whatman filter paper grade no. 1, 70 mm (Fisher Scientific); sterile 2 % collodion monomer in amyl acetate (Electron Microscopy Sciences); disposable glass pipettes (Fisher Scientific); clean, preferably deionized or distilled, water; forceps for holding slides; and fine-tip forceps for handling "mini-substrates."

2.3 Mini-substrates

Appropriate mini-substrates may be positioned upon the slide before the collodion membrane is applied. The substrates will be held in place on the slide by the thin collodion membrane and thus must be relatively small, lightweight, and flat.

2.3.1 Permanent Mini-substrates for TEM, SEM, AFM

For TEM

For grid cleaning: 50 ml glass beaker, 10 ml 95 % acetone, 100 mm plastic petri dish and lid, 9 cm diameter Whatman filter paper #1.

Copper TEM grids, 400 mesh recommended (Electron Microscopy Sciences); wooden applicator sticks (Fisher Scientific); regular brown paper towels.

For SEM and AFM

Appropriate discs or rectangles of heavy-duty aluminum foil or slide coverslip for particular SEM and AFM stages, each no more than about 36 mm^2 in area.

2.3.2 Removable Mini-substrates for Other Reactions

Collected collodion/biofilm for molecular analysis and for sectionable preparations, etc., may be removed from rigid substrates for chemical processing if underlain by absorbent paper mini-substrates.

Discs or rectangles of Whatman #1 filter paper, each no more than about 36 mm^2 in area.

3 Methods

3.1 Reusable Apparatus: The Petri Dish Strainer

1. In a fume hood or well-ventilated area, place a number of plastic petri dish bottoms upside down on paper towel padding (*see* **Note 1**).
2. Use the hot soldering iron to melt about 10-well-spaced drain holes through each petri dish bottom, quickly pushing the iron straight down and then back up (*see* **Note 2**).
3. Make one hole right at the periphery of the rim to facilitate rapid draining and drying, as in Fig. 1.

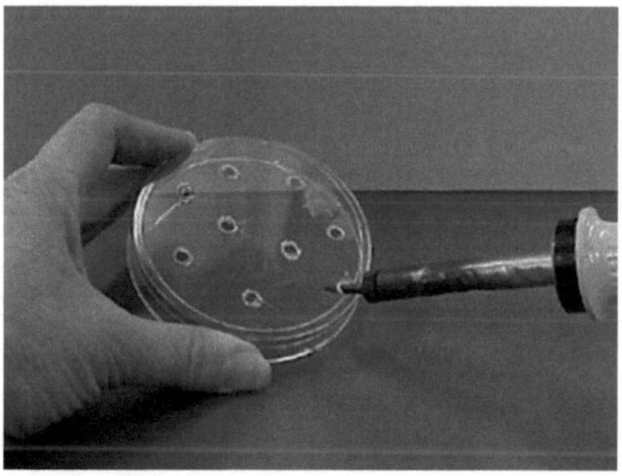

Fig. 1 Melting holes into the "strainer." Reproduced from [2] with permission

3.2 Preparing Collodion-Membraned Collection Slides

1. In a well-lighted and well-ventilated room, prepare an undisturbed work area (*see* **Note 3**).
2. Use a pencil to label a frosted-end microscope slide (*see* **Note 4**).
3. Lay the slide on a paper towel and thoroughly scrub the "good" side (label up) with a dry lab wipe. Blow off any dust, and avoid touching the good side (*see* **Note 5**).
4. Invert the lid of the petri dish on a paper towel, place the perforated "strainer" into it, center a 7 mm filter paper disk in the dish, and place the clean slide on top of the filter paper, label side up, as in Fig. 2.
5. Add about 50 ml clean water to the dish, making sure the slide is completely immersed (*see* **Note 6**). Excessive water will simply overflow the level of the inverted lid onto the paper towels.

   (If the collection slide will be used for purposes other than or in addition to light microscopy, *see* Subheading 3.3 below before proceeding. Mini-substrates may be positioned at this point.)
6. Use a disposable glass pipette (not a micropipettor) to draw up about 1 ml of collodion solution for about 20 slides (*see* **Note 7**).
7. From about 2 in. above the water surface, drop 2–3 drops onto the center of the water surface in the dish, as in Fig. 3. The floating liquid should spread out almost to the periphery of the strainer (*see* **Note 8**).
8. Move to a good angle above the dish to be able to see rainbow interference colors form as the liquid evaporates (*see* **Note 9**).
9. Allow the collodion membrane to dry completely. It may take up to 5 min. The resulting floating membrane should be silver/gold colored and perhaps wrinkling at the periphery. *See* Fig. 4.

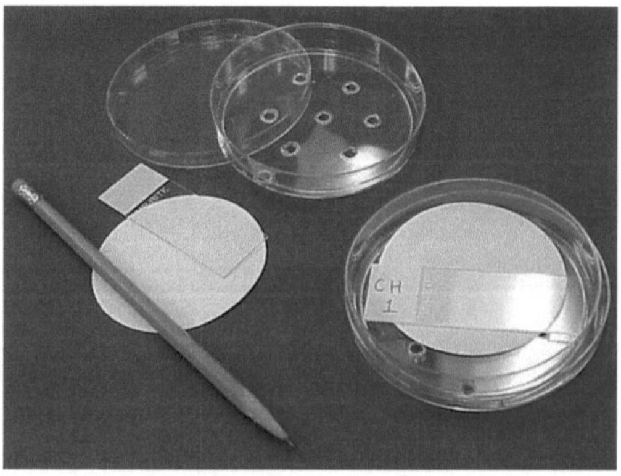

Fig. 2 Components and arrangement of the collodion membrane application apparatus. Reproduced from [2] with permission

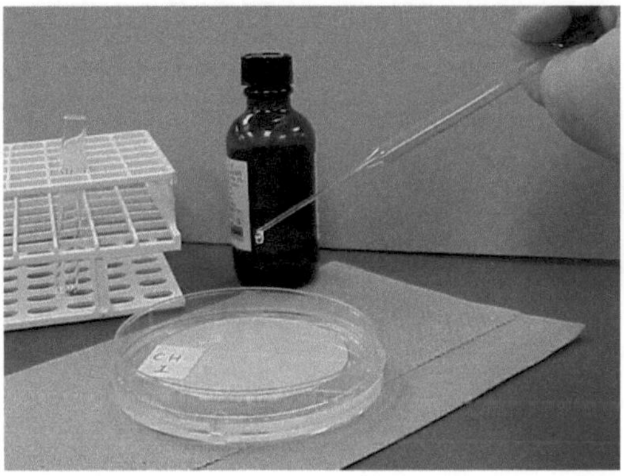

Fig. 3 Dropping the collodion on the water surface. Reproduced from [2] with permission

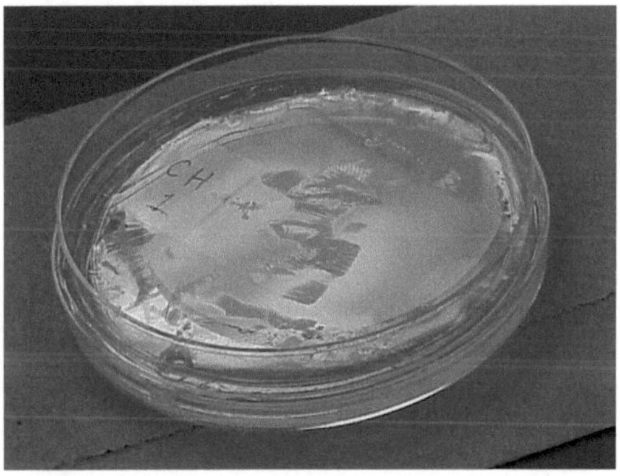

Fig. 4 The floating collodion membrane after the solvent has evaporated. Reproduced from [2] with permission

10. It may be necessary to nudge the membrane so that a smooth area is centered above the slide. To do so, use a wooden stick, and touch the top surface of the membrane near the periphery, not the edge, or the membrane may adhere to the stick.
11. Gently lift the strainer straight up out of the water, keeping it horizontal so that the membrane stays smooth and in place while the water drains out. The film should settle and make contact with the slide and the filter paper underlay (*see* **Note 10**).
12. Lay the strainer on a paper towel, being careful not to disturb the slide. The membrane should cover most of the slide and filter paper.
13. Empty the water from the petri dish lid and invert the lid on a dry area of paper towel.

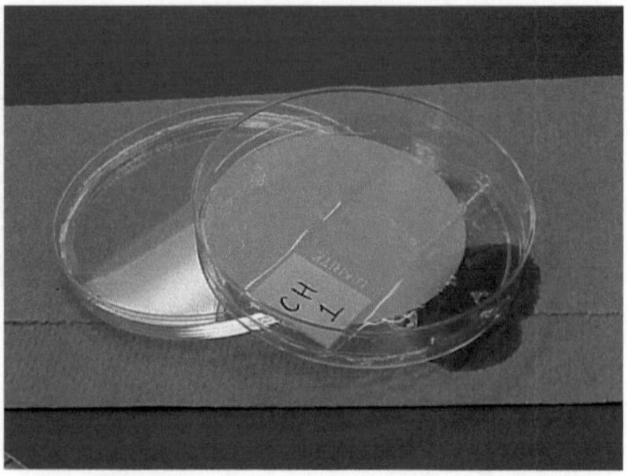

Fig. 5 Draining the freshly prepared slide. Reproduced from [2] with permission

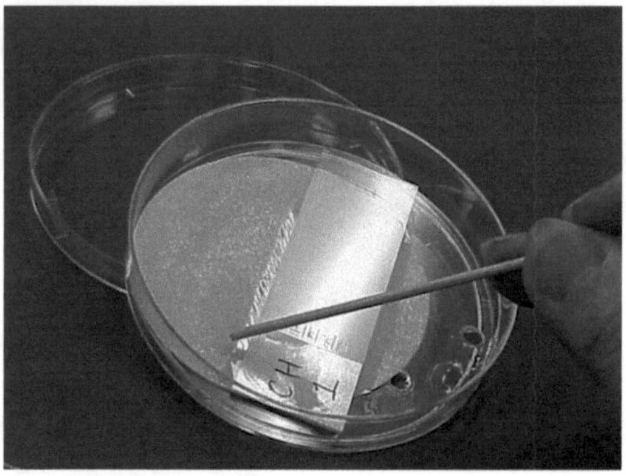

Fig. 6 Removing the slide from the underlying paper. Reproduced from [2] with permission

14. Carefully pick up the strainer with its slide still in position, locate the peripheral hole in the strainer, and prop the assembly against the edge of the lid so that the strainer is slightly tilted and the peripheral hole is in contact with the paper towel, allowing for quick drainage, as in Fig. 5.
15. Air-dry overnight. The filter paper and slide must be completely dry before a biofilm can be collected (*see* **Note 11**).
16. When the filter paper under the collecting slide looks completely dry, carefully run a wooden applicator stick along the edge of the slide, scoring the membrane, so that the slide can be removed from the filter paper without tearing the thin membrane. *See* Fig. 6.
17. The collodion-membraned capture slide is now ready for use. Protect the collodion (upper) surface of the slide until the collection is actually made. The slides may be stored in dry, covered slide boxes or other containers for many months without deleterious effects (*see* **Note 12**).

3.3 Applying Mini-substrates to the Collection Slide

Any or all of a number of "mini-substrates" may be positioned upon a given capture slide after **step 5** Subheading 3.2 above, but before the collodion membrane is formed and applied. After the desired substrates have been positioned, proceed with **step 6** Subheading 3.2 above.

3.3.1 Permanent SEM and AFM Mini-substrates

If SEM and AFM are to be performed on the collected biofilm, small pieces of appropriate examination substrate may be positioned near the end of the slide, leaving space for light microscopy near the center. Keep in mind that the weight and surface area must not put too great a strain on the very thin collodion membrane and cause it to break during biofilm capture.

1. Prepare pieces of clean heavy-duty aluminum foil, glass coverslip, carbon sheeting, or the like no more that about 36 mm^2 in surface area, following guidelines for the equipment to be used. All pieces should be as flat as possible (*see* **Note 13**).
2. Use clean fine-tip forceps (wipe occasionally with ethanol) to position 1–3 mini-substrate pieces upon the immersed slide. The very thin collodion membrane will later sandwich the pieces in place against the glass slide surface. Make sure that there is plenty of contact area for the membrane and slide between multiple mini-substrates, and leave a good space near the center of the slide for regular microscopic examination, as in Fig. 7.
3. Return to **step 6** Subheading 3.2 above.

3.3.2 Permanent TEM Mini-substrate

If TEM is to be done in the usual fashion for whole mounts on coated grids, several clean small-mesh grids may be incorporated into the collection slide.

1. Clean fine-mesh copper grids (size 400 is recommended) immediately before use by placing them in a small glass beaker, adding sufficient 95 % acetone to cover them, and swishing

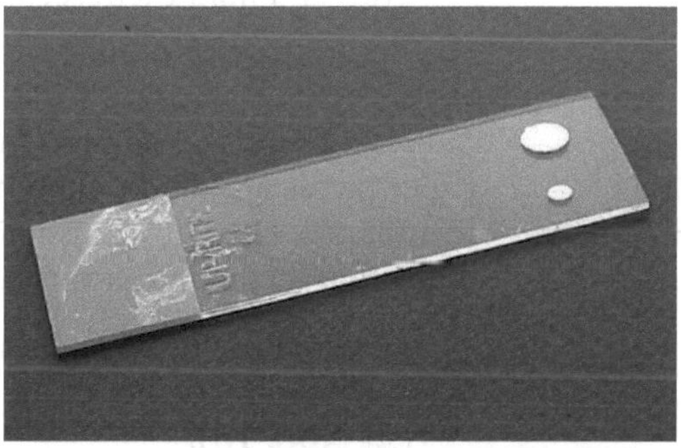

Fig. 7 The appearance of a slide with SEM and AFM mini-substrates sandwiched under the collodion membrane. Reproduced from [2] with permission

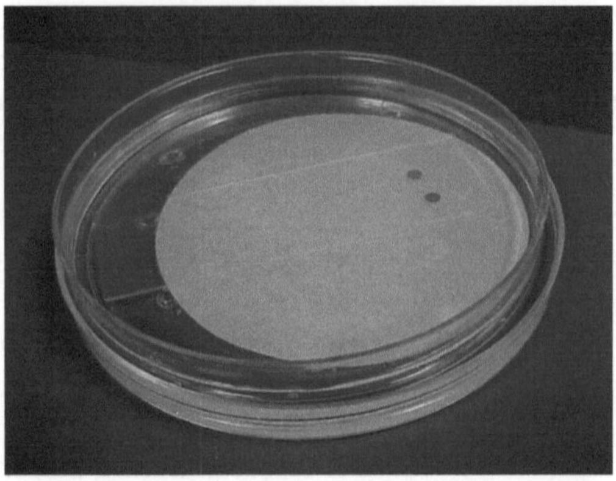

Fig. 8 A capture slide with 2 TEM grids positioned near the end. Reproduced from [2] with permission

them around for several seconds. Decant off the acetone, invert the beaker onto clean absorbent filter paper, and allow the grids to dry and fall onto the paper.

2. Transfer and disperse the cleaned grids onto a regular folded brown paper towel. Use fine-tip forceps to turn them dull side up, so that later you will be able to tell which side has the biofilm on it.
3. Dip the flat end of a wooden applicator stick into clean water, and touch it straight down to the dull surface of a grid on the brown paper towel. The grid should temporarily adhere to the stick (*see* **Note 14**).
4. Plunge the stick straight down through the water surface above the immersed slide to a location near the end. Use the stick to nudge it slightly if necessary. Care must be taken not to disturb the grid once it is in position.
5. Several grids may be positioned on the same slide, but make sure there is enough space between them so that the collodion membrane will be able to bond to the glass in between the grids and hold them securely in place. *See* Fig. 8.
6. Return to **step 6** Subheading 3.2 above.

3.3.3 Removable Collodion/Biofilm Mini-substrates

Subsets of captured biofilms may be removed from the slide if filter paper pieces are sandwiched beneath small areas of the collodion membrane. Thus, it is possible to carry out molecular analyses or TEM thin sectioning of the captured biofilm.

1. Prepare discs or rectangles of Whatman #1 filter paper (preferably sterile) of no more than about 1 cm^2 in surface area (*see* **Note 15**).
2. Use clean forceps to position 1–3 pieces of paper toward the end of the slide, leaving space in between for the membrane to

contact the slide and plenty of space toward the center of the slide for light microscopy.

3. Return to **step 6** Subheading 3.2 above.

3.4 Capturing the AWI Biofilm

Biofilms may be found in nature or grown in the lab (*see* **Notes 16** and **17**). The ideal biofilm for light microscopy observation is almost invisible on the water surface. It may look like a thin oil film with typical refraction colors if the observation angle is correct. Thicker biofilms, although more easily visible to the naked eye, will be difficult to analyze with light microscopy because the microbes and other biofilm constituents are heaped together too thickly to discriminate between individual cells. Other forms of microscopic or molecular analysis will still be possible, however, even on a routinely air-dried slide (*see* **Note 18**).

1. Have on hand a protective storage container that will not contact the captured biofilm (e.g., a slide box or tube), clean water in a container to a depth of about 3 in., and regular lab wipes.
2. Note the label before making the collection.
3. Hold the label end of the slide by hand or with tongs, and in one smooth movement, (a) touch the collodion membrane side to the biofilm as horizontally as you can; (b) immerse the slide completely except for the label, rotating it to a vertical position while doing so; and (c) immediately draw it out of the water vertically. This process should take about 1 s. *See* Figs. 9 and 10.
4. Rinse the slide by completely immersing it (except for the label) into a container of clean water for 1–2 s, and immediately draw it out vertically, as in Fig. 11.
5. Holding the slide by the edges, wipe off ONLY the bottom (the non-labeled side) with a lab wipe, as in Fig. 12. Do not wipe off the captured biofilm on the top!

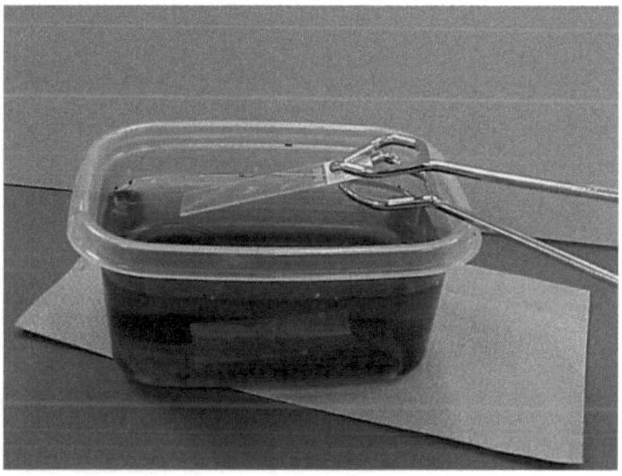

Fig. 9 Touching the capture slide horizontally to the water surface—capturing the AWI biofilm. Reproduced from [2] with permission

Fig. 10 Removing the captured biofilm vertically, without tearing it. Reproduced from [2] with permission

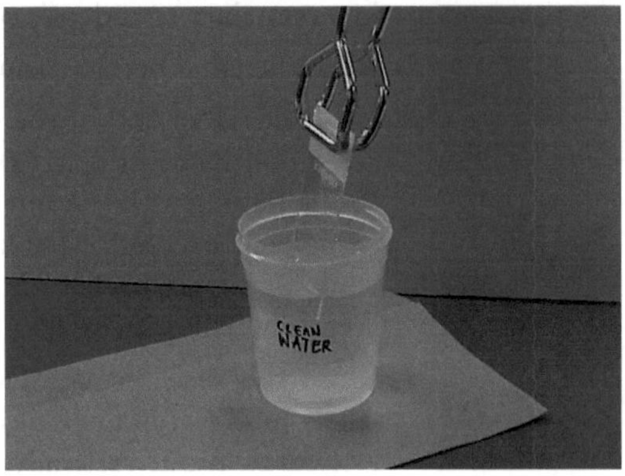

Fig. 11 Rinsing non-AWI microbes away from the biofilm. Reproduced from [2] with permission

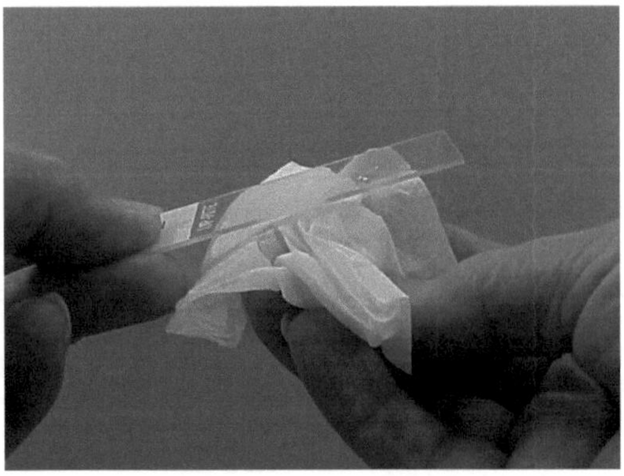

Fig. 12 Removing water and debris from the back of the slide to keep the microscope clean. Reproduced from [2] with permission

6. Place the slide in an airy protective container so that the captured biofilm can air-dry in situ, unless it is to be observed alive. (*see* **Note 18**). The dried biofilm can be stored for days to months in a suitable container which will protect the exposed surface until it is analyzed with microscopy or other methods (*see* **Note 19**).
7. Before any light microscopy is done on the captured biofilm, remove the mini-substrates as directed in the appropriate section below.

3.5 Light Microscopy

The thin AWI biofilm is an optimal preparation for introductory microscopy, and the same slide can be explored with several optical systems, ranging from basic student microscopes to more sophisticated fluorescence optics, and including laser dissection microscopy.

3.5.1 Observing Living Biofilms

Use collections that are no more than about 12 h old to observe living organisms in good condition positioned as they were in situ (*see* **Note 20**). Figure 13 shows a biofilm from nature observed within 2 h of capture.

1. Holding a clean coverslip with tweezers or by the edges, place it on a paper towel. Make sure two are not stuck together.
2. Put a drop of water about the diameter of a pencil eraser (~50 μl) on the center of the coverslip, and then carefully lower the slide, biofilm side down, toward it, without laying the slide completely down on top of the coverslip. Watch instead for the waterdrop to make contact with the slide, and immediately, the coverslip will adhere to the slide, with very few air bubbles.

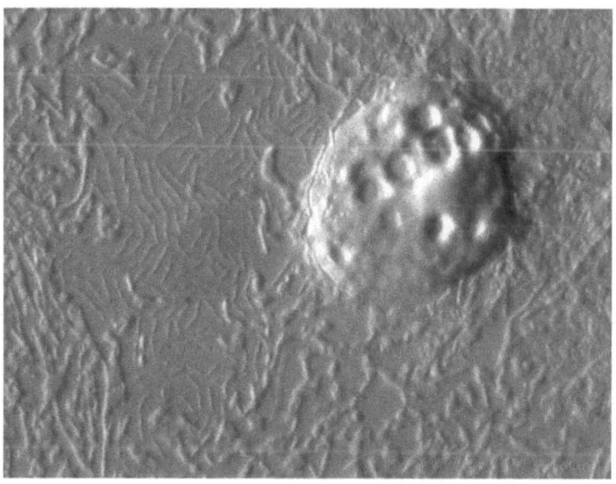

Fig. 13 A live AWI biofilm from nature, showing several varieties of bacteria and an amoeba which may be preying on them. Nomarski differential contrast optics were used

3. Tilt the slide on edge vertically against the paper towel to drain off excess water. Lay the slide coverslip side up on the paper towel to dry off the back of the slide before placing it on the microscope stage.
4. Observe the slide with a properly adjusted microscope (*see* **Note 21**). Notice especially any movement patterns of microbes, any close associations among two or more types of morphologically different microbes, and, in a multilayered biofilm, structural and organismal differences at different focal planes, etc. With time, some microbes may free themselves from the biofilm, sometimes assuming different morphologies.

3.5.2 Gram Staining

This method differs from routine Gram staining mainly in that it does not require the use of heat. Some results may differ slightly from classical results due to components of the biofilm or altered characteristics of microbes living as members of a biofilm community, but the biofilm itself will be arrayed as it was in nature, and additional details may be observed. *See* Fig. 14 for a typical Gram stain.

1. Have on hand the routine Gram stain components, except that no flame is necessary.
2. Place the dried slide, biofilm up, on a paper towel (*see* **Note 22**).
3. Place a flat drop of Crystal Violet stain about 1 cm in diameter on the center of the slide, directly upon a portion of the collected biofilm, avoiding spilling over the side.
4. Stain for about 30 s.
5. Holding the slide by the label end, gently dip it up and down into a cup of clean water for 5–10 s to wash out the excess stain (*see* **Note 23**).

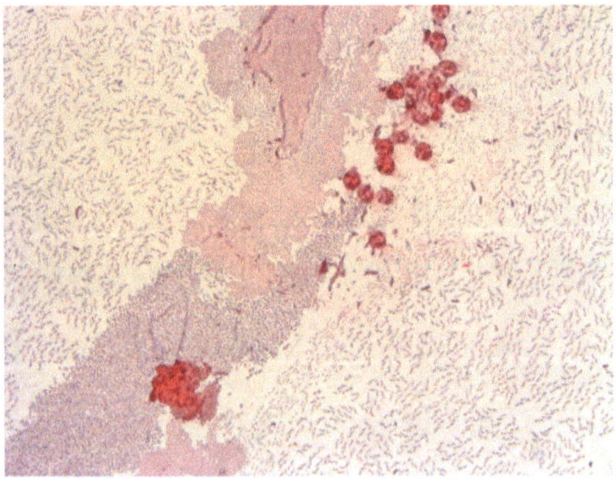

Fig. 14 A Gram-stained natural biofilm. *Larger red microbes* are algae. Copyright © 2004, American Society for Microbiology. All Rights Reserved. Reproduced by permission from Henk 2004. Applied and Environmental Microbiology, Apr. 2004, p. 2486–2493

6. Place a similar-sized drop of Gram's iodine on the same spot as before.
7. Stain for about 60 s.
8. Gently dip the slide up and down in a cup of 95 % ethyl alcohol for 3–5 s to decolorize it. (The same alcohol can be used for several slides.)
9. Dip the slide very briefly into a cup of clean water for about 3–5 s to rinse off the alcohol.
10. Place a flat drop of Safranin stain on the same spot as before.
11. Stain for about 60 s.
12. Dip the slide very briefly into a cup of clean water again to rinse off the alcohol.
13. Remove and tilt the slide on edge vertically against the paper towel to drain off excess water, and then dry off only the bottom of the slide (not the biofilm side) with a lab wipe.
14. Let it dry on the paper towel, biofilm side up, for at least 10 min.
    (a) In some labs, it is preferred that a coverslip be mounted to the slide with an oil drop before observing the Gram stain with the 100× oil immersion lens. In this case, after the slide has dried for at least 10 min, hold a clean coverslip with tweezers or by the edges, and place it on a paper towel. Make sure two are not stuck together. Put a drop of immersion oil about 3 mm in diameter on the center of the coverslip, and then carefully lower the slide, biofilm side down, toward it. When the oil drop makes contact with the slide, the coverslip will adhere to the slide, with very few air bubbles.
15. The AWI biofilm should be readily visible with 10, 20, and 40× objectives, but observation with a 100× oil immersion lens is optimal. Place an oil drop on the top of the slide (or cover slip) using the method prescribed for the particular microscope (*see* **Note 24**).

    Gram-positive bacteria usually stain purple, Gram-negative ones red. However, this is not always the case even under highly controlled conditions (*see* **Note 25**).

3.6 Fluorescence Microscopy

Many fluorescent dyes including immunostains and various indicator stains can be used on captured AWI biofilms.

3.6.1 DAPI Staining

DAPI is a very easy fluorescent dye to use with AWI biofilm collections. Something to keep in mind, however, is that unknown components or features of the biofilm may cause quenching or otherwise change staining efficiency from normal lab expectations. In addition, different organisms may have quite different optimal stain concentrations. Therefore, use of a concentrated stock

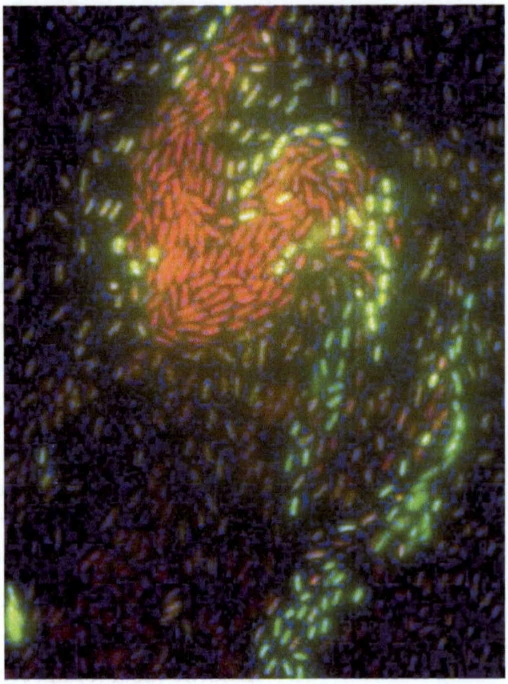

Fig. 15 A natural biofilm captured and stained with fluorescent dyes (DAPI and Molecular Probes Live/Dead BacLight stains) demonstrating the location of DNA in the nucleoids and the viability of the microbes. *Green fluorescence* indicates the bacteria are alive, and *red* indicates dead or deactivated bacteria

solution (10 μg/ml water) is recommended in this diffusion staining technique, especially for preliminary observations.

1. Apply a coverslip to a living or dry biofilm collection slide, using water only. *See* Subheading 3.5.1 **steps 1** and **2** above.
2. Using a micropipettor, apply a 10 μl drop of DAPI stock solution to the edge of the coverslip nearest the end of the slide.
3. Cut a small wedge from filter paper and place the point against the opposite side of the coverslip to draw liquid slowly across between the slide and coverslip via capillary action.
4. Leaving the paper in place, put the slide in a moist chamber, e.g., an inverted petri dish with a moist filter paper in the lid, and leave it in a dim area for about 10 min.
5. Remove the slide, dry the bottom, and observe with UV fluorescence (*see* **Note 26**). Figure 15 shows DAPI staining (nucleoids look blue) and a "live/dead" stain in which living cells appear green and dead or dying ones appear red.

3.6.2 Other Fluorescent Dyes

Many commercially fluorescent dyes (such as immunofluorescent stains or live/dead indicator stains) may be successfully applied to captured AWI biofilm samples. Long incubation periods and several solution changes may cause the collodion membrane to separate from the slide, however, so the prescribed protocols may need to be slightly modified in order to minimize mechanical disturbance. A small flow-through chamber may be devised using a product such as Liquid Blocker–Super PAP Pen (Electron Microscopy Sciences) as per included instructions. The field seen in Fig. 15 was prepared in this way.

3.6.3 Laser Capture for Selected Cell Analysis

Preliminary studies have been done using laser capture microscopes to select and remove single cells from thin captured AWI biofilms or fields of cells from thicker ones for PCR and further analysis. Although the collodion membrane, which is in direct contact with the glass slide, is not the recommended, hydrophilic substrate for laser dissection, selected cells were successfully catapulted into PCR tube receptacles, but in some cases, dissection was incomplete and shreds of collodion anchored the selected cells to the slide. Dissection and capture to an overlain membrane may facilitate the laser capture process. Particular laser capture devices will have varying requirements.

3.7 Use of Mini-substrates

Additional substrates that may have been added during the preparation of the collodion capture slide provide for ways of analyzing collected biofilms by methods other than light microscopy. The collodion (and therefore the captured biofilm) remains attached to the metal substrates that are used for TEM, SEM, and AFM, but it can be removed from paper substrates and processed for chemical treatments, such as TEM fixation and embedding, or for PCR and further molecular analysis.

3.7.1 TEM Grid

Addition of a TEM grid during preparation of the collection slide provides possibilities of many sorts of sample preparation and observation techniques. After the biofilm has dried on the collection slide, use fine-tipped TEM forceps to score around the periphery of the grid and remove it from the slide.

Direct Observation in the TEM

The grid may be observed as is in the TEM to detect high intrinsic contrast of any metallic components of a thin biofilm. Cubic metallic nanoparticles are seen in Fig. 16.

Negative Staining

The grid may be negatively stained [3] to view bacterial and archaeal ultrastructure, including morphological features specific to AWI biofilms, such as flotation morphologies. The method below makes special considerations for the collodion/biofilm sample. Figure 17b shows ends of 2 negative-stained microbes "velcroed" together in a raft-like biofilm.

1. Have on hand 9 cm Whatman #1 filter paper, a 100 mm diameter plastic petri dish, regular matte finish cellophane tape, fine-tip forceps, Parafilm (EMS), a 15 μl micropipettor, negative stain such as 2 % aqueous uranyl acetate (EMS), and distilled water.

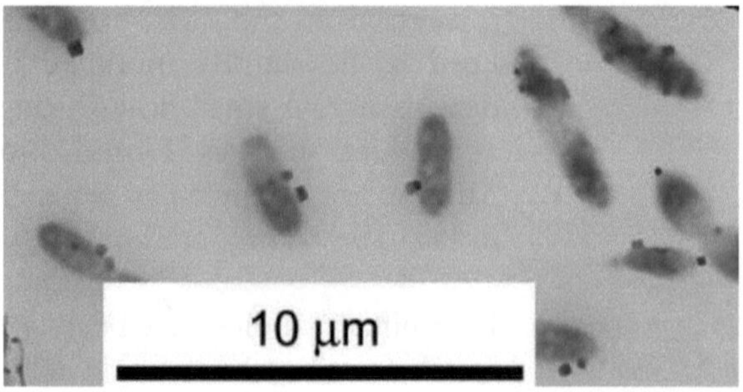

Fig. 16 An unstained TEM preparation from a copper grid mini-substrate. *Tiny dark squares* appear to be metallic crystals that have nucleated on the periphery of this particular bacterium

2. Prepare a storage receptacle. Place an absorbent filter paper circle in an inverted petri dish lid so that it lies flat, and tape down a strip of tape across the center of the paper extending upward against the sides of the lid so that the paper will not fall out. One edge of the tape should lie near the circle diameter. When the plastic dish bottom is inverted over the paper, this upside-down dish will be the protective grid storage receptacle and will provide attachment and labelable sites for a number of grids, as well as drainage for excess liquid.
3. With fine forceps, hold a grid that has been removed from the collection slide so that the biofilm is on top, and lightly touch the edge of the grid to the very edge of the tape, where a little of the adhesive is exposed. The grid should be held in place. Use pencil to label the tape beside each grid.
4. Place a small Parafilm (EMS) square in a protected area.
5. For each biofilm grid, pipette a fresh ~15 μl drop of negative stain onto the Parafilm, and nearby place a similar drop of distilled water.
6. To stain the grid, and without releasing the grid from the forceps, touch the biofilm side to a drop of stain, and immediately touch the edge of the grid vertically to absorbent filter paper to drain off the stain.
7. Continuing to hold the grid, again touch the biofilm side to a clean drop of water, and then drain against the filter paper as before.
8. Return the grid to its position against the tape edge, making sure that the biofilm side is up and that the grid contacts the adhesive and is not drawn back against the forceps. The subtending filter paper should immediately absorb excess liquid.
9. Observe with TEM.

Metal Shadowing

Very fine surface details of bacteria and viruses, or even nanoparticles and molecules, may be seen with TEM when the grid is metal shadowed by stationary low-angle evaporation of gold or tungsten, e.g., each evaporation device will have a protocol, but general methods are outlined in good electron microscopy textbooks [3]. In Fig. 17a, *see* a gold-shadowed bacterium that hangs from its "halo" which floats at the AWI.

Sputter Coating

Since the TEM grid is composed of a good conductor, portions of a captured AWI biofilm may also be easily observed with SEM. After the collodion membrane-supported area between grid bars

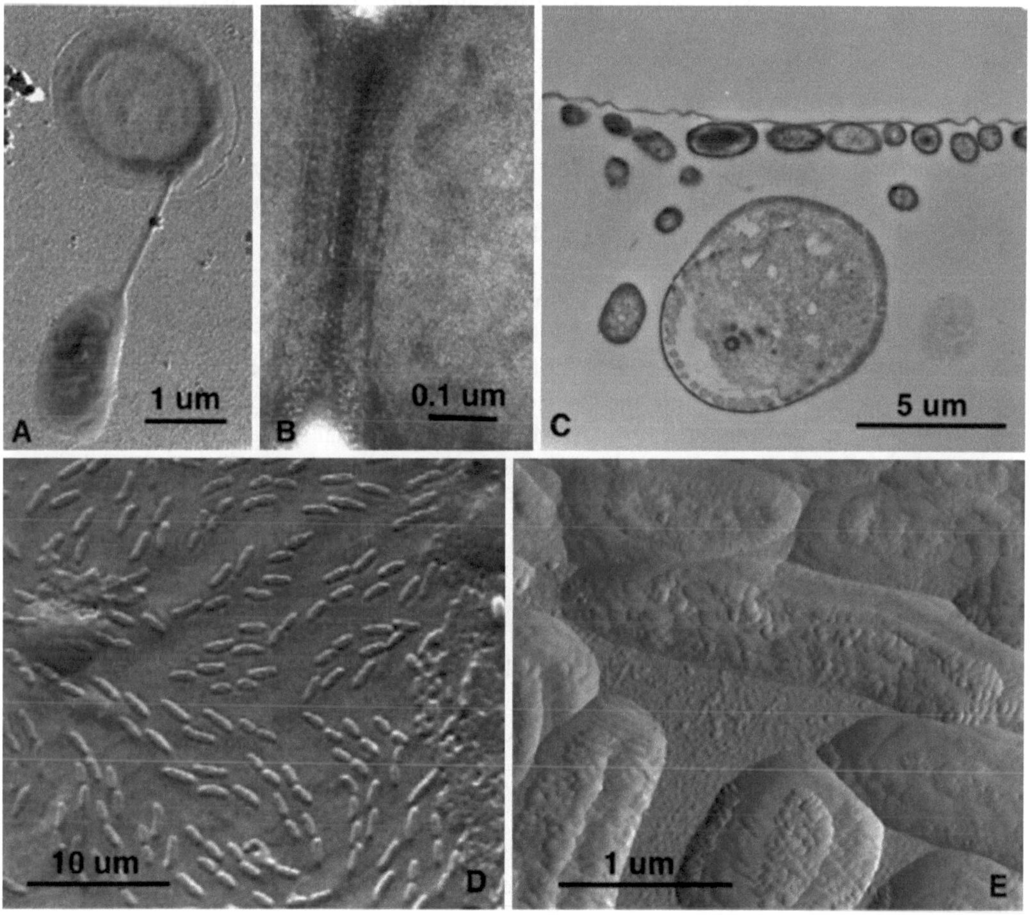

Fig. 17 Preparations made from mini-substrate supported AWI biofilms. (**a**) TEM of a gold-shadowed bacterium with flotation morphology. (**b**) TEM of the close contact between S-layers of adjacent AWI bacteria. (**c**) TEM of embedded and sectioned AWI microbes. The *linear array* is AWI biofilm bacteria. The *large round structure* is a protozoan, perhaps a bacterial predator, with basal body visible. The *thin gray line* at the top is the collodion membrane to which the AWI biofilm is attached. (**d**) A sputter-coated SEM preparation showing the arrangement of bacteria and some particulates at the undersurface of an AWI biofilm. (**e**) An AFM preparation showing details of the bacterial surface. Reproduced by permission from Henk 2004. Applied and Environmental Microbiology, Apr. 2004, p. 2486–2493

has been observed with TEM, the grid may be sputter coated with gold/palladium and the areas underlain by bars can be observed with good SEM contrast. *See* Fig. 17d.

Procedure for sputter coating is instrument dependent, but an overview is outlined in ref. 3. The grid may be attached to the SEM specimen stub with tiny dabs of conductive carbon or silver paint, which may be applied with a splinter formed when an applicator stick is broken in half.

3.7.2 Aluminum Foil, Coverslip Bits, Thin Carbon Wafer Bits

Biofilms underlain by substrates suitable for SEM and AFM, e.g., metal, carbon, and glass, may be suitable for observation and analysis with these instruments. Choices for appropriate substrates would depend on the particular requirements of the equipment. As with metal TEM grids, fine-tip forceps are used to score around and remove the mini-substrates from the collection slide, and they may be stored in a modified plastic petri dish storage container as in **step 2** in 3.6.2 above.

Sputter Coating for SEM Mini-substrates

See 3.7.1 "Sputter Coating" above. The relatively larger details visible with SEM include cell arrangement and appendages comprising the biofilm, as in Fig. 17d.

Preparation for AFM

Sample preparation of the captured AWI biofilm is instrument and purpose specific. Details visible with AFM would be on the nanoscale and could include atomic analysis of organic and inorganic biofilm components. Figure 17e shows a relatively low-magnification AFM image of bacterial surfaces that were captured on a carbon mini-substrate as simply air-dried.

3.7.3 Removable Biofilm Samples

If a portion of the collected biofilm needs to be removed from the capture slide for chemical or other analysis, a filter paper mini-substrate should have been applied when the collection slide was being prepared. Biofilm released from the collection slide can be chemically fixed, embedded, and thin sectioned for TEM as in Fig. 17c or processed for molecular analysis such as PCR and subsequent rDNA DGGE, for example. *See* Fig. 18d.

1. Before removing the filter paper mini-substrate, a small dot may be drawn on the upper surface of the paper-underlain biofilm with an indelible marker in order to see where the biofilm is after it is released from the paper.
2. As with the previous examples of mini-substrates, to remove the filter paper pieces from the dry biofilm collection on the slide, score around the periphery with fine-tip forceps (*see* **Note 27**).
3. A dry filter paper piece for molecular analysis may be stored dry in a regular labeled plastic microtube.

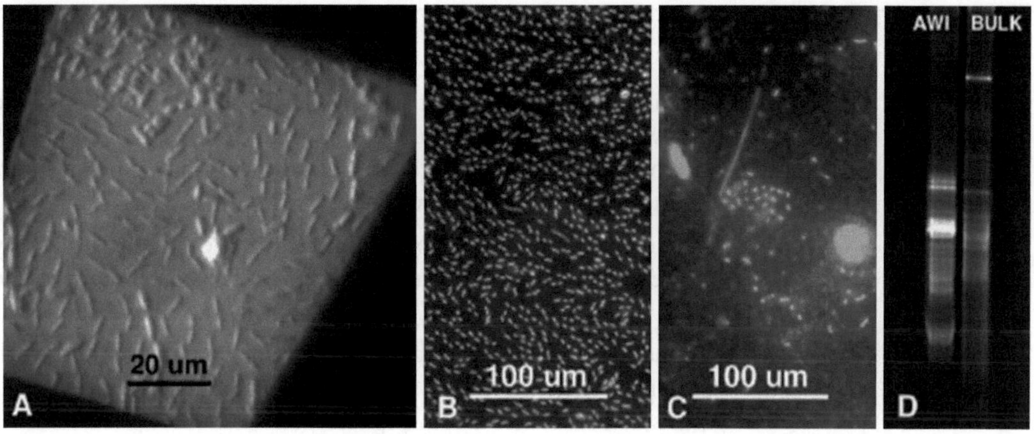

Fig. 18 Comparison of AWI and bulk characteristics from the same collection site. (**a**) The appearance of the captured natural AWI biofilm as seen by Nomarski light optics. (**b**) DAPI staining of that AWI biofilm. (**c**) DAPI staining of concentrated microbes from the water subtending the captured AWI. (**d**) Side-by-side comparison of DGGE of rDNA populations collected from the AWI organisms and from bulk water organisms sampled from water directly beneath the AWI. Indications are that two distinct communities exist in the same body of water. Copyright © 2004, American Society for Microbiology. All Rights Reserved. Reproduced by permission from Henk 2004. Applied and Environmental Microbiology, Apr. 2004, p. 2486–2493

4. When ready to remove the captured AWI biofilm from the filter paper, in a well-lighted area, fill a small clear container (such as a glass vial) to a depth of about 1 cm with sterile distilled water or the desired processing solution.
5. If the processing solution is aquatic and allows for good visualization of the released biofilm, the filter paper/biofilm may be added directly to the solution. When the collodion/biofilm is released, the filter paper can be removed and discarded.
   (a) If the processing solution is not of a suitable volume, or otherwise inappropriate, place the filter paper piece in water and carefully watch to see the collodion/biofilm membrane float away from the paper. Use clean fine-tip forceps quickly to remove the biofilm, immersing it into the appropriate solution for further processing for the particular purpose (*see* **Note 28**).

3.8 Comparison of AWI and Subsurface Communities

After an AWI biofilm has been collected on a collodion capture slide, a sample of the bulk water may be collected as well. Simple to complex microscopy and molecular analyses can be carried out on both samples using standard methods for the bulk sample and the above-described methodologies for the AWI sample. The results often yield striking differences between the two communities. *See* Fig. 18a–d.

4 Notes

1. Each strainer and its petri dish lid will accommodate 1 slide and will be in use for about 24 h, so make enough for the number of slides you will need on the following day. The strainers and their petri dish lid are reusable after 24 h.
2. The melted plastic will form a raised rim around the hole. Try to keep this rim to a minimum to avoid displacement of the slide later on.
3. It may be convenient to use cafeteria-type trays lined with paper towels, which will each hold six strainer assemblies and will stack crosswise to save space.
4. Super Up-Rite slides are suitable straight from the box; others may require special cleaning or the collodion membrane will not adhere properly.
5. You may carefully make a very small dot with a permanent marker near the center of the good side so that later the microscope may be focused there at high magnifications, very close to the plane of an almost invisible thin biofilm.
6. Thick rims of melted plastic around the strainer holes will elevate the filter paper and slide, perhaps preventing total immersion of the slide. Discard this strainer.
7. Locate a resting place for the pipette so that no liquid gets into the bulb while it is used.
8. If it does not, the dish assembly may have oily spots on its surface. It should be washed thoroughly with regular lab detergent, rinsed well, and dried thoroughly before use. Washing will usually be necessary after three to four uses.
9. The solvent smells something like bananas and has been used as a flavoring agent. However, avoid excessive inhalation. The ventilation in a regular laboratory area should be suitable.
10. While it is still wet, the membrane can slide around easily; do not rush.
11. A 60 °C drying oven may be used if the filter paper is not completely dry in appearance.
12. *See* ref. 2 for directions for making simple field collection kits and protective containers.
13. Appropriately sized discs may be prepared in many cases by using a sharp paper hole punch. Avoid burrs or wrinkles.
14. Grids have a tendency to be caught in the surface tension and to float on the water surface or to adhere to water on forceps, so avoid using forceps for this procedure.
15. A sharp paper hole punch may provide suitable discs. If desired, shapes may be coded for different samples or purposes.

16. *See* ref. 2 for instructions on growing "natural" AWI biofilms in the lab.
17. If known microbes are cultured in the lab for use with this method, access to the surface area must be taken into consideration. Culture tubes must be wide enough to accommodate a microscope slide and, for best results, should be maintained on a slant so that the slide can capture the surface biofilm from a horizontal angle. Cultures may also be grown in wide mouth containers that are deep enough to accommodate the vertical length of the slide.
18. Captured biofilm organisms may be kept alive on the slide for a maximum of about 12 h if the slide is maintained in a sealed chamber such as a lidded specimen cup or a 50 cc capped plastic centrifuge tube. The living sample must be removed from the paper mini-substrate as soon as possible for good sample preservation if TEM fixation, embedding, and sectioning are to be done.
19. It may be useful to collect a regular aquatic sample from the collection site for comparison or additional analysis.
20. Some stains kill microorganisms, so if microbial activities such as cell division, motility, predation, or other microbial interactions are to be observed, it is best to avoid stains and use some sort of differential interference contrast microscopy, e.g., Nomarski optics. However, a microscope with phase-contrast optics or even a basic light microscope with a properly adjusted condenser lens will also be useful.
21. If you placed a small mark in the center of the slide in **step 3** Subheading 3.2, move that dot into the light spot and bring the dot into focus. The biofilm will not be precisely at that plane, but a very slight adjustment of the fine focus knob then should bring the thin biofilm into focus.
22. If desired, a circle about 1 cm in diameter can be drawn on the back of the slide with a marker to designate the area that will be stained with the repeated additions of stains.
23. In order only to visualize AWI biofilm microbes, especially with student microscopes, the staining process may be stopped here. The slide should be dried for at least 10 min and then a coverslip mounted with water (as in Subheading 3.5.1 **steps 1** and **2**). Quite reasonable observations can be made with the 20× and 40× dry objectives.
24. *See* ref. 2 for student microscope tips.
25. Archaea may also be found in AWI biofilms. A simple negative-stain TEM preparation (Fig. 17b) from the same collection slide may help visualize the crystalline proteins characteristic of the archaeal S-layer which comprises the archaeal cell wall.

26. There will probably be a continuum of staining intensity: too concentrated near the edge of the coverslip where the DAPI was applied and too dilute at the far side. Somewhere in the middle, however, various optima will be seen for the various microbes. In general, the optima will show bluish-white DNA fluorescence at bacterial nucleoids as well as at eukaryote nuclei and sometimes at cilia basal bodies. Engulfed bacterial nucleoids may sometimes be seen within protozoan food vacuoles. Where DAPI concentration is too great, nuclear disruption often produces yellow fluorescence. If DAPI concentration is below optimal, nucleoid DNA may emit a dim dark blue. Some red chlorophyll fluorescence is also visible at UV excitation wavelengths. Other biofilm components, e.g., oils and various mineral particles, may also autofluoresce at UV excitation wavelengths to produce assorted emission colors.
27. If TEM fixation is to be done, do not let the collected biofilm dry completely. Have the fixation chemicals ready in advance, and remove the filter paper/biofilm directly to the fixative without disturbing the remaining biofilm collection, which should then be left to dry securely on the glass slide. The filter paper can be fished out of the fixation vial later in the process.
28. The presence of the collodion membrane has not been seen to be detrimental in molecular analyses and is beneficial in TEM fixation processing.

Acknowledgement

This work was supported by the Department of Biological Sciences at Louisiana State University. Thanks to many students, starting with Bret Territo.

References

1. Henk MC (2004) Method for collecting air-water interface microbes suitable for subsequent microscopy and molecular analysis in both research and teaching laboratories. Appl Environ Microbiol 70:2486–2493
2. Henk MC (2002) Microscopy of natural and experimental biofilms in the undergraduate laboratory: a simple collection kit with multiple applications. In: 2002 Proceedings of the association for biology laboratory education. http://www.ableweb.org/volumes/vol-24/5-henk.pdf
3. Bozzola JJ, Russell LD (1992) Electron microscopy principles and techniques for biologists, 2nd edn. Jones and Bartlett, Sudbury, MA

Chapter 23

Biofilm-Growing Bacteria Involved in the Corrosion of Concrete Wastewater Pipes: Protocols for Comparative Metagenomic Analyses

Vicente Gomez-Alvarez

Abstract

Advances in high-throughput next-generation sequencing (NGS) technology for direct sequencing of environmental DNA (i.e., shotgun metagenomics) are transforming the field of microbiology. NGS technologies are now regularly being applied in comparative metagenomic studies, which provide the data for functional annotations, taxonomic comparisons, community profile, and metabolic reconstructions. For example, comparative metagenomic analysis of corroded pipes unveiled novel insights on the bacterial populations associated with the sulfur and nitrogen cycle, which may be directly or indirectly implicated in concrete wastewater pipe corrosion. The objective of this chapter is to describe the steps involved in the taxonomic and functional analysis of metagenome datasets from biofilm involved in microbial-induced concrete corrosion (MICC).

Key words Metagenome, 454 Pyrosequencing, Next-generation sequencing (NGS), Metabolic pathways, Function annotation, Taxonomic classification

1 Introduction

Microbial-induced concrete corrosion (MICC) is a significant cause of deterioration and premature failure for the estimated 800,000 miles of wastewater collection infrastructure [1]. Failure to adequately address the deteriorating infrastructure networks threatens our environment, public health, and safety and in particular represents a significant economic burden to local governments due to the maintenance and replacement costs [2]. Metabolic processes within the sulfur biogeochemical cycle are known to play an important role in MICC [3–5]. In wastewater systems, these processes include the formation of sulfide, volatilization of hydrogen sulfide (H_2S), biological oxidation of sulfide, and precipitation of metal sulfides (Fig. 1). The primary source of sulfur is sulfate (SO_4^{2-}) which can be reduced by sulfate-reducing bacteria (SRB)

Gianfranco Donelli (ed.), *Microbial Biofilms: Methods and Protocols*, Methods in Molecular Biology, vol. 1147,
DOI 10.1007/978-1-4939-0467-9_23,

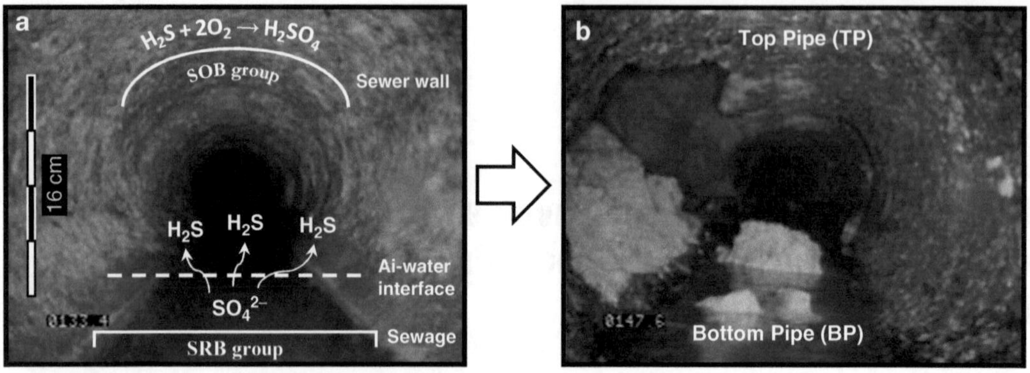

Fig. 1 The corrosion process occurs by the reaction of the biogenic sulfuric acid (H_2SO_4) with the cementitious material of the concrete. (**a**) In wastewater, sulfate ($SO4_2^-$) is reduced by sulfate-reducing bacteria (SRB) to sulfide (H_2S). H_2S is transferred across the air-water interface to the sewer atmosphere where sulfur/sulfide-oxidizing bacteria (SOB) on the pipe surface oxidize the H_2S to H_2SO_4 and (**b**) results in the eventual structural failure of sewers. Photo courtesy of the Metropolitan Sewer District of Greater Cincinnati

to H_2S under anaerobic conditions. H_2S is transferred across the air-water interface to the sewer atmosphere and is converted on the pipe surface to biogenic sulfuric acid (H_2SO_4) by sulfide-oxidizing bacteria (SOB). The corrosion process occurs by the reaction of H_2SO_4 with the cementitious material of the concrete and results in the eventual structural failure of sewers (Fig. 1).

Capillary or "Sanger" sequencing of 16S rRNA and functional genes has been conducted to characterize the microbial community of wastewater systems [6–10]. Although those approaches provided useful but limited information, they still may not integrally reflect the whole metabolic processes and functional capabilities of microbial communities. Since then, development of high-throughput next-generation sequencing (NGS) technology has fundamentally revolutionized microbiology with a powerful tool for metagenomic studies [11, 12]. NGS technologies provide the capability to sequencing millions of different molecules at once, hence the direct genetic analysis of genomes (metagenome) contained in a particular environmental samples [11–14]. Commercially available NGS platforms (454, Illumina, SOLiD, Ion Torrent, and PacBio) utilize diverse chemistry and base detection tools [15]. Each instrument performed well with respect to maximize sequence reads; however, the quality of the data and potential sources of errors, including sequencing artifacts, varies by model. Therefore, the choice of a particular NGS platform should be carefully considered.

Regardless of the sequencing technology used to generate metagenomes, there are many challenges involving the analysis of the data, including a considerable amount of data and the computational resources needed for complex analyses. Recently, cyber infrastructure resources (Table 1) have been developed for

Table 1
Publicly available open source tools for the analyses of metagenomic data

Tool	Description	URL	References
MG-RAST	Fully automated service for quality control, annotation, and analysis for metagenomic shotgun samples	http://metagenomics.anl.gov	[16]
CAMERA	Platform for depositing, analyzing, visualizing, and sharing metagenomic data	http://camera.calit2.net	[17]
IMG/M	Community resource for analysis and annotation of metagenome datasets and microbial genomes	http://img.jgi.doe.gov/cgi-bin/m/main.cgi	[18]
WebCARMA	Taxonomic and functional classification of unassembled short reads from metagenomic communities	http://webcarma.cebitec.uni-bielefeld.de/cgi-bin/webcarma.cgi	[19]
MEGAN	Integrated approach to the taxonomic and functional analysis of metagenomic data	http://www-ab.informatik.uni-tuebingen.de/software/megan	[20]
Cytoscape	Platform for visualizing interaction networks/biological pathways and integrating with annotations and metadata	http://www.cytoscape.org	[21]
STAMP	Software for analyzing metagenomic profiles and graphical environment for performing statistical analyses	http://kiwi.cs.dal.ca/Software/STAMP	[22]

researchers interested in analyzing and depositing metagenome datasets [16–22]. Still, the interpretation of sequence data (i.e., protein function) by an open source system currently relies on homology searches against annotated sequence databases. Nevertheless, these resources provide the basis for functional annotation, taxonomic comparison, phylogenomic profile, and metabolic reconstruction [14].

Analysis of metagenomes provides access to the functional gene composition of microbial communities and when appropriately applied could provide a better insight of the microbial structure in response to environmental factors or biotic interactions [13, 14]. The goal is to generate functional profiles and comparisons across multiple metagenomes to identify functional differences between microbial communities [23]. This chapter gives an overview in the study of microbial communities in concrete wastewater biofilms, with particular emphasis on the steps involved in the functional analysis of metagenomes (Fig. 2). Our approach [24] utilized the 454 FLX Titanium platform (Roche, 454 Life Sciences, Branford, CT) that generated an average of 350–450 bp (after trimming) and offered multiplexing, allowing for multiple sample analysis (i.e., barcoding) in a single run.

2 Materials

1. Computer with internet connection and internet browser (*see* **Note 1**).
2. High capacity data storage.
3. Biological sequence editor software.
4. Image analysis software.
5. A spreadsheet program for organization and analysis of data in tabular form.
6. Software(s) able to perform statistical data analysis.

3 Methods

3.1 Sampling Biofilms

1. Collect biofilm samples from predefined sections of a corroded concrete sewer pipe (*see* **Note 2**).
2. Remove the biomass from the surface with sterile metal spatula by scraping the surface area (*see* **Note 3**).
3. Transfer the biomass into a sterile tube with an additional sterile metal spatula.
4. Transport on ice and stored at −20 °C until DNA extraction (*see* **Note 4**).

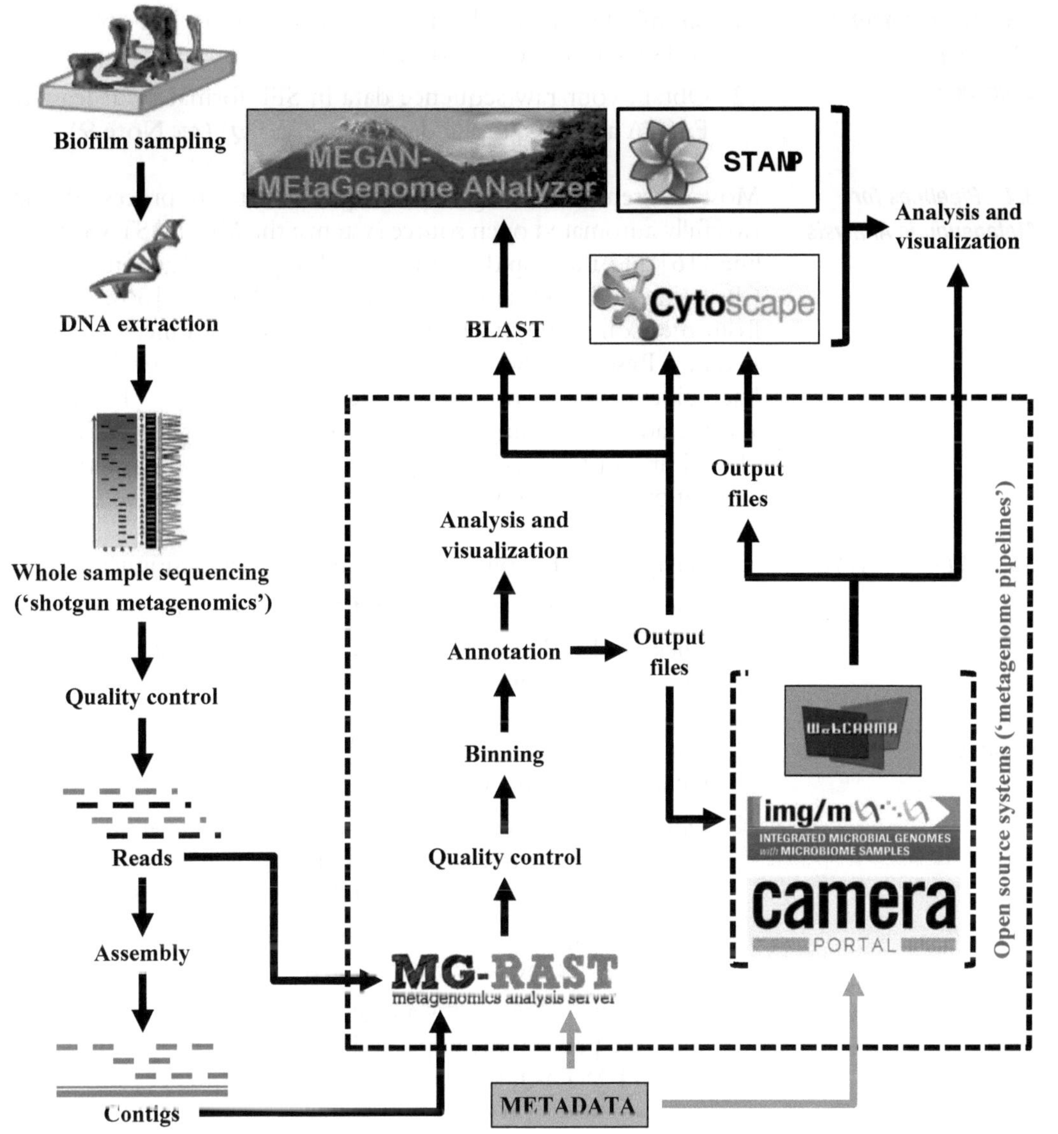

Fig. 2 Schematic overview of the major steps of a comparative biofilm metagenomic study. A list and description of open source system for comparative analysis of metagenomes is provided in Table 1

3.2 Extraction of Total DNA from Biofilms

1. Transfer the recommended sample size of biofilm to a new sterile tube (*see* **Note 5**).
2. Extract total DNA with the appropriate protocol designed by the investigator (*see* **Note 6**).
3. Measure the concentration of the genomic extract with a Quant-iT™ PicoGreen® dsDNA Assay Kit assay (Invitrogen™, Molecular Probes®, Inc, Eugene, OR) (*see* **Note 7**).

3.3 Environmental DNA Direct Sequencing

1. Submit DNA samples for sequencing using the appropriate NGS technology (*see* **Note 8**).
2. Obtain your raw sequence data in SFF format or at least the FASTA and QUAL file data (i.e., FASTQ) (*see* **Note 9**).

3.4 Pipelines for Metagenomic Analysis

Most unassembled metagenome sequence data are processed using two fully automated open source systems: the MG-RAST v3.0 pipeline [16] and the Rapid Analysis of Multiple Metagenomes with a Clustering and Annotation Pipeline (RAMMCAP) [30], available from the Community Cyberinfrastructure for Advanced Microbial Ecology Research and Analysis (CAMERA) [17] (*see* **Note 10**). Each pipeline (i.e., workflow) includes multiple steps: (1) sequence and metadata uploading; (2) quality control; (3) gene calling, i.e., identification of protein coding regions; (4) protein clustering; (5) annotation against known databases; and (6) comparative analysis.

3.4.1 MG-RAST Pipeline

1. Navigate to the following link: http://metagenomics.anl.gov.
2. Log in or create a new account for the MG-RAST server.
3. Click on the "Up Arrow" (data upload) link to start uploading data to the server.

Prepare Data

1. Use the "Prepare Data" section to upload any sequence and metadata files into your inbox.
2. Optional: download the blank metadata template and fill out the 12 mandatory terms (ftp://ftp.metagenomics.anl.gov/data/misc/metadata/MGRAST_MetaData_template_1.4.xlsx).
3. Upload sequence, barcode (if necessary), and metadata (optional) files to your inbox folder (*see* **Note 11**).
4. Use the menu entry "Manage Inbox" to calculate sequence and base pair count, unpacked compressed files, convert SFF files to FASTAQ, and demultiplex your sequence file.

Data Submission

1. Optional: select a spreadsheet with metadata for the project you want to submit (*see* **Note 12**).
2. Create a project to upload a job to the MG-RAST server.
3. Select the sequence file(s) from your inbox to be included in your project.
4. Specify the pipeline options for the analysis of your sequences (*see* **Note 13**).
5. Submit data to the MG-RAST pipeline. An accession numbers (MG-RAST ID's) will be automatically assigned to your datasets.

MG-RAST: Analysis Modules

1. Click on the "Earth" (browse) link to access the "Metagenome Project List" page.
2. Select and inspect the sequencing quality of the uploaded metagenome in the "Metagenome Overview" page (*see* **Note 14**).

Taxonomy: 16S rRNA-Based Phylogeny

1. Download the FASTA-formatted file containing the predicted ribosomal sequences via the menu entry "Metagenome Overview" → "Metagenome Download" (*see* **Note 15**).
2. Retrieve the 16S rRNA gene sequences (*see* **Note 16**).
3. Re-examine the 16S rRNA gene sequences using a database of known 16S rRNA gene sequences.
4. Construct a phylogenetic tree following the methodology described elsewhere [10, 24, 35].

Taxonomy: Protein-Derived Annotation

1. Taxonomic analyses of protein coding sequences can be accessed via the menu entry "Metagenome Overview" → "Analyze" → "Best Hit Classification" (*see* **Note 17**).
2. Select an expected *e*-value cutoff of $1e^{-05}$.
3. Download the taxonomic classification count tables (*see* **Note 18**).
4. Open table with a spreadsheet program (e.g., EXCEL).

Functional Role: Protein Annotation

1. Functional analyses of protein coding sequences can be accessed via the menu entry "Metagenome Overview" → "An alyze" → "Hierarchical Classification" (*see* **Note 19**).
2. Select an expected *e*-value cutoff of $1e^{-05}$.
3. Download the functional annotation count tables (*see* **Note 18**).
4. Open table with a spreadsheet program (e.g., EXCEL).
5. Identify proteins and assign their respective enzyme commission number (EC).
6. Assign reads to gene families for metabolic pathways reconstruction (Fig. 3a).

Viewing Metabolic Pathway: KEGG Map

1. Explore the whole metabolic pathway using the KEGG Mapper, an internal tool of MG-RAST based on the KEGG pathway mapping system [38].
2. The KEGG Mapper can be accessed via the menu entry "Metagenome Overview" → "Analyze" → "Hierarchical Classification" → "open KEGG Mapper".

Output Files

1. Download the nucleotide sequence FASTA-formatted file containing the predicted coding regions via the menu entry "Metagenome Overview" → "Metagenome Download".

3.4.2 CAMERA Pipeline

1. Navigate to the following link: https://portal.camera.calit2.net/gridsphere/gridsphere.
2. Log in or create a new account for the CAMERA server.
3. Click on the "Workflows" link, select "Metagenomic Data Annotation and Clustering" in the workflow list, and press "Start" to start uploading data to the server.

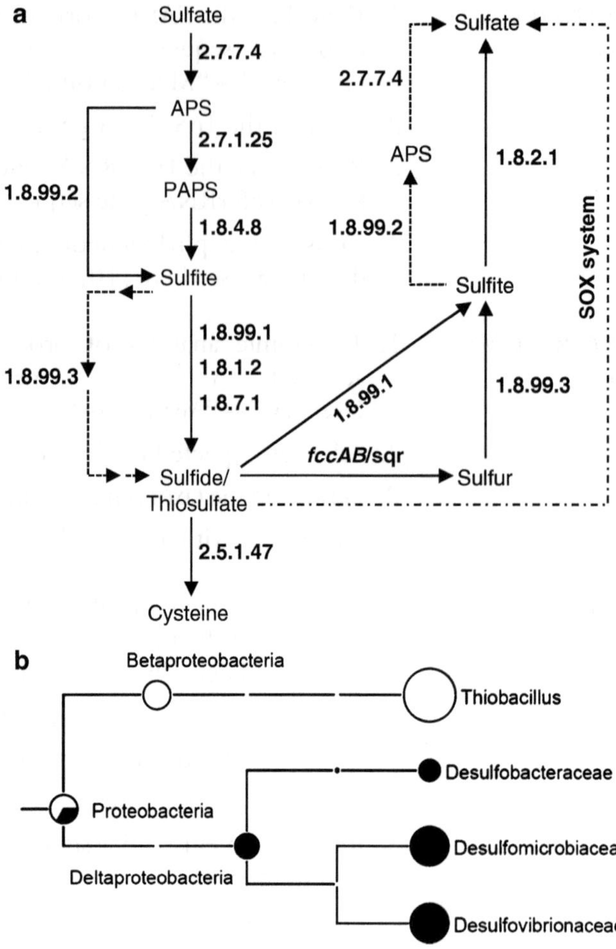

Fig. 3 Sample outputs from a descriptive analysis of the sulfur metabolic pathway. (**a**) Diagram with identified components (i.e., enzyme commission number) of the metabolic pathway. (**b**) Classification and relative abundance of taxonomic groups computed by MEGAN [20] of sulfur-related proteins. *Each circle* is scaled logarithmically to represent the number of reads that were assigned to each taxonomic group. Pipe sections: top (*white*) and bottom (*black*). Adapted from a comparative metagenomic study of a corroded concrete pipe [24]

Data Submission

1. Type a project name.
2. Specify the pipeline options for the analysis of your sequences (*see* **Note 20**).
3. Upload your FASTA-formatted sequence file (*see* "Output Files").
4. Submit data to the RAMMCAP pipeline (*see* **Note 21**).

CAMERA: Analysis Modules

1. Click on the "Workflow Status/Results" link to access the RAMMCAP Workflow Results Summary page.

Functional Role: Protein Annotation

1. Download the compressed ZIP file containing the functional annotation count tables (*see* **Note 18**).
2. Open the tables (Pfam, TIGRfam, and COG) with a spreadsheet program (e.g., EXCEL).
3. Identify proteins and assign their respective enzyme commission number (EC).
4. Assign reads to gene families for metabolic pathways reconstruction (Fig. 3a).

Output Files

1. Download the nucleotide sequence FASTA-formatted file containing the predicted coding regions.

3.5 Data Normalization

1. Normalize the counts (relative abundance) of each downloaded table from MG-RAST and RAMMCAP against the total number of hits in their respective database (e.g., SEED, COG, etc.) using effective sequence counts (ESC). ESC is a composite measure of sequence number and average genome size (AGS) of the metagenome [39].

$$\mathrm{ESC} = N_s\left(\frac{G}{G_s}\right)$$

where N is the total gene count in the x dataset, G is the average of G_s values across all samples (in Mb), and G_s is the average genome size (AGS) in the x dataset. AGS is the number of megabases per genome present in the sample [40]:

$$\mathrm{AGS}(\text{in Mb}) = \frac{a + b \times L^{-c}}{x}$$

where x is the marker gene density (i.e., # of matches to marker genes divided by the total # of sequenced base pairs [megabases surveyed] or count per megabase [Mb]), L is the read length, and the constant parameters a=18.26, b=3,650, and c=0.733 were chosen such as to minimize the weighted sum of squares [40, 41] (*see* **Note 22**).

3.6 Comparative Metagenomic Analysis

3.6.1 Integrative Taxonomic Analysis of Functional Genes

1. Identify sequences assigned to functional groups (e.g., sulfur and nitrogen) from MG-RAST and RAMMCAP output table files (*see* **Note 23**).
2. Retrieve sequences from the nucleotide sequence FASTA-formatted file (*see* "Output Files" sections).
3. Analyze selected functional gene sequences with BLASTX against the NCBI non-redundant protein sequence (nr) database using the CAMERA server.
4. Import the BLASTX results into the software MEGAN [20] for assignment and comparison of taxonomic groups [42] (*see* **Note 24**).

5. Generate a dendrogram (i.e., tree representation of the NCBI taxonomy) to display the taxonomic affiliation of selected sequences via the menu entry "Compare" (Fig. 3b).

3.6.2 Statistical Data Analysis

Univariate Analyses

Richness and Diversity of Functional Genes

1. Calculate rarefaction curves for richness (e.g., ChaoI, ACE) and diversity (e.g., Shannon, Simpson indices) estimators (*see* **Note 25**).

Proportion of Functional Genes

1. Calculate and compare the proportion of each functional group or specific genes (*see* **Note 26**).
2. Identify functional groups with statistically significant difference between metagenomes (*see* **Note 27**) (Fig. 4a).
3. Calculate the proportion of genomes in the metagenome that is capable of that function (*see* **Note 28**).

Enrichment of Functional Genes

1. Calculate the odds ratio or the relative risk of observing a given group in the sample relative to the comparison dataset [43]:

$$\text{Odd ratio} = \left(\frac{A}{B} \Big/ \frac{C}{D} \right)$$

where A is the number of hits to a given category in the x dataset, B is the number of hits to all other categories in the x metagenome, C is the number of hits to a given category in the y dataset, and D is the number of hits to all other categories in the y dataset.

2. An odds ratio of 1 indicates that the community DNA has the same proportion of hits to a given category as the comparison dataset (Fig. 4b).

Multivariate Analyses

Data Transformation and Distance Matrix

1. Log(x+1)-transform the counts of each downloaded table (*see* **Note 29**).
2. Create a distance matrix using the Bray-Curtis similarity coefficient of the transformed data (*see* **Note 30**).

Exploratory Data Analysis

1. Apply hierarchical clustering (e.g., Unweighted Pair Group Method with Arithmetic Mean [UPGMA]) and reduction (e.g., nonmetric multidimensional scaling [nMDS]) methods to reduce and visualize the complex collection of data points.
2. Explore the relationship of metagenomes using nMDS ordination plots and cluster dendrograms.
3. Test the robustness of the ordination plot and dendrogram (*see* **Note 31**).

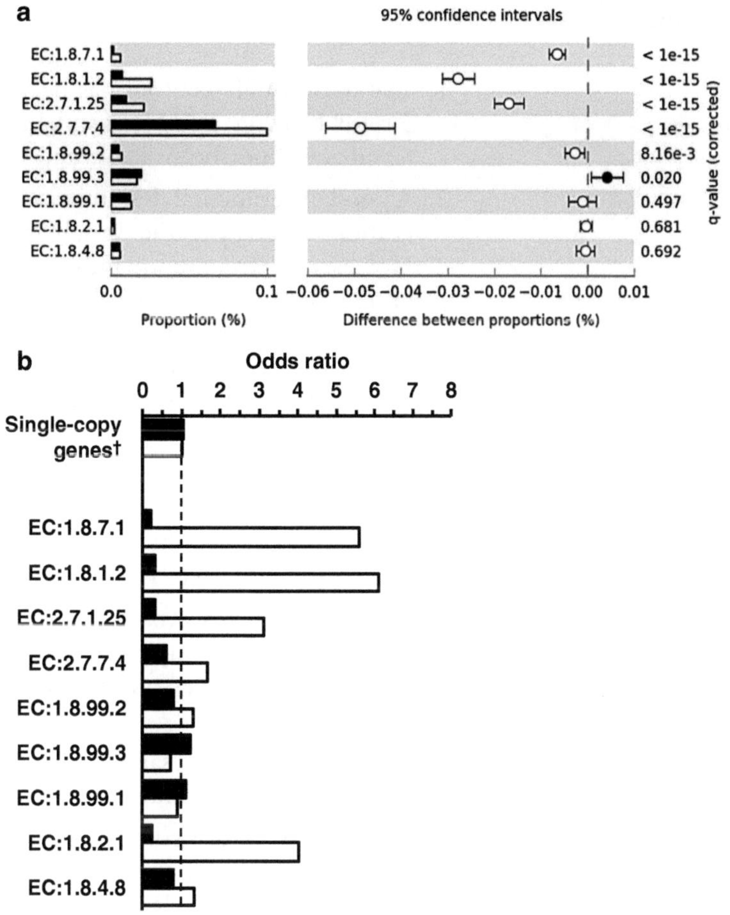

Fig. 4 Sample outputs from a univariate analysis of the sulfur metabolic pathway. (**a**) Box plot with the mean proportion of identified components (i.e., enzyme commission number) of the metabolic pathway. Statistical differences in mean proportions between two sections of a corroded pipe were based on the Fisher's exact test using corrected *q*-values (Storey's FDR multiple test correction approach) using the STAMP software [22]. (**b**) Bar chart shows the odds ratio values for each functional group. †Single-copy genes' (*gyrA*, *gyrB*, *recA*, *rpoA*, and *rpoB*) odds ratio value of 1 indicated a similar proportion of genomes. Error bars represent the standard error of the mean. Pipe sections: top (*white*) and bottom (*black*). Adapted from a comparative metagenomic study of a corroded concrete pipe [24]

4. Use the results of the exploratory data analysis to create hypotheses (*see* **Note 32**).

Network-Based Analysis

1. Import gene functional annotations and construct molecular interaction networks using the software package Cytoscape [21].
2. Display the network and filter to select subsets of nodes and/ or interactions that share a particular function (e.g., sulfur metabolism) (*see* **Note 33**).

4 Notes

1. The MG-RAST webtool has been optimized for the Mozilla Firefox browser. For Microsoft Internet Explorer users, many features will not be available and/or function correctly.
2. A good experimental design in microbial ecology includes replication to analyze the variability within one sample [25]. However, there are quantitative approaches for analyzing unreplicated datasets [22, 26]. For example, our study [24] used randomization procedures (e.g., Fisher's exact test with Storey's FDR multiple test correction approach) to statistically test differences in functional groups between two unreplicated sections of a corroded pipe.
3. Approximately 4 cm^2 of surface area was removed from the corroded pipe [24]. Carefully avoid concrete particles and corroded material which may introduce inhibitor substances and possibly interfere with downstream assays (e.g., PCR).
4. For RNA-based downstream methods, avoid RNA degradation by freezing samples in liquid nitrogen immediately and transporting samples on dry ice. Store at −80 °C until RNA extraction.
5. The recommended sample weight is 0.1–0.3 g and varies according to the manufacturer's recommendations.
6. At the time of the study [24], we used the UltraClean Soil DNA kit following the manufacturer's instructions (MoBio Laboratories Inc., Solana Beach, CA). For samples that have a high presence of humic and inhibitor substances or to increase the yield of DNA, the investigator can use a modified protocol previously described in Gomez-Alvarez et al. [27]. Briefly, 5 g of sample was ground with sterile sand and liquid nitrogen with a mortar and pestle, and then DNA was extracted using a CTAB-based method [28] with subsequent cesium-chloride gradient purification.
7. NanoDropUV and NanoDrop Fluorometry (Thermo Fisher Scientific, Wilmington, DE) measurements do not accurately measure the DNA concentration and are not compatible with 454 pyrosequencing protocols.
8. Commercially available NGS platforms for metagenomic applications include the following: the 454 pyrosequencing technology (Roche, 454 Life Sciences, Branford, CT), Illumina (Illumina, Inc., San Diego, CA), SOLiD (Life Technologies™, Applied Biosystems®, Carlsbad, CA), Ion Torrent (Life Technologies™, Carlsbad, CA), and PacBio (Pacific Biosciences, Menlo Park, CA). Of the NGS technologies, both the 454 and Illumina have been used extensively in

metagenomic studies. The 454 applies emulsion polymerase chain reaction (ePCR) in beads to clonally amplify random DNA fragments, while Illumina performs solid-surface PCR amplification on a surface [14]. For the rest of the technologies, the read length and/or error rate make those platforms at this moment unusable for metagenomic studies.

9. CAMERA and MG-RAST performs the quality control step using raw sequence data from FASTQ or SFF format data files. It is also critical to know the barcode sequence used for each sample if multiplexing was used.
10. Although a number of cyberinfrastructure resources have been developed for researchers [29], the MG-RAST server is the most widely used tool for the analysis of shotgun metagenomics [16].
11. The sequence files must be FASTA, FASTQ, or SFF format, and files larger than 50 MB should be compressed (e.g., GZIP, ZIP). All files should be named without spaces using alphanumeric and only with the following three characters: ., –, and _. The barcode file should be plain text ASCII containing lines with a barcode sequence followed by a unique filename separated by a tab.
12. The names of the sequence files must exactly match the library file_name fields or match the library metagenome_name fields minus the file extension in the metadata file.
13. MG-RAST recommendation is to use quality filtering with the default parameters:
    (a) Select if your sequence file(s) contain assembled data and include the coverage information (unlikely). Assembly has the potential to introduce biases and will likely only assemble organisms that are relatively abundant.
    (b) Select removal of artificial replicate sequences, i.e., reads that began at the same position but varied in length or contained a sequencing discrepancy produced by sequencing artifacts [27]. In our study [24], prior to sequence upload, we implemented a dereplication pipeline (http://microbiomes.msu.edu/replicates) to identify and remove clusters of artificially replicated sequences. Filter parameters included a cutoff value of 0.9, no length difference requirement, and an initial base pair match of 3 bp.
    (c) Screen for the removal of any host-specific species sequences (e.g., *H. sapiens*, NCBI v36) using DNA level matching with bowtie [31].
    (d) Select removal of low-quality sequences using a modified DynamicTrim algorithm [32]. Use values 15 and 5 as the lowest Phred score that will be counted as a high-quality

base and the minimum amount of bases below the above-specified quality contain in a sequence, respectively.

(e) Use a value of 2.0 as multiplicator of standard deviation for length cutoff to discard sequences whose length differs from the average sequence length.

(f) Specify a value of 5 for the maximum allowed number of ambiguous (non-ACGT) base pairs allowed in a particular sequence [33].

14. Sequence quality is a primary concern to any metagenomic analysis, and MG-RAST incorporates the tool DRISEE to measure sequencing error for whole-genome shotgun metagenomic data. DRISEE can detect error introduced during sample collection, genetic purification/amplification, and sequencing [34].

15. MG-RAST identifies ribosomal sequences and annotates them in addition to providing protein annotation. Sequences are prescreened and identified using QIIME-UCLUST for at least 70 % identity to ribosomal sequences against RNA databases (Greengenes, LSU and SSU SILVA database project, and Ribosomal Database Project II).

16. In addition to the presence of the important 16S rRNA gene marker, the files can contain fragmentary and nonoverlapping regions of small and large subunit ribosomal RNA from archaeal, bacterial, and eukaryotic and nonfunctional internal transcribed spacer (ITS) regions. Open source programs such as Megraft [36] and Metaxa [37] are able to identify and exclude the non-16S rRNA sequences in metagenomic libraries.

17. Putative open reading frames (ORF) are identified and their corresponding protein sequences annotated using BLAT against the M5NR protein databases for taxonomy classification. The M5NR is an integration of many sequence databases into one single, searchable database [16].

18. The tables provide information of protein assignments with the caveat that a protein sequence could be assigned to more than one closely related organism (i.e., taxonomy) or subsystem (i.e., function). There are cases that one read map to multiple annotations. In addition, long reads and contigs can contain sections with more than two predictable genes, which are annotated separately. The "abundance" and "hit" columns in the tables represent the number of sequences that contain a given annotation and the number of unique sequences (i.e., clustering) that were found in a particular annotation, respectively.

19. Putative ORF are identified and their corresponding protein sequences annotated using BLAT against the SEED subsystem for function classification [16].
20. Use the default parameters with the exception of:
    (a) Select the method "HHM based rRNA finding" from the drop down "rRNA prediction method" menu.
    (b) Specify an expected *e*-value cutoff of $1e^{-05}$ for Pfam, TIGRfam, and COG in the "Advanced Parameters" panel.
21. The RAMMCAP pipeline (i.e., CAMERA) assigns functions by comparison to Pfam, TIGRfam, and COG databases [30].
22. AGS is defined as an ecological measure of genome size that also includes multiple plasmid copies, inserted sequences, and associated phages and viruses [40]. This is particularly important since comparable average genome size permitted us to quantitatively compare the metagenomic data [41].
23. Use Level 2 SEED subsystem in MG-RAST to identify functional genes (e.g., "Nitrogen Metabolisms" subsystem).
24. Compare the metagenomes at the genus level (when available) using absolute read counts with default parameters for the lowest common ancestor (LCA) algorithm of min-score of 35, a top-percent value of 10 %, and min-support of 5 [42].
25. Rarefaction curves are used to estimate the relative richness within a sample or to compare the richness between metagenomes. For example, the ChaoI estimator of COG richness is calculated using the number of individual COGs per unique COG function.
26. Direct comparison of the distribution of different functions (i.e., genes) is not possible within the metagenome or between samples, since length and copy number of the genes was not incorporated in the formula.
27. For metagenomic studies with one representative, the statistical difference of functional genes between two samples is based on the Fisher's exact test with corrected *q*-values (Storey's FDR multiple test correction approach) using the software package STAMP [22].
28. Divide the AGS to the amount of DNA (in kb) per function-specific gene [41].
29. Transformation is strictly a numerical approach to decrease the possibility of distortion, particularly since the Bray-Curtis distance measure is very sensitive to outlying values [44].
30. The Bray-Curtis distance measure accounts for the abundance distributions of attributes (i.e., genes, functions).

31. The robustness of the nMDS ordination plot is evaluated with a Shepard diagram (i.e., stress test), which measures the goodness of fit of an nMDS plot, e.g., <0.1 indicates a good fitting solution [45]. Clustering is evaluated using the bootstrapping method, e.g., the percentage of replicates where each node is still supported is given on the dendrogram [46].
32. Avoid using the same data to test these hypotheses [47].
33. In the network graph, the nodes represent the genes (i.e., biological entities) and the edges between nodes represent the biological interactions. Data are mapped to nodes or edges using attribute values (e.g., protein functions), which can be used to perform network searches and filtering and to control visual aspects of nodes and edges [48].

Acknowledgements

The United States Environmental Protection Agency (USEPA) through the Office of Research and Development funded this research. R.P. Revetta and J.W. Santo Domingo of the USEPA participated in design and coordination of the study. It has been subjected to the Agency's peer and administrative review and has been approved for external publication. Any opinions expressed in this manuscript are of the authors and do not necessarily reflect the official positions and policies of USEPA. Any mention of trade names or commercial products does not constitute endorsement or recommendation.

References

1. USEPA (United States Environmental Protection Agency) (2009) State of technology review report on rehabilitation of wastewater collection and water distribution systems. Office of Research and Development, Cincinnati, OH, EPA/600/R-09/048
2. USEPA (United States Environmental Protection Agency) (2002) State of technology review report on rehabilitation of wastewater collection and water distribution systems. USEPA Urban Watershed Management Branch, Edison, NJ, EPA/600/JA-02/226
3. Mori T, Nonaka T, Tazaki K et al (1992) Interactions of nutrients, moisture, and pH on microbial corrosion of concrete sewer pipes. Water Res 26:29–37
4. Vollertsen J, Nielsen AH, Jensen HS et al (2008) Corrosion of concrete sewers—the kinetics of hydrogen sulfide oxidation. Sci Total Environ 394:162–170
5. Zhang L, De Schryver P, De Gusseme B et al (2008) Chemical and biological technologies for hydrogen sulfide emission control in sewer systems: a review. Water Res 42: 1–12
6. Vincke E, Boon N, Verstraete W (2001) Analysis of the microbial communities on corroded concrete sewer pipes—a case study. Appl Microbiol Biotechnol 57:776–785
7. Okabe S, Ito T, Satoh H (2003) Sulfate-reducing bacterial community structure and their contribution to carbon mineralization in a wastewater biofilm growing under microaerophilic conditions. Appl Microbiol Biotechnol 63:322–334
8. Okabe S, Odagiri M, Ito T, Satoh H (2007) Succession of sulfur-oxidizing bacteria in the microbial community on corroding concrete in sewer systems. Appl Environ Microbiol 73:971–980

9. Satoh H, Odagiri M, Ito T, Okabe S (2009) Microbial community structures and in situ sulfate-reducing and sulfur-oxidizing activities in biofilms developed on mortar specimens in a corroded sewer system. Water Res 43:4729–4739
10. Santo Domingo JW, Revetta RP, Iker B et al (2011) Molecular survey of concrete sewer biofilm microbial communities. Biofouling 27:993–1001
11. Jones WJ (2010) High-throughput sequencing and metagenomics. Estuaries Coasts 33:944–952
12. Simon C, Daniel R (2009) Achievements and new knowledge unraveled by metagenomic approaches. Appl Microbiol Biotechnol 85:265–276
13. Gilbert JA, Meyer F, Bailey MJ (2011) The future of microbial metagenomics (or is ignorance bliss?). ISME J 5:777–779
14. Thomas T, Gilbert J, Meyer F (2012) Metagenomics—a guide from sampling to data analysis. Microb Inform Exp 2:3
15. Shokralla S, Spall JL, Gibson JF, Hajibabaei M (2012) Next-generation sequencing technologies for environmental DNA research. Mol Ecol 21:1794–1805
16. Meyer F, Paarmann D, D'Souza M et al (2008) The metagenomics RAST server—a public resource for the automatic phylogenetic and functional analysis of metagenomes. BMC Bioinformatics 9:386
17. Sun S, Chen J, Li W et al (2011) Community cyberinfrastructure for advanced microbial ecology research and analysis: the CAMERA resource. Nucleic Acids Res 39:D546–D551
18. Markowitz VM, Chen I-MA, Chu K et al (2012) IMG/M: the integrated metagenome data management and comparative analysis system. Nucleic Acids Res 40:D123–D129
19. Gerlach W, Jünemann S, Tille F et al (2009) WebCARMA: a web application for the functional and taxonomic classification of unassembled metagenomic reads. BMC Bioinformatics 10:430
20. Huson DH, Mitra S, Ruscheweyh H-J et al (2011) Integrative analysis of environmental sequences using MEGAN4. Genome Res 21:1552–1560
21. Shannon P, Markiel A, Ozier O et al (2003) Cytoscape: a software environment for integrated models of biomolecular interaction networks. Genome Res 13:2498–2504
22. Parks DH, Beiko RG (2010) Identifying biologically relevant differences between metagenomic communities. Bioinformatics 26:715–721
23. Prakash T, Taylor TD (2012) Functional assignment of metagenomic data: challenges and applications. Brief Bioinform 13:711–727
24. Gomez-Alvarez V, Revetta RP, Santo Domingo JW (2012) Metagenome analyses of corroded concrete wastewater pipe biofilms reveal a complex microbial system. BMC Microbiol 12:122
25. Prosser JI (2010) Replicate or lie. Environ Microbiol 12:1806–1810
26. Lennon JT (2011) Replication, lies and lesser-known truths regarding experimental design in environmental microbiology. Environ Microbiol 13:1383–1386
27. Gomez-Alvarez V, Teal TK, Schmidt TM (2009) Systematic artifacts in metagenomes from complex microbial communities. ISME J 3:1314–1317
28. Zhou J, Bruns MA, Tiedje JM (1996) DNA recovery from soils of diverse composition. Appl Environ Microbiol 62:316–322
29. De Filippo C, Ramazzotti M, Fontana P, Cavalieri D (2012) Bioinformatic approaches for functional annotation and pathway inference in metagenomics data. Brief Bioinform 13:696–710
30. Li W (2009) Analysis and comparison of very large metagenomes with fast clustering and functional annotation. BMC Bioinformatics 10:359–367
31. Langmead B, Trapnell C, Pop M, Salzberg SL (2009) Ultrafast and memory-efficient alignment of short DNA sequences to the human genome. Genome Biol 10:R25
32. Cox MP, Peterson DA, Biggs PJ (2010) SolexaQA: at-a-glance quality assessment of Illumina second-generation sequencing data. BMC Bioinformatics 11:485
33. Huse SM, Huber JA, Morrison HG et al (2007) Accuracy and quality of massively parallel DNA pyrosequencing. Genome Biol 8:R143
34. Keegan KP, Trimble WL, Wilkening J et al (2012) A platform-independent method for detecting errors in metagenomic sequencing data: DRISEE. PLoS Comput Biol 8:e1002541
35. Gomez-Alvarez V, Revetta RP, Santo Domingo JW (2012) Metagenomic analyses of drinking water receiving different disinfection treatments. Appl Environ Microbiol 78: 6095–6102
36. Bengtsson J, Hartmann M, Unterseher M et al (2012) Megraft: a software package to graft ribosomal small subunit (16S/18S) fragments onto full-length sequences for accurate species richness and sequencing depth analysis in pyrosequencing-length metagenomes and similar environmental datasets. Res Microbiol 163:407–412

37. Bengtsson J, Eriksson KM, Hartmann M et al (2011) Metaxa: a software tool for automated detection and discrimination among ribosomal small subunit (12S/16S/18S) sequences of archaea, bacteria, eukaryotes, mitochondria, and chloroplasts in metagenomes and environmental sequencing datasets. Antonie Van Leeuwenhoek 100:471–475
38. Kanehisa M, Goto S (2000) KEGG: Kyoto encyclopedia of genes and genomes. Nucleic Acids Res 28:27–30
39. Beszteri B, Temperton B, Frickenhaus S, Giovannoni SJ (2010) Average genome size: a potential source of bias in comparative metagenomics. ISME J 4:1075–1077
40. Raes J, Korbel JO, Lercher MJ et al (2007) Prediction of effective genome size in metagenomic samples. Genome Biol 8:R10
41. Frank JA, Sørensen SJ (2011) Quantitative metagenomic analyses based on average genome size normalization. Appl Environ Microbiol 77:2513–2521
42. Mitra S, Rupek P, Richter DC et al (2011) Functional analysis of metagenomes and metatranscriptomes using SEED and KEGG. BMC Bioinformatics 12:S21
43. Gill SR, Pop M, Deboy RT et al (2006) Metagenomic analysis of the human distal gut microbiome. Science 312:1355–1359
44. Michie MG (1982) Use of the Bray-Curtis similarity measure in cluster analysis of foraminiferal data. Math Geol 14:661–667
45. Wickelmaier F (2003) An introduction to MDS: reports from the Sound Quality Research Unit (SQRU) No. 7, 1–26. http://homepages.uni-tuebingen.de/florian.wickelmaier/pubs/Wickelmaier2003SQRU.pdf
46. McKenna JE (2003) An enhanced cluster analysis program with bootstrap significance testing for ecological community analysis. Environ Model Software 18:205–220
47. Zuur AF, Ieno EN, Elphick CS (2010) A protocol for data exploration to avoid common statistical problems. Methods Ecol Evol 1:3–14
48. Smoot ME, Ono K, Ruscheinski J et al (2011) Cytoscape 2.8: new features for data integration and network visualization. Bioinformatics 27:431–432

Chapter 24

Culture-Independent Methods to Study Subaerial Biofilm Growing on Biodeteriorated Surfaces of Stone Cultural Heritage and Frescoes

Francesca Cappitelli, Federica Villa, and Andrea Polo

Abstract

Actinobacteria, cyanobacteria, algae, and fungi form subaerial biofilm (SAB) that can lead to material deterioration on artistic stone and frescoes. In studying SAB on cultural heritage surfaces, a general approach is to combine microscopy observations and molecular analyses. Sampling of biofilm is performed using specific adhesive tape and sampling of SAB and the substrate with sterile scalpels and chisels. Biofilm observations are carried out using optical and scanning electron microscopy. Specific taxa and EPS in biofilm can be readily visualized by fluorochrome staining and subsequent observation using fluorescence or confocal laser scanning microscopy. The observation of cross sections containing both SAB and the substrate shows if biofilm has developed not only on the surface but also underneath. Following nucleic acid extraction, 16S rRNA gene sequencing is used to identify bacterial taxa, while 18S rRNA gene and internal transcribed spacer (ITS) sequence analysis is used to study eukaryotic groups. In this chapter, we illustrate the protocols related to fluorescence in situ hybridization (FISH), scanning electron microscopy (SEM), and denaturing gradient gel electrophoresis (DGGE).

Key words Optical, Fluorescence and electron microscopy, Fluorescence in situ hybridization (FISH), Scanning electron microscopy (SEM), Denaturing gradient gel electrophoresis (DGGE), Adhesive tape strip, Sterile scalpel

1 Introduction

According to the International Council on Monuments and Sites (ICOMOS) [1] biofilm is "a mono- to multilayered microbial colony attached to surfaces with varying thickness of up to 2 mm. Often a biofilm consists of very few cells of different microorganisms embedded in large amounts of extracellular slime. These cohesive often sticky layers may shrink and expand according to the supply of water. Biofilms often create multicoloured biopatina by production of colouring agents."

Gianfranco Donelli (ed.), *Microbial Biofilms: Methods and Protocols*, Methods in Molecular Biology, vol. 1147,
DOI 10.1007/978-1-4939-0467-9_24,

Biofilm on stone heritage and frescoes is called "subaerial biofilm" (SAB) as the microbial communities develop at the atmosphere–stone interface [2]. SAB can cause damage to cultural heritage surfaces through a variety of mechanisms, including chemical reactions with materials (e.g., microbial excretion of aggressive acids), physical disruption (e.g., extracellular polymeric substances (EPS) production that can cause mechanical stress on the mineral structure), and esthetic alterations (e.g., pigment production).

In the study of SAB on cultural heritage surfaces, a general approach is to combine microscopy observations and molecular analyses. Direct biofilm observation is carried out using optical and scanning electron microscopy. Specific taxa and EPS in biofilm can be readily visualized due to autofluorescence or by staining with fluorochromes and subsequent observation using fluorescence or confocal laser scanning microscopy. The observation of cross sections containing both SAB and the substrate shows if biofilm has developed not only on the surface but also underneath. Following nucleic acid extraction, 16S rRNA gene sequencing is used to identify bacterial taxa, while 18S rRNA gene and internal transcribed spacer (ITS) sequence analysis is used to study eukaryotic groups. Bacterial (particularly actinobacteria and cyanobacteria), algal, and fungal communities are the most deteriorating biological agents growing as biofilm on stone material and frescoes [2].

Sampling areas are selected after visual inspection of putative microbiological alterations. Two sampling methods are generally used: (1) the noninvasive adhesive tape strip [3], by which only the biofilm is sampled—not the underlying substrate—for microscopy analyses and in situ visualization techniques, and (2) the sterile scalpel or chisel, for microscopy observations and molecular analyses following nucleic acid extraction. The collected samples, their number and dimension, should be representative of the overall substrate's characteristics and state of conservation. When dealing with cultural heritage, it is not frequent for microinvasive sampling to be permitted, and when it is, the samples are very few and extremely small, putting severe constraints on the microbiologist's work.

2 Materials

2.1 Fluorescence In Situ Hybridization

Prepare all solutions using ultrapure water (prepared by purifying deionized water to attain a sensitivity of 18 MΩ cm at 25 °C) sterilized by autoclave and analytical grade reagents in aseptic conditions. Prepare and store all reagents at room temperature (unless indicated otherwise). Diligently follow all waste disposal regulations when disposing of waste materials.

2.1.1 Formaldehyde Fixative and Permeabilization Solution

1. Paraformaldehyde stock solution: 8 % (wt/v) solution of paraformaldehyde in water (*see* **Note 1**). Add 8 g of paraformaldehyde to 90 mL of distilled water in a beaker with a magnetic stirrer (*see* **Note 2**). Mix and make up to 100 mL with water. Store at 4 °C for up to 1 month.
2. Paraformaldehyde working solution: 4 % (wt/v) solution of paraformaldehyde by diluting the stock solution with an equal volume of phosphate buffered saline (PBS, 130 mM NaCl, 10 mM sodium phosphate, pH 7.2). Store at 4 °C. The solution should be made up and used the same day, it must not be stored.
3. Ethanol/water series: 50 % v/v, 80 % v/v, and 100 % ethanol (*see* **Note 3**).

2.1.2 Hybridization and Washing Solutions

1. 5 M NaCl: 29.2 g NaCl in 100 mL of distilled water. Autoclave the solution and store at room temperature.
2. 1 M Tris–HCl pH 7.2–7.6: 15.8 g Tris–HCl in 100 mL of distilled water. Adjust pH with HCl. Autoclave the solution and store at room temperature.
3. 10 % wt/v sodium dodecyl sulfate (SDS): 10 g SDS in 100 mL of distilled water. Filter the solution and store at room temperature.
4. Hybridization buffer: 0.9 M NaCl, 20 mM Tris–HCl pH 7.2, 0.01 % wt/v SDS (*see* **Note 4**), 40 % v/v deionized formamide (*see* **Notes 5** and **6**). To prepare 2 mL of hybridization buffer, add-in the following order: 360 μL 5 M NaCl, 40 μL 1 M TRIS–HCl, 800 μL formamide, 800 μL Milli-Q (autoclaved, cell free) water, and 2 μL 10 % SDS. Mix compounds well. Freshly prepare the solution before use.
5. Hybridization mixture: mix 50 μL of the hybridization buffer with 50 ng of the selected probe (*see* **Note 7**). Make up fresh immediately prior to use, and store under dark conditions.
6. Washing buffer: to prepare 50 mL of washing buffer, add-in the following order: 460 μL 5 M NaCl (for 40 % stringency, *see* **Note 8**) (Table 1), 1 mL 1 M TRIS–HCl, 48.5 mL water, and 50 μL 10 % SDS. Mix components well.

2.1.3 EPS Matrix Staining and Counterstaining Solutions

1. ConA Stock solution: 5 mg concanavalin A (Con A, *see* **Note 9**) in 2 mL 0.1 M sodium bicarbonate, pH 8.3, and briefly vortex (*see* **Note 10**).
2. DAPI Stock solution: 5 mg/mL 4′,6-diamidino-2-phenylindole (DAPI) dihydrochloride stock solution by dissolving 10 mg in 2 mL of deionized water (*see* **Note 11**).
3. EPS staining and counterstaining working solution: 200 μg/mL ConA (*see* **Note 12**), 10 μg/mL DAPI (*see* **Note 13**).

Table 1
NaCl concentration in the washing buffer according to % formamide of the hybridization buffer (SILVA rRNA database project, available at www.arb-silva.de/)

% Formamide	NaCl (M concentration)	μL 5 M NaCl in 50 mL
0	0.900	9,000
5	0.636	6,300
10	0.450	4,500
15	0.318	3,180
20	0.225	2,150
25	0.159	1,490
30	0.112	1,020
35	0.080	700
40	0.056	460
45	0.040	300
50	0.028	180
55	0.020	100
60	0.014	40

To prepare 1 mL of dye solution, dilute 40 μL ConA stock solution and 2 μL DAPI stock solution to a final volume of 1 mL PBS. Mix compounds well. Freshly prepare the solution before use (*see* **Note 14**).

2.1.4 Equipment

1. Adhesive tape strips (Fungi tape™).
2. Microscope slides.
3. Frame-seal incubation chambers (1.0 cm^2 area).
4. Tube heater, thermomixer, or water bath.
5. In situ PCR block set at 45 °C (*see* **Note 15**).
6. Confocal microscope equipped with suitable lasers.
7. Software for image analyses (e.g., Imaris and ImageJ).

2.2 Scanning Electron Microscopy

2.2.1 Samples Preparation and Analysis

1. 0.1 M sodium cacodylate buffer solution: dissolve 21.4 g of sodium cacodylate [$Na(CH_3)_2\ AsO_2 \cdot 3H_2O$] in 900 mL of demineralized water. Adjust pH to 7.2 with HCl. Add demineralized water to a volume of 1,000 mL (*see* **Note 16**).
2. 2 % Glutaraldehyde in 0.1 M sodium cacodylate buffer: add 8 mL of concentrated glutaraldehyde solution (25 %) in 92 mL of 0.1 M sodium cacodylate buffer (pH 7.2).

3. 2 % Osmium tetroxide in 0.1 M sodium cacodylate buffer: add 5 mL of concentrated osmium tetroxide solution (4 %, Sigma) in 5 mL of 0.1 M sodium cacodylate buffer (pH 7.2) (*see* **Note 17**).
4. Ethanol/water series: 50, 60, 70, 80, 90 % v/v, and absolute ethanol solutions.
5. Polyester or a bicomponent epoxy resin.
6. Demountable molds (Teflon, silicone, polyethylene molds, etc.).
7. Precision cutter with diamond cut-off wheel.
8. SiC grinding and polishing papers: 600, 1,200, 2,400, 4,000 grit sizes.
9. Conductive adhesive tape (i.e., C tape, Cu tape, C tabs, etc.) or conductive paint (i.e., colloidal silver paint, conductive carbon cement, etc.).
10. Pure gold or other conductive metals (Au/Pd, Pt, Cr) or carbon for coating (*see* **Note 18**).
11. Equipment for metallization of samples by sputtering or evaporation (*see* **Note 19**).

2.3 Denaturing Gradient Gel Electrophoresis

Prepare all solutions using ultrapure water (prepared by purifying deionized water to attain a sensitivity of 18 MΩ cm at 25 °C) sterilized by autoclave and analytical grade reagents in aseptic conditions. Prepare and store all reagents at room temperature (unless indicated otherwise). Diligently follow all waste disposal regulations when disposing of waste materials.

2.3.1 DNA Extraction Equipment and Solutions

1. Sterilize as many mortars and pestles as samples to be pulverized.
2. 2 mL tubes sterilized by autoclave.
3. Thermomixer.
4. Freezer −80 °C.
5. 0.5 M ethylenediaminetetraacetic acid (EDTA) solution: add 186.1 g disodium ethylenediamine tetraacetate·$2H_2O$ into 800 mL of distilled water and adjust the pH to 8.0 with NaOH (about 18–20 g of NaOH solid pellets). Dilute the solution to 1 L with distilled water. Filter the solution through a 0.5 μm filter and sterilize in an autoclave.
6. 1 M tris(hydroxymethyl)aminomethane–HCl (Tris–HCl) buffer: 121.1 g of tris(hydroxymethyl)aminomethane (Tris base) in 800 mL of demineralized water and adjust the pH to 8.0 with concentrated HCl. Adjust the volume to 1,000 mL with demineralized water and sterilize in an autoclave.
7. Tris–EDTA (TE) buffer: 10 mM Tris–HCl solution (pH 8.0), 1 mM EDTA. To prepare 100 mL of TE buffer, add-in the

following order: 1 mL of 1 M Tris–HCl (pH 8.0) and 0.2 mL of 0.5 M EDTA and made up with double distilled water up to 100 mL. Autoclave the solution. Leave aliquots at 4 °C for current use (550 μL for each sample) and store remaining aliquots at −20 °C.

8. 35 mg/mL lysozyme solution in TE buffer: for 20 mL place 0.7 g of lysozyme in 20 mL of TE buffer. Aliquot and freeze at −20 °C.
9. 20 mg/mL proteinase K solution in TE buffer: 100 mg of proteinase K in 5 mL of demineralized, DNAse, and proteinase-free water. Aliquot and freeze at −20 °C (*see* **Note 20**).
10. 10 % sodium dodecyl sulfate (SDS) solution in water: 10 g of SDS in 80 mL of deionized/Milli-Q water and mix it (*see* **Note 21**). Heat to 68 °C if necessary. Adjust the volume to 100 mL with deionized/Milli-Q water and mix it again. Filter the solution to remove any undissolved material if necessary.
11. 5 M NaCl solution in water: add 29.2 g NaCl in 100 mL of distilled water. Autoclave the solution and store at room temperature.
12. 10 % (v/v) cetyltrimethylammonium bromide (CTAB) solution in 0.7 M NaCl (CTAB/NaCl). To prepare 100 mL of CTAB/NaCl 10 % (v/v), dissolve 4.1 g of NaCl in 80 mL of demineralized water and slowly add 10 g of CTAB while heating the solution to 55 °C and stirring. Adjust the final volume to 100 mL and store above 15 °C to prevent CTAB from precipitation.
13. Phenol–chloroform–isoamyl alcohol (25:24:1, v/v) for molecular biology (*see* **Notes 22** and **23**).
14. Chloroform:isoamyl alcohol (24:1 v/v) suitable for nucleic acid purification (*see* **Note 23**).
15. Isopropyl alcohol (2-propanol) (*see* **Note 23**).
16. Seventy percent ethanol solution in water.
17. A NanoDrop spectrophotometer (*see* **Note 24**).

2.3.2 DNA Amplification Equipment and Solutions

1. A laminar flow cabinet.
2. Thermocycler to perform the polymerase chain reactions (PCR).
3. PCR Buffers, dNTPs, $MgCl_2$, and General Reagents: 10× PCR buffer (200 mM Tris–HCl solution at pH 8.4, 500 mM KCl solution), 50 mM (or 25 mM) $MgCl_2$ solution, 10 mM (or 20 mM) dNTP mix, PCR grade dimethyl sulfoxide (DMSO), and PCR grade bovine serum albumin solution (BSA) (*see* **Note 25**). Store at −20 °C.
4. 5 U/μL taq DNA Polymerase GoTaq®. Store at −20 °C.

5. Olygonucleotide primers for bacterial community, cyanobacterial communities, algal community, and fungal communities. Forward primers are needed with a GC clamp (Table 2) (*see* **Note 26**) [4–8]. Store at −20 °C.
6. Agarose for molecular biology.
7. Tris/Borate/EDTA (TBE) buffer stock solution: make a 5× solution of TBE in water. Add 54 g of Tris base, 27.5 g of boric acid, and 20 mL of 0.5 M EDTA (pH 8) in a 1 L graduated cylinder. Add demineralized water to a volume of 1 L. Mix to dissolve the components. Store at room temperature.
8. TBE working solution: make a 0.5× solution. Dilute 1–10 of the TBE stock solution in demineralized water. Store at room temperature.
9. Glass Erlenmeyer flasks.
10. Microwave.
11. A quantitative and qualitative marker suitable for segments between 400 and 1,000 bp.
12. A horizontal electrophoresis apparatus: tank, lid, cables, a power supply, gel trays, and multi-well combs.
13. Gel loading solution: saccharose 40 % wt/v, bromophenol blue 0.05 % wt/v, EDTA 0.1 mM pH 8, and SDS 0.5 % (wt/v). To prepare 100 mL of gel loading solution, add 40 g of saccharose, 0.05 g of bromophenol blue, 20 mL of 0.5 M EDTA, and 0.5 g of SDS; adjust volume to 100 mL with demineralized water.
14. 0.1 % ethidium bromide solution in water: 100 μL of concentrated ethidium bromide solution (1 %) in 1,000 mL of demineralized water (*see* **Notes 23** and **27**).
15. A gel documentation system (Gel Doc).

2.3.3 8 % Polyacrylamide Denaturing Gradient Gel Electrophoresis (DGGE)

1. A universal mutation detection system for 16 cm gels.
2. A gradient maker.
3. A magnetic stirrer.
4. A power supply.
5. A peristaltic pump.
6. Seventy percent ethanol solution in demineralized water.
7. Staining boxes: plastic container to stain the gels (*see* **Note 28**).
8. Tris–acetate-EDTA (TAE) buffer stock solution: make a 50× solution of TAE in water. Add 242 g of Tris base, 57.1 mL of glacial acetic acid, and 100 mL of 0.5 M EDTA (pH 8) with demineralized water to a volume of 1 L. Store at room temperature.

Table 2
Most popular primers used in PCR-DGGE analysis on artwork

Primer	Target taxa	Sequence
357F-GC	Bacteria	CGCCCGCCGCGCGCGGCGGGCGGGGCGGGGGCACGGGGGGCCTACGGGAGGCAGCAG
907 R		CCGTCAATTCCTTTGAGTTT
CYA359-GC	Cyanobacteria and plastids	CGCCCGCCGCGCCCCGCGCCCGTCCCGCCGCCCCCGCCCGGGGGAATTTTCCGCAATGGG
CYA781Ra		GACTACTGGGGTATCTAATCCCATT
CYA781Rb		GACTACAGGGGTATCTAATCCCTTT
NS1-GC	Algae	CGCCCGCCGCGCGCGGCGGGCGGGGCGGGGGCACGGGCCAGTAGTCATATGCTTGTC
NS2		GAATTACCGCGGCTGCTGGC
NS5	Fungi	AACTTAAAGGAATTGACGGAAG
ITS4		TCCTCCGCTTATTGATATGC
ITS1F-GC		CGCCCGGGGCGCGCCCCGGGCGGGGCGGGGGCCGGGGGGCTTGGTCATTTAGAGGAAGTAA

9. TAE working solution: make a 1× TAE solution. Dilute 1–50 of the 50× TAE stock solution in demineralized water. Store at room temperature.
10. 100 % Denaturing solution: to prepare 200 mL of solution, add-in a graduated cylinder 40 mL of 40 % acrylamide/bisacrylamide 37.5:1, 84 g of urea, 80 mL of 100 % formamide, and 4 mL of 50× TAE. Add water to a volume of 200 mL. Mix for 3 h with mild heat. Mix and store at −20 °C (*see* **Note 23**).
11. 0 % Denaturing solution: to prepare 200 mL of solution, add-in a graduated cylinder 40 mL of 40 % acrylamide/bisacrylamide 37.5:1 and 2 mL of 50× TAE. Add water to a volume of 200 mL. Mix and store at −20 °C (*see* **Note 23**).
12. 10 % Ammonium persulfate (APS) solution: to prepare 1 mL of solution, add 0.1 g of APS in 1 mL Milli-Q water. Store at 4 °C (*see* **Note 29**).
13. Tetramethylethylenediamine (TEMED) (*see* **Note 23**).
14. Denaturing loading blue: to prepare 50 mL of denaturing loading blue, mix 0.25 g of bromophenol blue, 0.25 g of xylene cyanol, 0.2 mL of 0.5 M EDTA, 30 mL of 100 % glycerol, and demineralized water to a volume of 50 mL. Adjust pH to 8.0.
15. SYBR Green solution: add 80 μL of concentrated SYBR Green in 800 mL of TAE 1×. Store at 4 °C at the dark (*see* **Notes 23** and **30**).
16. A gel documentation system (Gel Doc).

3 Methods

3.1 Fluorescence In Situ Hybridization

Fluorescence in situ hybridization (FISH) is a culture-independent method for mapping the spatial arrangement of microorganisms at the single-cell level and for assessing the composition of microbial communities and their in situ dynamics [9]. The protocol applies FISH to three-dimensionally (3D) biofilm-growing microbial community on biodeteriorated cultural heritage surfaces and entails an adhesive tape sampling procedure that reproduces a mirror image of the biofilm present on the selected area. It is not in any way invasive for the substrate [2]. The sample is then subjected to hybridization by fluorescently labeled, rRNA-targeted oligonucleotide probes, extracellular polymeric matrix staining, and signal detection. This approach has been successfully used to map microbial colonization on stone and fresco materials.

3.1.1 Pretreatment of the Slides and Fixation

1. Immobilize the adhesive tape strip on a glass slide and wash twice by dipping the slide in sterile PBS at room temperature (*see* **Note 31**). Air-dry the slide at any step.

Table 3
Most popular probes used in FISH analysis on artwork

Target group	Probe name	Probe sequence (5′–3′)	Group coverage %
Bacteria	EUB338	GCTGCCTCCCGTAGGAGT	94
Eukarya	EUK516	ACCAGACTTGCCCTCC	90
Archaea	ARCH915	GTGCTCCCCCGCCAATTCCT	90
Alphaproteobacteria	ALF1B	CGT TCG YTC TGA GCC AG	68
Betaproteobacteria	BET42a	GCCTTCCCACTTCGTTT	86
Gammaproteobacteria	GAM42a	GCCTTCCCACATCGTTT	76
Actinobacteria	HGC69a	TATAGTTACCACCGCCGT	93
Geodermatophilus/Blastococcus group	Geo/Blasto	CCATCCCCAGCCGGAAACC	
Modestobacter genus	Modesto	TTCGCCGCTAGGGCA	
Cyanobacteria	CYA361	CCCATTGCGGAAAATTCC	91

2. Dip the slide in a fresh solution fixation buffer and incubate for 4 h at 4 °C.
3. After the fixation, wash the samples in PBS at room temperature to remove any remaining fixative and store in 1:1 (v/v) PBS/ethanol at −20 °C (*see* **Note 32**). Samples can be stored for at least several months.
4. Drain slide and allow it to air-dry. Confine a selected area with a frame-seal incubation chamber.
5. Dehydrate slides through an ascending ice-cold ethanol/water series of 50, 80, and 100 % for 3 min each at room temperature. Air-dry the slide at any step (*see* **Note 33**).

3.1.2 Probe Hybridization

1. Add 50 ng of the selected probe/probes (*see* **Notes 34** and **35**) (Table 3) [4, 10–18] to 50 μL of pre-warmed hybridization buffer. Protect the hybridization mixture in the dark at 4 °C.
2. Add 50 μL of the hybridization mixture in the square area, remove the upper sticky tape, and carefully seal the slide with a plastic sealer (*see* **Note 36**).
3. Place the slides with samples to be hybridized in an in situ PCR block at 46 °C for at least 3 h (*see* **Note 37**).
4. Carefully remove the plastic sealer from the slide with forceps.
5. Prepare 50 mL of washing buffer in a polyethylene tube and preheat in a 48 °C water bath (*see* **Note 38**).

6. Quickly rinse off the hybridization mixture with a few milliliters of preheated washing buffer (*see* **Note 39**).
7. Immediately add 100 μL of the preheated washing buffer to the square area, making certain that the cells are completely covered, and incubate at 48 °C. Replace the buffer every 5 min for 15 min (*see* **Note 40**).
8. Gently rinse the slide with distilled water and air-dry it (*see* **Note 41**).

3.1.3 EPS Staining, Counterstaining, and Signal Detection

1. Add 50 μL of the EPS staining and counterstaining working solution to the square area, making certain that the cells are completely covered, and incubate for 10 min at room temperature in the dark (*see* **Note 42**).
2. Rinse slide with distilled water to ensure the removal of excess stain, and let air-dry.
3. Mount the slides with the appropriate antifading mounting medium (*see* **Note 43**), and view the samples using confocal laser microscopy (*see* **Note 44**). The sample can be stored for a few days to many months in the dark at −20 or −70 °C (*see* **Note 45**).
4. Confocal microscope equipped with suitable lasers (major providers: Leica Microsystems, Carl Zeiss, Nikon, and Olympus Corporation) (*see* **Note 46**) (Table 4). For the best optical resolution, use an oil-immersion objective of high numerical aperture (NA) (1.30) with a ×63 or ×100 magnification. Good results can also be obtained with high NA (1.25) water-immersion objectives. The topic of image acquisition is a large

Table 4
Most frequently used fluorochromes in FISH analysis [24]

	Wavelengths		
Fluorochromes	**Excitation (nm)**	**Emission (nm)**	**Color**
AMCA	351	450	Blue
FITC	492	528	Green
FluoX	488	520	Green
TRITC	557	576	Red
Texas Red	578	600	Red
Cy3	550	570	Orange
Cy5	651	674	Infrared

one, and the reader is referred to the extensive literature on this topic [19, 20].

5. Process images in either Imaris software for qualitative presentation (Bitplane Inc.) or ImageJ software (Java software for image-processing analysis; freely available at http://rsbweb.nih.gov/ij/ for quantitative analysis.

3.2 Scanning Electron Microscopy

Scanning electron microscopy (SEM) allows high-magnification visualization of biofilms on frescoes and stone substrates. In particular, the biofilm structure, the superficial morphology of biofilm, and the endolithic growth can be studied. Moreover, as biofilms are able to develop within the substrates discontinuities (i.e., porosity, cracks or fissure) and to cause disruption of the mineral matrix, the observation of polished cross sections can provide further insight into the damaging process [5].

3.2.1 Analysis of the Fragments

Carry out all procedures as reported by Gulotta et al. [21], Cappitelli et al. [22, 23], and Polo et al. [5].

1. On the same day as the sampling by means of steel scalpel or sharp chisel (*see* **Note 47**), fix samples overnight in 2 % glutaraldehyde solution.
2. Post-fix samples with 2 % osmium tetroxide solution.
3. Dehydrate samples in serial higher concentrations of ethanol (50, 60, 70, 80, 90 %, and absolute).
4. Dry the samples to a critical point.
5. Mount the samples on a stub using conductive adhesive tape and coat them with a nanometric layer of pure gold or other conductive metals or carbon (*see* **Notes 18** and **48**) by sputtering or evaporation method (*see* **Note 19**).
6. Analyze the superficial morphology of the sample with biofilm by SEM in the secondary electrons mode.

3.2.2 Analysis of Polished Cross Section

1. Embed the samples into polyester or bicomponent epoxy resin by means of demountable molds.
2. Cut the prepared samples using a precision cutter in order to expose to the surface the cross section.
3. Grind and polish the cross section with SiC paper with progressively growing grit sizes. As the grit size increases, the section achieves a more polished appearance, thus allowing better microscopic observations.
4. Analyze the polished cross section by SEM in backscattered electrons mode to study the sample stratigraphy, the endolithic growth, and the damage extent. The coupled use of an Energy Dispersive X-Ray Spectrometer (EDS) can also provide elemental information of the sample.

3.3 Denaturing Gradient Gel Electrophoresis

Community-level studies in SAB on artistic stone and frescoes are relying more and more on culture-independent methods based on the analysis of DNA (or RNA). The amount of sample needed to study microbial communities using such molecular techniques is smaller than for cultivation efforts (only a few mg of sample are required), and this is a major advantage for studies in the cultural heritage field. In addition, culture-independent molecular methods allow the study of a much broader spectrum of microorganisms residing in SAB as both the cultivable and non-cultivable fractions of the microbial community are identified. Here, the PCR-denaturing gradient gel electrophoresis (PCR-DGGE) protocol is illustrated (*see* **Note 49**).

3.3.1 Total DNA Extraction from Environmental Samples

Carry out all procedures in aseptic conditions (*see* **Note 50**).

1. Preferably on the same day as the sampling by sterile scalpel (*see* **Notes 51–53**), pulverize the samples by mortar and pestle (*see* **Note 54**) in sterile conditions, and put the powder into 2.0 mL tubes with 550 mL of TE. Store the samples at –20 °C.
2. Subject samples to three thermal cycles –80/+70 °C for 2–3 min each cycle using the thermomixer set at 70 °C and –80 °C for the freezer cycle (*see* **Note 55**).
3. Add 50 μL of lysozyme solution (35 mg/mL) and incubate samples at 37 °C for 30 min (*see* **Note 56**).
4. Add 53 μL of proteinase K (20 mg/mL) and incubate samples at 37 °C for 60 min (*see* **Note 57**).
5. Add 200 μL of the 10 % SDS solution (*see* **Note 58**) and incubate at 37 °C for 30 min; at the same time, incubate the CTAB at 37 °C for 10 min.
6. Add 100 μL of 5 M NaCl and 80 μL of CTAB/NaCl solution (*see* **Note 59**). Mix carefully and incubate at 65 °C for 10 min.
7. Add an equal volume of phenol–chloroform–isoamyl alcohol (25:24:1). Mix and centrifuge 10 min at 16,000 × *g*. Transfer the supernatant to new sterile 2 mL tubes (*see* **Note 60**).
8. Add an equal volume of chloroform–isoamyl alcohol (24:1). Mix and centrifuge 10 min at 16,000 × *g*. Transfer the supernatant to new sterile 2 mL tubes (*see* **Note 60**).
9. Add 0.6 volume of isopropyl alcohol. Leave the samples 1 h at room temperature (*see* **Note 61**).
10. Centrifuge at 13,000 × *g* for 5 min and carefully remove the supernatant (*see* **Note 62**).
11. Wash the pellet of DNA with 500 μL of 70 % ethanol.
12. Centrifuge at 16,000 × *g* for 5 min and carefully remove the ethanol. Leave the pellet to dry in thermomixer at 37 °C for 15 min.

13. Add 50 μL of TE buffer pH 8 (*see* **Note 63**). Store at −20 °C.
14. Measure the DNA concentration by a NanoDrop spectrophotometer (*see* **Note 64**).

3.3.2 Amplification of Microbial Communities

Analyze the bacterial communities, particularly the cyanobacterial and the actinobacterial communities, by amplifying 16S rRNA gene fragments; the algal communities, by amplifying 18S rRNA gene fragments; and fungal communities, by amplifying the internal transcribed spacer (ITS) region fragments by PCR. Chemical and thermal conditions of PCRs depend on the used primers and must be optimized case-by-case (*see* **Note 65**). Carry out all the amplification reactions in a laminar flow cabinet with sterilized tools. Verify the amplification reaction results by agarose gel electrophoresis, performed as follows:

1. In an Erlenmeyer flask add an appropriate volume of 0.5× TBE and 1.2 % (wt/v) of agarose. Close the Erlenmeyer flask using a sheet of waxed paper and a rubber band (*see* **Note 66**).
2. Place flask in a microwave oven and heat to obtain an agarose solution (*see* **Note 67**).
3. Transfer the agarose solution to a gel tray (*see* **Note 68**) equipped with a multi-well comb and wait for the polymerization of the gel.
4. Remove the comb and transfer the gel to the tank submerged in 0.5× TBE.
5. Load 2 μL of PCR product together with 2 μL of gel loading solution, and 2 μL of marker, into the gel wells (*see* **Note 69**).
6. Set the power supply at 100 V and start the run, stopping it when the blue loading solution is almost at the end of the gel (*see* **Note 70**).
7. Carefully remove the gel and stain it in the 0.1 % ethidium bromide solution for 20 min (*see* **Note 71**).
8. Wash the gel in demineralized water.
9. Take pictures of the stained gel directly on a UV transilluminator by the gel documentation system.

3.3.3 Analysis of Obtained Amplicons by DGGE

1. Clean and degrease the glass plates with 70 % ethanol solution.
2. Place the glass plates and spacers together with the sandwich clamps in the casting stand, where a rubber strip placed at the bottom prevents leakage.
3. Assemble the gel sandwich: place the 16 cm glass plate on top of the large plate, taking care to correctly place the spacers along each edge of the plate assembly. Attach the plate clamps (tight enough to hold everything together) and place the

entire assembly into the rear slot of the pouring stand. Loosen the clamps slightly and use the spacing card to ensure proper spacer alignment. Fix the plate clamps (tight, to prevent "leakage") and remove the plate assembly from the pouring stand. Inspect the plate assembly to ensure that the two glass plates and the spacers form a flush surface along the bottom of the assembly. If the surface is not flush, reset the plate assembly, as breaches in the seal of the plate assembly with the bottom of the pouring stand will result in leaking gel solutions. Place the glass plates in the casting stand, where a rubber strip lies along the bottom to prevent leakage (*see* **Note 72**).

4. Place the gradient maker on a magnetic stirrer, with a magnetic stir bar in each well to mix the solutions during gel preparation. Place the far end of the small (10 cm) delivery tubing between two glass plates.
5. Create 11.5 mL of the two proper solutions with the highest denaturant concentration and with the lowest denaturant concentration by mixing the 0 % and the 100 % denaturing solutions (*see* **Note 73**).
6. Add 50 μL APS and 15 μL TEMED to the proper high- and low-denaturing solutions and mix well immediately (*see* **Note 74**).
7. Quickly pour the high-denaturing solution into the right leg of the gradient maker (on the pump side) and the low-denaturing solution into the left leg. Make sure the pump is off and the gradient maker-channels are closed. Open the gradient maker-channels and simultaneously start the pump at 10 mL/min. When the solutions mix and reach the bottom of the glasses, decrease the speed of the pump to 2.5 mL/min. Stop the flow when the solutions are finished (*see* **Note 75**).
8. Remove the needle and rinse the gradient maker and tubing with Milli-Q water with the pump at the max flow speed.
9. Carefully, without disturbing the gel solution, add approximately 2 mL of 1× TAE buffer to the gel solution to form a layer on top of the gel solution, approximately 5 mm thick. This layer will help the top boundary of the gel to be smooth.
10. Leave the gel to polymerize 2–3 h.
11. Remove the 1× TEA buffer with a syringe or a sheet of blotting paper.
12. Clean a comb with 70 % ethanol solution and place the comb between the glass plates (*see* **Note 76**).
13. Make the staking gel solution: 3 mL of 0 % denaturing solution, 2 μL of TEMED, and 40 of APS.
14. Drop the staking gel solution between the glass plates by pipette. Be sure to avoid trapping any air bubbles (*see* **Note 77**). Let the gel polymerize from at least 1 h to overnight (*see* **Note 78**).

15. Carefully remove the comb from the glass plates.
16. Clean the wells immediately with 1× TAE.
17. Add approximately 7 L of 1× TAE and fill the buffer chamber, put about 0.5 L aside for later use.
18. Switch on the heater and recirculation pump. Preheat the buffer in the universal mutation detection apparatus to 60 °C; this will take about 1 h.
19. When the temperature is about 50 °C, interrupt the heating and attach the gel plates to the core assembly. Two sets of plates must be attached. If only one gel is to be run, the other set of plates will be an assembly of two plates with no spacers or gel. Be sure that the smallest glass plates and the rubber gasket of the core seal hermetically.
20. Place the core assembly in the buffer chamber, fill the top reservoir with remaining buffer, and check that the upper part does not leak (*see* **Note 79**). Check the buffer height: surface should reach at least "run," better between "run," and "maximum."
21. Flush each well with buffer by a syringe to remove any unpolymerized acrylamide. Failure to do this might result in uneven well floors and unresolved bands (*see* **Note 80**).
22. Continue heating to 60 °C. Do not add plate assembly when the buffer is too hot (>55 °C) as this will cause the plates to crack.
23. Make a prerun at 60 V for 1 h (*see* **Note 81**).
24. Load approximately 70 ng of DNA from the PCR product with 3 μL of denaturing loading blue to each sample (*see* **Note 82**).
25. Run at 200 V for 10 min before turning on the recirculation pump so as to minimize sample loss through washing sample out of the wells (*see* **Note 81**).
26. Run at 90 V for 16 h at 60 °C (*see* **Note 81**).
27. Remove the glass plates from the gel clamps. Carefully separate the plates, leaving the gel exposed on the small plate. Use a spacer to trim the well walls, but be sure to leave the left-most wall slightly higher than the others to use as a reference. For easy manipulation, the gel should be transferred to the staining box in which it is stained and transported.
28. Stain the gel for 20 min in 400 mL of SYBR Green solution at the dark.
29. Remove the SYBR Green from the staining box and wash the gel in 400 mL of demineralized water for 10 min in the dark. Repeat the washing another two times for 5 and 1 min, respectively.
30. Slide the gel off the glass plate on to a UV transilluminator and photograph the gel by the Gel Doc apparatus.

31. Excise all bands and transfer them in a 1.5 mL tube and elute the bands in 50 μL of TE. Store the bans in a thermomixer at 37 °C for 4 h and overnight at 4 °C. Store the bands at −20 °C.
32. Amplify again the DNA as reported in Subheading 3.3.2, with primers without GC clamp.
33. Sequence the amplified bands and analyze the sequences using the sequences databases.

4 Notes

1. It is a hazardous solution and can cause cross-linking. Gloves and safety glasses must always be worn, and the solution must be made under a chemical fume hood.
2. While stirring the cloudy suspension on a hotplate at 60 °C, add 1–2 drops of 1 M NaOH. The solution should clear immediately. Take care the solution does not boil. Once the powder has dissolved and the solution clears, turn off the heat and equilibrate to pH 7 with diluted HCl. Filter while still warm (37–40 °C) through a 0.45 μm filter membrane.
3. The success of the FISH technique depends on the accessibility of the intracellular target site. The complicated cell wall structures of many rock-inhabiting bacteria and fungi (mostly melanized) can hinder the entry of the probe and the emission of the fluorescent signal [24]. Alternative effective permeabilization methods employing enzymes (e.g., lysozyme and glucanase for Gram positive and fungi, respectively), surfactants, osmotic, and microwave irradiation strategies can be used [25–28]. There is a trade-off between permeability and cell integrity that must be taken into account to optimize results for individual sample types.
4. Add sodium dodecyl sulfate last to avoid precipitates. During the preparation of the hybridization buffer, place SDS in the lid of the 2 mL centrifuge tube so it is added only when the lid is closed and shaken.
5. Deionized formamide is a toxic solution and should be handled with care, avoiding contact with eyes and skin and also avoiding inhalation and ingestion. Wear protective gloves and handle under the fume hood. Store formamide at 4 °C.
6. The formamide concentration is dependent on the probe used and determines the stringency of the hybridization. Stringency is a characteristic of the hybridization buffer that denotes the degree of homology between the probe and its rRNA target site. Adjustment of the salt concentration and the temperature at which the hybridization reaction is conducted may also modulate hybridization stringency. While formamide concentration

and temperature are directly proportional to stringency, salt concentration is inversely correlated to stringency. We find that it is more convenient to keep the incubator set at one temperature and to modulate the stringency by adding formamide. The hybridization solution containing 40 % formamide is optimized for the FISH probes, which are ~50-nt long with a GC content of 50 %. Lower formamide concentration will favor nonspecific probe binding, whereas higher formamide concentration will prevent even the specific binding of probes to target rRNA.

7. The probe stock usually comes dried and needs to be diluted in sterile water according to manufacturer instructions. Alternatively, measure a small aliquot in the photometer and use the following formula: 1 OD_{260} ~ 20 ng/μL. Keep the probe solution on ice under dark conditions. Labeled probes can be stored at −20 °C for several months.

8. The washing buffer is used to remove the unbound probe and minimize background fluorescence and should be prepared at the same stringency as the hybridization buffer. To avoid using excess amounts of formamide, thus minimizing operator risks and hazardous waste, washing buffer stringency is achieved by decreasing the salt concentration (Table 1).

9. Concanavalin A selectively binds to α-mannopyranosyl and α-glucopyranosyl residues. It is used to visualize the polysaccharidic component of the extracellular polymeric matrix as polysaccharides are a major fraction of the EPS matrix [29]. A broad selection of Concanavalin A conjugates is commercially available. Select appropriate fluorochromes for multicolor analysis.

10. In most cases a small percentage of the conjugate will remain as a visible aggregate in solution. For longer storage, divide the solution into aliquots and freeze at −20 °C. Protect from light and avoid repeated freezing and thawing.

11. DAPI is a known mutagen and should be handled with care. For long-term storage, the stock solution can be aliquoted and stored at −20 °C. For short-term storage, the solution can be kept at 2–6 °C, protected from light. When handled properly DAPI solutions are stable for at least 6 months. The excitation maximum for DAPI bound to dsDNA is 358 nm, and the emission maximum is 461 nm. DAPI can be excited with a xenon or mercury-arc lamp or with a UV laser.

12. The biofilm matrix can contain considerable protein amounts. In order to visualize the proteinaceous component of the extracellular polymeric matrix, 10 μg/mL of the amine-reactive dye BODIPY 630/650-X SE can be used.

13. Although polysaccharides and proteins are the major fractions of the EPS matrix, biofilms of various origins have been found

to contain extracellular DNA (eDNA). eDNA can be detected with DAPI.

14. Centrifuge the dye solution briefly in a microcentrifuge before use; only the supernatant should then be added to the experiment. For short-term storage, the solution can be kept on ice, protected from light.
15. A simple homemade humidity chamber can be made as follows: place filter papers saturated in water at the bottom of a plastic box. Put two glass rods on the top of the filter paper and place the slides on top of these; then put the lid on the box and place the box in an incubator at 45 °C. This forms a chamber with relative humidity of 60–70 %.
16. Use caution when working with the solid sodium cacodylate and the buffer solutions: the compounds are toxic and contain arsenic.
17. The solution must be prepared and used under a chemical fume hood.
18. To make the samples conductive to the electron beam. Conductive materials give a high secondary electron yield. A higher secondary electron signal allows better images. Also in backscattered electrons mode, conductive materials give a high backscattered emission.
19. The sputtering method is fast and easy but can cause a slight heating of the sample. The thickness of the sputtered layer depends on the current, sputtering time, target to sample distance, and material being sputtered. The evaporation method allows a lower heating of the sample, but it is usually slower than sputtering and less controllable as for the thickness of the coating layer. It needs a big vacuum chamber with long distance between the target and sample to minimize sample heating.
20. Alternatively in TE buffer added with 50 % glycerol to be stored for longer times.
21. Do not dissolve in 100 mL of deionized/Milli-Q water. In most cases, solution volume increases when the large amount of solute is dissolved in solvent.
22. To be stored in cold room at 2–8 °C sealed with parafilm.
23. It is a hazardous solution. Gloves and safety glasses must always be worn, and the solution must be manipulated under a chemical fume hood.
24. Alternatively, an agarose gel electrophoresis equipment.
25. PCR reagents can be purchased ready.
26. The CYA781Ra and CYA1781Rb primers (Table 2) are specific for the amplification of 16S rRNA gene segments from cyanobacteria and plastids [6].

27. Ethidium bromide is one of the most widely used nucleic acid stains in molecular biology laboratories. However, being both highly toxic and a mutagen, some safer alternatives are available as SYBR Safe™ DNA gel stain.
28. The gels must stay horizontally in the staining boxes immersed in the dye solution. The staining boxes are equipped with a hole to drain the dye.
29. Make fresh when required or aliquot 0.5 mL into 1.5 mL tubes and freeze until needed.
30. Make fresh SYBR Green every 7–8 colorations or 2 weeks.
31. Inorganic material collected with the adhesive tape and surrounding the microbial community can also be fluorescent. The washing step aims to remove residual salts, reducing sample autofluorescence. However, the geometric shape of rock-salt crystal structures can be easily recognized and discerned from the fluorescent signal associated to the biomass.
32. Residual salt crystals can remain after evaporation when air-drying the sample, causing background fluorescence. Stock samples in 1:1 (v/v) water/ethanol for immediate use or in PBS/ethanol for long storage.
33. Do not store slides after this pre-hybridization treatment. Proceed with the in situ hybridization within hours.
34. Comprehensive resources for >600 probe sequences (e.g., probeBase and SILVA dataset) that provide various online tools for checking specificity and coverage are available online [30–32]. A list of the most popular group-specific probes is provided in Table 3.
35. Specificity of the rRNA hybridization should be checked by incubating an additional area of the fixed samples for 1 h at 37 °C in 0.5× PBS, pH 7.4, with and without 0.5 mg/mL RNase A and 83 U/mL RNase I. This sample represents the negative control.
36. Avoid trapping air bubbles between the plastic sealer and the slide. Air bubbles can hamper hybridization. To avoid these, pipette the volume of hybridization mixture into the frame at one end and roll the flexible plastic sealer over the frame, beginning at the end where the mixture is added.
37. Overnight hybridization at 37 °C can also be performed for convenience. However, long incubation times might increase background fluorescence noise.
38. During the washing step, increase the water bath temperature by 2 °C, with respect to the hybridization temperature, to prevent nonspecific probe binding and to promote the diffusion of unbound probe outside the cells.

39. Avoid the slide cooling down as temperature is a critical parameter in defining stringency. During this and subsequent steps, protect the sample with aluminum foil to prevent bleaching.
40. Slight background staining of the adhesive tape may occur. In this case, improve the washing step by putting the slide into the 50-mL falcon tube containing the preheated washing buffer, and incubate at 48 °C for at least 20 min.
41. The stream of rinse water is aimed above the selected square area and allowed to cascade over it and off the slide. Do not point the wash bottle stream directly at the chamber with sample. Also, rinse the back of the slide to remove any salts present.
42. Air bubbles can hamper staining.
43. The use of the antifading reagent is very important for the imaging of fluorescent dyes. The choice of a specific antifading reagent depends on the label used. Citifluor AFI (Citifluor, London, UK) and Slow Fade antifade kit (Molecular Probes, Eugene, USA) can be used to protect a wide range of fluorophores from photobleaching.
44. Probe-conferred fluorescence fades much more rapidly than DAPI fluorescence in the microscopic image, and UV excitation will also bleach the probe signal. Thus, it is safer to visualize probe-stained cells first, then, subsequently, all cells from the same field of vision in UV excitation.
45. Microorganisms (e.g., photosynthetic bacteria, eukaryotic microalgae) frequently exhibit a natural autofluorescence from chlorophyll or other pigments. Autofluorescence decay gradually follows cell death. Thus, autofluorescence measurements are best conducted on freshly collected samples. However, autofluorescence phenomena should be taken into consideration in the choice of the fluorochrome.
46. In alternative to the CLSM, a standard wide-field epifluorescence microscope can be used to acquire FISH images. Use an upright microscope equipped with epifluorescence illumination and an appropriate combination of filters (Table 4). A microscope objective with a high numerical aperture (NA) is desirable to collect the maximum number of photons. The fluorescence filter sets are optimized to maximize emission collection and minimize the spectral overlap in case of multicolor FISH.
47. Use a sheet of weighing or other smooth paper to facilitate the collection of fragments and powder material from the substrate during sampling.
48. Make the gold layer as thin as possible (nanometric scale) and keep the coating layer particles as small as possible to obtain well-defined images of the sample's surface.

49. The PCR-DGGE approach is the most frequently used culture-independent method based on the analysis of DNA or RNA for studies in the cultural heritage field. However, the composition and structure of microbial communities living on the surface of stone have also been studied recently by pyrosequencing [33].
50. In a laminar flow cabinet or near a Bunsen burner.
51. Collect less than 15 mg of material.
52. Use a new sterile scalpel for every sample.
53. Use a sheet of paper to gather the sample powder and to put the sample in the tube.
54. If samples need to be pulverized, otherwise neglect.
55. The transfer during the thermal cycles must be fast to ensure strong thermal shock to break the cellular walls.
56. To break down the cell wall for lysing bacteria.
57. To digest protein and remove contamination from preparations of nucleic acid. Addition of proteinase K to nucleic acid preparations rapidly inactivates nucleases that might otherwise degrade the DNA or RNA during purification.
58. SDS is a strong anionic detergent that can solubilize the proteins and lipids that form the membranes. This will help the cell membranes and nuclear envelopes to break down. In addition to removing the membrane barriers, SDS helps release the DNA from histones and other DNA-binding proteins by denaturing them.
59. To remove residues of cellular walls, polysaccharides and unscathed proteins from the solution containing DNA.
60. Be careful to transfer the supernatant without including the substances at the interface.
61. To allow the DNA precipitation.
62. Use 100 μL tips to remove carefully the supernatant. In alternative, use firstly 1,000 μL tips and complete the removal with 100 μL tips.
63. To obtain more concentrated DNA, add 25 or 10 μL of TE buffer.
64. Alternatively, the DNA extraction can be verified by an electrophoresis run with 1 % agarose gel.
65. Bacterial community can be analyzed with primers GC-357 F and 907 R and with the following chemical conditions: 1× of PCR buffer, 1.8 mM of $MgCl_2$, 0.2 mM of dNTP mix, 0.3 μM of each primer, 2 μg/μL of BSA, and 1.25 U of Taq DNA polymerase. Carry out the reactions in a total volume of 25 μL containing 2.5 μL of template DNA (if necessary dilute the

template before the PCR). Thermal cycling program: initial denaturation at 94 °C for 4 min, 10 cycles consisting of denaturation at 94 °C for 30 s, annealing at 61 °C for 1 min and extension at 72 °C for 1 min, 20 cycles consisting of denaturation at 94 °C for 30 s, annealing at 56 °C for 1 min and extension at 72 °C for 1 min, and a final extension step at 72 °C for 10 min [5].

Cyanobacterial communities can be analyzed using the primers CYA359-GC and CYA781R (equimolar mixture between primers CYA781Ra and CYA781Rb) according to Nubel et al. [6] (*see* **Note 26**) and with the following chemical conditions: 1× of PCR buffer, 1.8 mM of $MgCl_2$, 0.2 mM of dNTP mix, 0.5 μM of each primer, 5 % of DMSO, and 0.63 U of Taq DNA polymerase. Carry out the reactions in a total volume of 25 μL containing 3 μL of template DNA (if necessary dilute the template before the PCR). Thermal cycling program: initial denaturation at 94 °C for 5 min, 35 cycles consisting of denaturation at 94 °C for 45 s, annealing at 60 °C for 45 s and extension at 72 °C for 2 min, and a final extension step at 72 °C for 10 min.

Algal communities can be studied with primers NS1-GC and NS2 and with the following chemical conditions: 1× of PCR buffer, 1.8 mM of $MgCl_2$, 0.2 mM of dNTP mix, 0.5 μM of each primer, 5 % of DMSO, and 0.63 U of Taq DNA polymerase. Carry out the reactions in a total volume of 25 μL containing 3 μL of template DNA (if necessary dilute the template before the PCR). Thermal cycling program: initial denaturation at 94 °C for 5 min, 35 cycles consisting of denaturation at 94 °C for 45 s, annealing at 63 °C for 45 s and extension at 72 °C for 2 min, and a final extension step at 72 °C for 10 min [4, 5].

Fungal communities can be investigated by a semi-nested PCR performed with primer combinations NS5 and ITS4 (for the first PCR step) and ITS4 and GC-ITS1 (for the second PCR step). Perform the first amplification step with the following chemical conditions: 1× of PCR buffer, 1.8 mM of $MgCl_2$, 0.2 mM of dNTP mix, 0.5 μM of each primer, and 0.625 U of Taq DNA polymerase. Carry out the reactions in a total volume of 25 μL containing 2 μL of template DNA (if necessary dilute the template before the PCR). Thermal cycling program: initial denaturation at 95 °C for 3 min, 30 cycles consisting of denaturation at 95 °C for 45 s, annealing at 52 °C for 45 s and extension at 72 °C for 2 min, and a final extension step at 72 °C for 10 min. Use the first PCR product as template for the second amplification step. Perform the second amplification step with the following chemical conditions: 1× of PCR buffer, 1.8 mM of $MgCl_2$, 0.2 mM of dNTP mix, 0.5 μM of each primer, 5 % of DMSO, and 0.625 U of Taq DNA polymerase. Carry out the reactions in a total volume of 25 μL containing 2 μL of template DNA (if necessary dilute the template before the PCR). Thermal cycling program: initial denaturation at 94 °C for 5 min, 35 cycles consisting of denaturation at 94 °C for 45 s, annealing at 58 °C for 45 s and extension at 72 °C for 2 min, and a final extension step at 72 °C for 10 min [5, 34]. Alternatively, the fungi community can be studied by amplifying 18S rRNA gene fragments [4].

66. Pierce the waxed paper to allow air to escape during the heating.
67. After some seconds of boiling, check the agarose suspension until it is completely dissolved.
68. Gels may be cast with a UV-transparent plastic tray directly on the gel stage of the electrophoresis bases using the gel casting gates. Gels may also be cast on the removable plastic trays with the aid of the gel caster or with standard laboratory tape.
69. The samples composed by 2 μL of PCR product and 2 μL of gel loading solution can be prepared on a hydrophobic surface (for example parafilm) to be loaded in the agarose gel.
70. Or at least 3/4 of gel length.
71. The staining could require more than 20 min if the ethidium bromide solution is old.
72. If necessary, two rubber strips can also be used to prevent leakage.
73. The suitable denaturing gradient of proper solutions depends on samples; therefore, this must be set up case-by-case in order to obtain an optimal resolution of the different gene fragments. To optimize the analysis, perform a first test with a wide denaturing gradient for the gradient setup, and a second test with a more focused denaturing gradient. In general 40–60 and 30–50 % are wide denaturing gradients for bacteria and fungi, respectively.
74. These reagents begin the polymerization of the acrylamide.
75. Alternatively, stop the flow when the acryl is approximately 0.5 cm under the comb level (the comb will be placed later).
76. To obtain good wells ensure that the comb and spacers are of the same thickness.
77. If necessary, remove air bubbles moving the comb up and down.
78. In case of overnight polymerization, envelop the cast with wet blotting paper and plastic film, and conserve the gel at 4 °C to avoid gel dehydration.
79. In case of leakage from the upper part, ensure that the smaller glass plate and the rubber gasket of the core are tightly sealed.
80. To be sure bring a little bit of loading dye into the wells and flush the wells with 1× TAE by syringe until all dye is removed.

81. Check that the metal filament in the upper part of the core assembly is fully immersed by the buffer to avoid the run interruption. In case of leakage of buffer from the upper part (*see* **Note 79**), the metal filament could not be fully immerged.
82. The volume loaded depends on the success of the PCR and the number of expected products. For example, soil samples produce many products; therefore, the maximum volume should be loaded. Conversely, when running a single PCR product, a few microliters will be enough. In general, biofilm samples from fresco and stone materials do not yield many products. It might be useful to make a first test to calibrate the volume to be loaded.

References

1. ICOMOS International Scientific Committee for Stone (ISCS) (2008) ICOMOS-ISCS Illustrated glossary on stone deterioration patterns. Ateliers 30 Impression, Champigny/ Marne, France
2. Gorbushina AA (2007) Life on the rocks. Environ Microbiol 9:1613–1631
3. Urzì C, De Leo F (2001) Sampling with adhesive tape strips: an easy and rapid method to monitor microbial colonization on monument surfaces. J Microbiol Methods 44:1–11
4. Polo A, Cappitelli F, Brusetti L et al (2010) Feasibility of removing surface deposits on stone using biological and chemical remediation methods. Microb Ecol 60:1–14
5. Polo A, Gulotta D, Santo N et al (2012) Importance of subaerial biofilms and airborne microflora in the deterioration of stonework: a molecular study. Biofouling 28:1093–1106
6. Nubel U, Garcia-Pichel F, Muyzer G (1997) PCR primers to amplify 16S rRNA genes from cyanobacteria. Appl Environ Microbiol 63: 3327–3332
7. White TJ, Bruns T, Lee S, Taylor J (1990) Amplification and direct sequencing of fungal ribosomal RNA genes for phylogenetics. In: Innis MA, Gelfand DH, Sninsky JJ, White TJ (eds) PCR protocols, a guide to methods and applications. Academic, San Diego, pp 315–322
8. Gardes M, Bruns TD (1993) ITS primers with enhanced specificity for basidiomycetes—application to the identification of mycorrhizae and rusts. Mol Ecol 2:113–118
9. Urzi C (2008) Fluorescent in-situ hybridization (FISH) as molecular tool to study bacteria causing biodeterioration. In: May E, Jones M, Mitchell J (eds) Heritage microbiology and science: microbes, monuments and maritime materials. Royal Society of Chemistry, Cambridge, pp 143–150
10. Amann RI, Binder BJ, Olson RJ et al (1990) Combination of 16S rRNA-targeted oligonucleotide probes with flow cytometry for analyzing mixed microbial populations. Appl Environ Microbiol 56:1919–1925
11. Cappitelli F, Principi P, Pedrazzani R et al (2007) Bacterial and fungal deterioration of the Milan Cathedral marble treated with protective synthetic resins. Sci Total Environ 385: 172–181
12. Müller E, Drewello U, Drewello R et al (2001) In situ analysis of biofilms on historic window glass using confocal laser scanning microscopy. J Cult Herit 2:31–42
13. Stahl DA, Amann R (1991) Development and application of nucleic acid probes. In: Stackebrandt E, Goodfellow M (eds) Nucleic acid techniques in bacterial systematics. John Wiley & Sons Ltd, Chichester, pp 205–248
14. Piñar G, Gurtner C, Ramos C et al (2002) Identification of Archaea in deteriorated ancient wall paintings by DGGE and FISH analysis. In: Galan E, Zezza F (eds) Protection and conservation of the cultural heritage of the Mediterranean cities. Balkema, Lisse
15. Manz W, Amann R, Ludwig W et al (1992) Phylogenetic oligodeoxynucleotide probes for the major subclasses of Proteobacteria: problems and solutions. Syst Appl Microbiol 15: 593–600
16. Roller C, Wagner M, Amann R et al (1994) In situ probing of Gram-positive bacteria with high DNA G+C content using 23S rRNA-targeted oligonucleotides. Microbiology 140: 2849–2858
17. Urzi C, La Cono V, Stackebrandt E (2004) Design and application of two oligonucleotide probes for the identification of Geodermatophilaceae strains using fluorescence in situ hybridization (FISH). Environ Microbiol 6: 78–685

18. Schönhuber W, Zarda B, Eix S et al (1999) In situ identification of cyanobacteria with horseradish peroxidase-labeled, rRNA-targeted oligonucleotide probes. Appl Environ Microbiol 65:1259–1267
19. Pawley JB (1995) Handbook of biological confocal microscopy, 2nd edn. Springer, New York, pp 453–467
20. Pawley D, Flinchbaugh J (2006) The current state: progress starts here. Manuf Eng 137:71
21. Gulotta D, Goidanich S, Bertoldi M et al (2012) Gildings and false gildings of the baroque age: characterization and conservation problems. Archaeometry 54:940–954
22. Cappitelli F, Toniolo L, Sansonetti A et al (2007) Advantages of using microbial technology over traditional chemical technology in removal of black crusts from stone surfaces of historical monuments. Appl Environ Microbiol 17:5671–5675
23. Cappitelli F, Salvadori O, Albanese D et al (2012) Cyanobacteria cause black staining of the national museum of the American Indian building (Washington, D.C., USA). Biofouling 28:257–266
24. Urzì C, La Cono V, De Leo F, Donato P (2003) Fluorescent in situ hybridization (FISH) to study biodeterioration. In: Saiz-Jimenez C (ed) Molecular biology and cultural heritage. Balkema Publishers, Lisse, pp 55–60
25. de Vos MM, Nelis HJ (2003) Detection of *Aspergillus fumigatus* hyphae by solid phase cytometry. J Microbiol Methods 55:557–564
26. Teertstra WR, Lugones LG, Wosten HAB (2004) In situ hybridization in filamentous fungi using peptide nucleic acid probes. Fungal Genet Biol 41:1099–1103
27. Prigione V, Marchisio VF (2004) Methods to maximise the staining of fungal propagules with fluorescent dyes. J Microbiol Methods 59:371–379
28. Villa F, Cappitelli F, Principi P et al (2009) Permeabilization method for in-situ investigation of fungal conidia on surfaces. Lett Appl Microbiol 48:234–240
29. Flemming H-C, Wingender J (2010) The biofilm matrix. Nat Rev Microbiol 8:623–633
30. Loy A, Maixner F, Wagner M, Horn M (2007) ProbeBase—an online resource for rRNA-targeted oligonucleotide probes: new features 2007. Nucleic Acids Res 35: D800–D804
31. Pruesse E, Quast C, Knittel K et al (2007) SILVA: a comprehensive online resource for quality checked and aligned ribosomal RNA sequence data compatible with ARB. Nucleic Acids Res 35:7188–7196
32. Quast C, Pruesse E, Yilmaz P et al (2013) The SILVA ribosomal RNA gene database project: improved data processing and web-based tools. Nucleic Acids Res 41:590–596
33. Cutler NA, Oliver AE, Viles HA et al (2013) The characterisation of eukaryotic microbial communities on sandstone buildings in Belfast, UK, using TRFLP and 454 pyrosequencing. Int Biodeterior Biodegr 82:124–133
34. Giacomucci L, Bertoncello R, Salvadori O et al (2011) Microbial deterioration of artistic tiles from the façade of the Grande Albergo Ausonia & Hungaria (Venice, Italy). Microb Ecol 62: 287–298

Chapter 25

Biofilms of Thermophilic Bacilli Isolated from Dairy Processing Plants and Efficacy of Sanitizers

Sara A. Burgess, Denise Lindsay, and Steve H. Flint

Abstract

In many environments, bacteria can attach to a surface and grow into multicellular structures, otherwise known as biofilms. Many systems for studying these biofilms in the laboratory are available. To study biofilms of the thermophilic bacilli in milk powder-manufacturing plants, standard laboratory biofilm techniques need to be adapted. The focus of this chapter is on techniques that can be used for growing and analyzing biofilms of thermophilic bacilli that are isolated from dairy processing plants. These techniques include laboratory methods as well as how to set up a pilot-scale experiment. The laboratory methods consist of a microtiter plate assay, which is used for strain selection, and the CDC reactor, which is used for testing sanitizers and antimicrobial surfaces. In dairy processing, if a new sanitizer or antimicrobial surface appears to be promising, it is useful to carry out pilot-scale experiments before introducing it to a manufacturing plant. We describe how to set up a pilot-scale experiment for testing the efficacy of sanitizers against the thermophilic bacilli.

Key words Biofilm, Thermophiles, Milk, Dairy, *Anoxybacillus*, *Geobacillus*, Sanitizers

1 Introduction

Thermophilic bacilli are a key reason for poor quality in milk powder, caused by the release of bacteria from biofilms growing in preheaters and evaporators in the milk powder-manufacturing plant [1, 2]. Therefore, understanding how these bacteria grow and testing different sanitizers and antimicrobial surfaces against these bacteria are important. A number of different methods have been used for growing and analyzing biofilms of the thermophilic bacilli, including microtiter plate assays, recirculating flow systems [3, 4] and continuous flow-through systems [3, 5]. The method selected depends on what is being analyzed. The common components among all these systems are the use of sterile reconstituted milk as the growth medium and the use of stainless steel as the surface for attachment. We have found that the use of other growth media

Gianfranco Donelli (ed.), *Microbial Biofilms: Methods and Protocols*, Methods in Molecular Biology, vol. 1147,
DOI 10.1007/978-1-4939-0467-9_25, © Springer Science+Business Media New York 2014

such as trypticase soy broth (TSB) and the use of glass as the surface result in less robust biofilms.

Microtiter plate assays are useful for strain selection. Thermophilic bacilli do vary in their ability to form biofilms; therefore, it is important to select the most robust biofilm former before testing sanitizers and antimicrobial surfaces. Microtiter plates also have a higher throughput than flow-through systems. We outline a protocol for growing biofilms of the thermophilic bacilli on stainless steel coupons in microtiter plates. Removal of the attached cells from the coupons and subsequent plate counts can be used for comparing strains.

For testing the efficacy of sanitizers, we have found that the CDC reactor gives the most robust results. In general, biofilms formed under static conditions tend to be more susceptible to sanitizers than those formed using the CDC reactor [6]. As biofilms of thermophilic bacilli form under turbulent flow in milk powder-manufacturing plants, it is important that in the laboratory, the sanitizers being tested against these bacteria are used on a biofilm that has also been formed under turbulent flow.

If a new technology is to be introduced into a dairy processing plant to prevent the formation of biofilms, it is important to test it at a pilot-scale level. The results from a laboratory experiment can vary greatly from those at a pilot-scale level or in a manufacturing plant. We describe how a pilot-scale pasteurizer can be used to test the effect of sanitizers on biofilms of the thermophilic bacilli.

2 Materials

2.1 Microtiter Plate Assay

1. Strains of thermophilic bacilli.
2. Trypticase soya agar (TSA).
3. Milk plate count agar (MPCA).
4. Milk plate count agar + 0.2 % starch (MPCA + S).
5. Reconstituted skim milk (RSM): add 100 g of skim milk powder to 910 mL of water and autoclave at 115 °C for 15 min.
6. 0.1 % peptone.
7. Six-well microtiter plate.
8. 1 cm^2 stainless steel coupons with a grade 2B finish (*see* **Note 1**). Autoclave before use.
9. Sterile plastic tubes.
10. Glass beads.

2.2 CDC Biofilm Reactor

This method is used for testing the efficacy of sanitizers against thermophilic bacilli and is based on the method described by Luppens et al. [7]. It can also be used for testing antimicrobial surfaces.

1. Strains of thermophilic bacilli.
2. TSA.
3. TSB.
4. RSM.
5. 0.1 % peptone.
6. Potassium phosphate buffer.
7. Sanitizers.
8. Difco Neutralizing Buffer.
9. CDC reactor and magnetic plate stirrer (Biosurface Technologies Co., Bozeman, Montana).
10. Masterflex L/S digital economy drive multi-channel peristaltic pump (Cole-Parmer, Thermo Scientific, North Shore City, New Zealand), Masterflex Santoprene tubing (2.79 mm, catalogue number 643-48), and connectors.
11. Red natural rubber connecting tubing (Global Science, Albany, Auckland, New Zealand).
12. Grade 2B surface finish stainless steel coupons (Biosurface Technologies Co.).
13. Waste bucket.
14. Incubator or refrigerator at 4–5 °C.

2.3 Pilot-Scale Pasteurizer

This system incorporates the modified Robbins device, which is based on the Robbins device designed by McCoy et al. [8]. It is a stainless steel unit that can be installed in a section of manufacturing plant.

1. Pilot scale plate heat exchanger (PHE) (e.g., Sondex PHE).
2. Modified Robbins device (Fig. 1).
3. Sterile vacuum sampling tubes (Vacuette® Greiner Labortechnik, Thermo Science) and sterile vacutainer needles.
4. 1.5 % sodium hydroxide.
5. 1 % nitric acid.

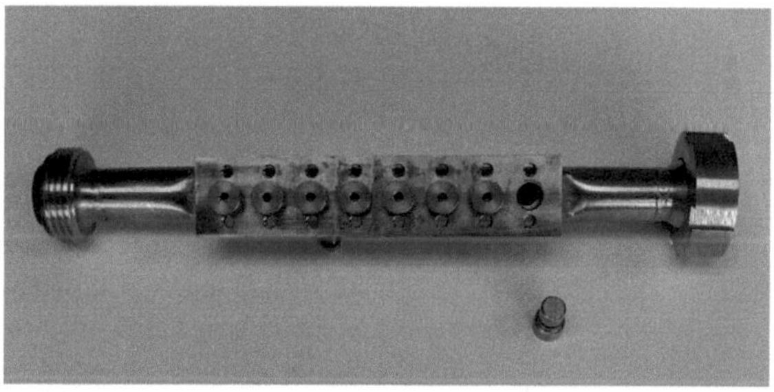

Fig. 1 The modified Robbins device

6. Approximately 60,000 L of raw or pasteurized milk (*see* **Note 2**).
7. Rinse water.
8. MPCA.
9. MPCA + S.
10. TSB.
11. 0.1 % peptone.
12. Sanitizer, e.g., Perform (Orica, New Market, Auckland, New Zealand).

3 Methods

3.1 Microtiter Plate Assay

1. Streak strains of interest on to TSA and incubate overnight at 55 °C.
2. Resuspend the culture in 0.1 % peptone and adjust to an absorbance of 0.1_{600nm} (*see* **Note 3**).
3. Set up a six-well microtiter plate by placing 2–4 stainless steel coupons into each well and add 5 mL of RSM.
4. Transfer 50 μL of the bacterial suspension into each well of the six-well microtiter plate. Incubate for 8 h at 55 °C (*see* **Note 4**).
5. Remove the RSM and replace with 5 mL of fresh RSM.
6. Incubate at 55 °C for a further 8 h.
7. Remove the stainless steel coupons and gently rinse with 10 mL of 0.1 % peptone. Repeat.
8. To remove the attached cells, place the coupon in a sterile plastic tube with 5 g of glass beads and 10 mL of 0.1 % peptone.
9. Mix by vortex for 1 min.
10. For total thermophile counts, carry out serial tenfold dilutions in 0.1 % peptone and plate on to MPCA (*see* **Note 5**).
11. For thermophilic spore counts, heat treat the attached cell suspension at 100 °C for 30 min and carry out serial tenfold dilutions and plate on to MPCA + S (*see* **Note 6**).
12. Incubate the plates at 55 °C for 2 days before counting (*see* **Note 7**).

3.2 CDC Biofilm Reactor

3.2.1 Reactor Setup

1. Autoclave the CDC reactor, tubing, and stainless steel coupons (*see* **Note 8**).
2. Connect the components aseptically as outlined in Fig. 2.
3. Immediately before starting the reactor, pour 100 mL of sterile RSM into the CDC reactor vessel and switch the magnetic stirrer on to a temperature of 55 °C (*see* **Note 9**) and a speed of 200 rev/min.

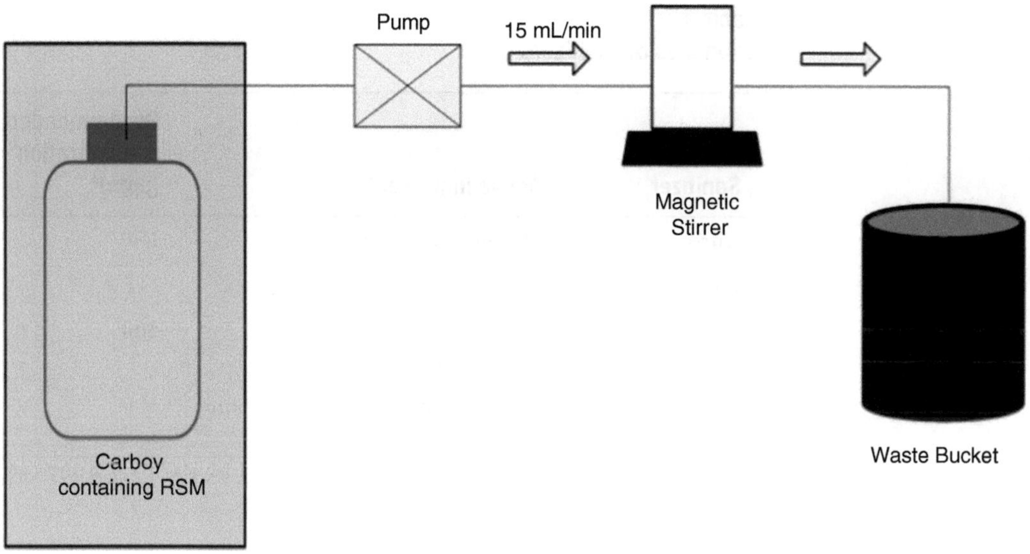

Fig. 2 Schematic diagram of the setup for the CDC reactor

3.2.2 Culture Preparation

1. Streak strains of interest on to TSA and incubate overnight at 55 °C.
2. Subsample one colony into 10 mL of TSB, incubate at 55 °C, and grow until mid-exponential phase (*see* **Note 10**).
3. Dilute the culture by 1/100 in 100 mL of TSB and grow for a further 6–8 h until mid-exponential phase.
4. Transfer the 100 mL of culture into 2 L of sterile RSM.
5. Pour the RSM culture into the sterile carboy and connect to the pump aseptically.

3.2.3 Operation of the Reactor

1. Switch on the peristaltic pump to a flowrate of 15 mL/min (*see* **Note 11**).
2. The RSM will pass through rubber tubing from a storage container, located in the 4 °C incubator, into the CDC reactor.
3. Run for 24 h.

3.2.4 Testing the Efficacy of the Sanitizer

1. Remove the coupons, rinse gently in 10 mL of 0.1 % peptone, and repeat (*see* **Note 12**).
2. For sanitizer exposure, place the coupons in a volume (between 3 and 5 mL) of sanitizer or potassium phosphate buffer in a closed 50 mL tube. Refer to Table 1 for examples of sanitizers [9].
3. After exposure at room temperature for the time chosen (usually 1–5 min), remove the sanitizer solution and add 10 mL of neutralizing buffer.

Table 1
Examples of sanitizers

Sanitizer	Active ingredient	Recommended concentration (ppm)[a]
Citrox	Flavonoids from citrus fruits	150
Iodophor multi	Iodine and iodophor complexes	15
Oxonia active	Hydrogen peroxide and peroxyacetic acid	400
Ster-Bac	n-Alkyldimethylbenzylammonium chlorides	400

[a]These concentrations are based on British Standard BS EN 1276:1997 [8]

4. Place glass beads in the tube and mix the tube by vortex for 1 min to remove the attached cells from the coupon.
5. Prepare plate counts as described in **steps 8–12** in Subheading 3.1.
6. Compare the plate counts from the treated coupons with those from the untreated coupons to determine the efficacy of the sanitizer or antimicrobial surface.

3.3 Pilot-Scale Pasteurizer

3.3.1 Operation of the Pilot-Scale Pasteurizer

1. Set up the pilot-scale system so that the raw milk flows from a balance tank and is heated using a preheater PHE before it passes through a regenerative PHE, followed by a high heater in which it is pasteurized. Then cool the pasteurized milk using the cooling side of the regenerative PHE and an additional cooler (Fig. 3). The temperature profile of the pilot-scale system is illustrated in Fig. 4.
2. Install the modified Robbins device in the section of manufacturing plant post-pasteurization, where temperatures are optimal for the growth of thermophilic bacteria. In this example, it was installed between cooling pass 2 and cooling pass 3.
3. Install rubber septum sample points at various locations throughout the system for taking milk samples. At a minimum, they should be located ex (exiting) balance tank, exregenerative PHE—heating side, ex high heater, and ex regenerative PHE—cooling side.
4. Before starting each run, sanitize the system by circulating 0.2 % Perform at ambient temperature for 15 min; follow with a water rinse for 10 min (*see* **Note 13**).
5. Operate the pilot-scale system under turbulent flow (*see* **Note 14**). In this case, it was operated at a flowrate of 2,500 L/h for 24 h.

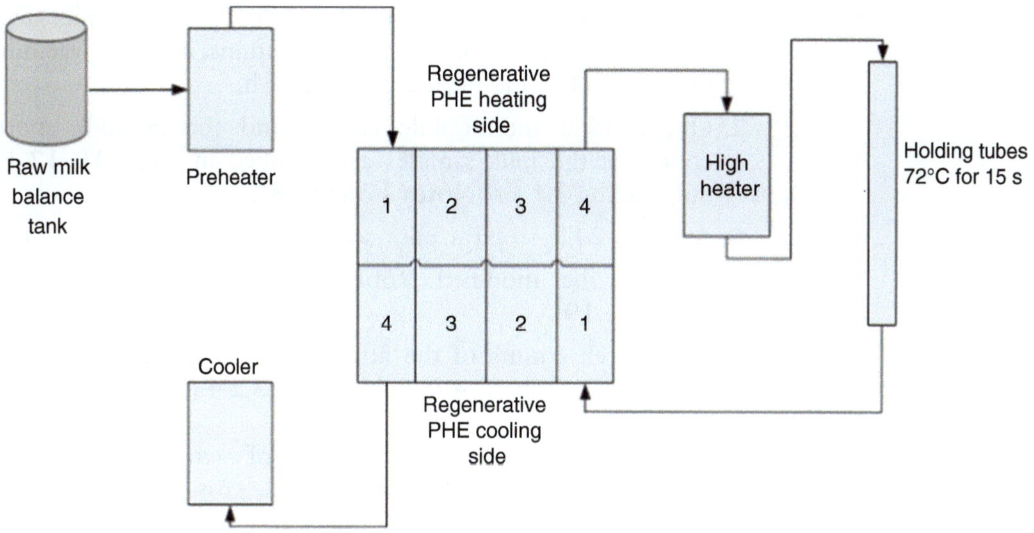

Fig. 3 Schematic diagram of the pilot plant setup

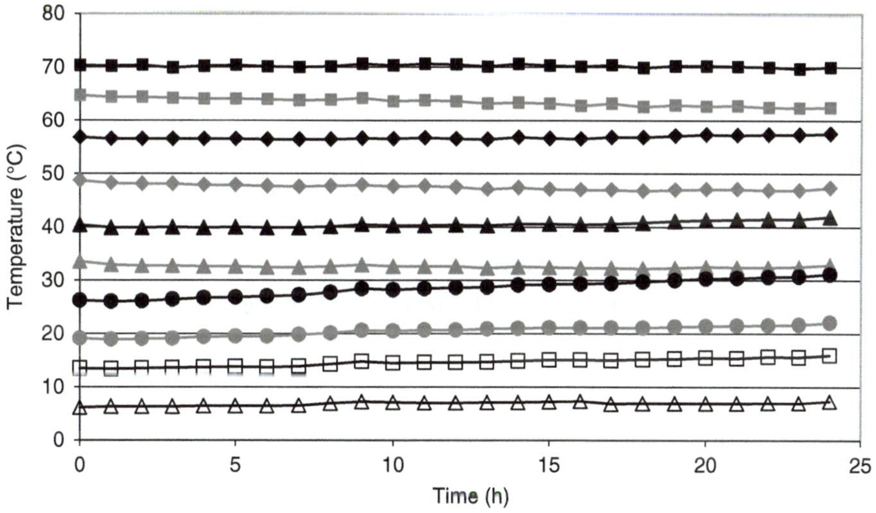

Fig. 4 Temperature profiles through sections of the regenerative PHE: exiting the preheater (*open triangle*), exiting section 1 of the heating side of the regenerative heater (*grey circle*), exiting section 2 of the heating side of the regenerative heater (*grey triangle*), exiting section 3 of the heating side of the regenerative heater (*grey diamond*), exiting section 4 of the heating side of the regenerative heater (*grey square*), exiting section 4 of the cooling side of the regenerative heater (*black square*), exiting section 3 of the cooling side of the regenerative heater (*black diamond*), exiting section 2 of the cooling side of the regenerative heater (*black triangle*), exiting section 1 of the cooling side of the regenerative heater (*black circle*), and exiting the cooler (*open square*)

3.3.2 Preparation of the Culture

1. Prepare the culture as described in **steps 1–3** of Subheading 3.2.2.
2. Transfer 10 mL of the culture into 2 L of TSB and incubate for a further 4–6 h until mid-exponential phase (*see* **Note 15**).
3. Pump the culture at a rate of 5 mL/min into the balance tank throughout the run (*see* **Note 16**).

3.3.3 Sampling Regime

1. To monitor the growth of thermophilic bacteria in milk over time, take samples from the rubber septums, using the vacuum sampling tubes and needles, every 2–4 h.
2. Obtain total thermophile counts and thermophilic spore counts of the milk samples as described in **steps 10–12** in Subheading 3.1 (*see* **Notes 17** and **18**).
3. After 18–24 h, stop the pilot-scale system and rinse with water.
4. Remove the modified Robbins device from the system (*see* **Note 19**).
5. Obtain cell counts of the attached cells on the coupons as described in **steps 1–4** in Subheading 3.2.4 (*see* **Note 20**).

3.3.4 Opening and Sampling of the PHE

6. Open the PHE and swab at the end of each run in the optimum temperature zone for thermophile growth (*see* **Note 21**).
7. After swabbing, remove the sponge end of the swab and place in a stomacher bag with 10 mL of 0.1 % peptone.
8. Stomach the swab for 2 min.
9. Carry out serial dilutions in 0.1 % peptone and prepare plate counts as described in **steps 10–12** in Subheading 3.1.

3.3.5 Clean-in-Place Regime (CIP)

10. For a standard CIP, flush the system with 1.5 % sodium hydroxide at 76 °C for 30 min, rinse with water at ambient temperature for 10 min, flush with 1 % nitric acid wash at 65 °C for 30 min, and then rinse again with water at ambient temperature (*see* **Note 22**).
11. Open the PHE to check that there is no visible foulant/biofilm. If there is, repeat the CIP.

4 Notes

1. Prior to use for the first time, passivate the stainless steel coupons in 50 % nitric acid at 70 °C for 30 min and rinse with water. For subsequent use, clean with 5 % Pyroneg (pyrogenically negative cleaner, Thermo Fisher Scientific.) and rinse with distilled water. Coupons composed of other materials, such as rubber, polyurethane, or polyvinylchloride, should be sterilized as appropriate.
2. This volume is based on operating the PHE for 24 h at a flowrate of 2,500 L/h.
3. A bent 200 μL pipette tip can be used to scrape off colonies. As colonies of the dairy thermophilic bacilli are generally quite small, scraping off half a plate into 3 mL of 0.1 % peptone should give an OD_{600nm} of approximately 0.2–0.5. Additional peptone can then be added to bring the OD_{600nm} down to 0.1.

4. Incubation for longer than 8 h can result in coagulation of the milk. To prevent coagulation, 100 mM MOPS (3-(N-morpholino) propanesulfonic acid) can be added to the medium.
5. Plate counts can be prepared using pour plates [2] or droplet plates [10]. If the droplet plate technique is used, the plates need to be prepared at least 2 days in advance to allow time for drying.
6. When spores are expected to be in low numbers, a 1 mL spread plate can be included in the test procedure.
7. If the droplet plate technique is used, the incubation time should be reduced to 1 day.
8. Stainless steel coupons can be autoclaved in the holders within the CDC reactor vessel. For permanent antimicrobial surfaces, sterilize as recommended by the manufacturer and place in the coupon holder aseptically. Untreated stainless steel coupons should also be included as a control. For temporary antimicrobial surfaces, dip sterile stainless steel coupons into the antimicrobial agent and place into the coupon holder. A separate control run should be carried out.
9. To ensure that the CDC reactor remains at a constant temperature of 55 °C, it can be placed in a small water bath on top of the magnetic stirrer.
10. Mid-exponential phase should be reached within 6–8 h of incubation. If the thermophilic bacilli are grown for too long, they will reach stationary phase and cells will start to die off.
11. Biofilms of *Geobacillus stearothermophilus* have a doubling time of approximately 25 min [3]; therefore, the CDC reactor must be run at a flowrate of greater than 14 mL/min to ensure that the residence time is less than the doubling time.
12. If antimicrobial surfaces are being tested instead of sanitizers, after rinsing the coupon, the attached cells can be removed and counted (Subheading 3.1, **steps 8–12**). Alternatively, surfaces can be swabbed to remove bacterial cells as described in ISO 18593:2004 [11].
13. If an antibacterial agent is being tested for its effect on extending the run length, replace the sanitizer with the antibacterial agent. Prepare the antibacterial agent according to the manufacturer's instructions. In this case, individual runs would need to be compared with and without the antibacterial agent and no sanitizer would be used at the end of the run.
14. The flowrate will depend on the system being used.
15. The thermophile count of the inoculum should be approximately 10^7 cfu/mL.
16. The thermophile count of the raw milk after the addition of the inoculum should be approximately 10^3 cfu/mL.

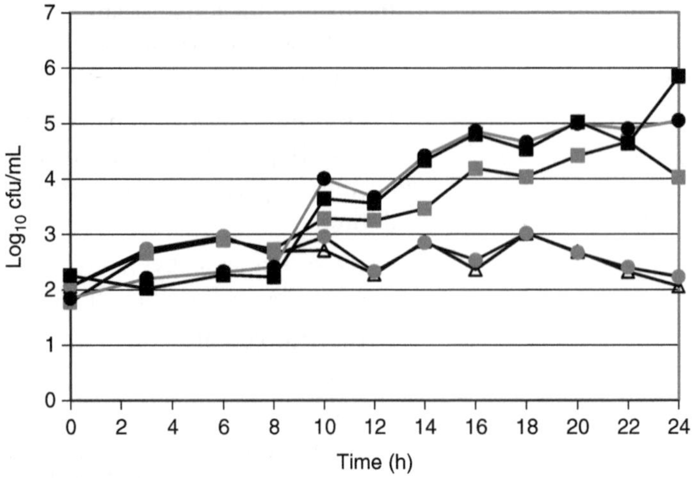

Fig. 5 Total thermophile counts in the milk from a pilot-plant-scale PHE inoculated with *G. stearothermophilus* throughout a 24-h run: exiting the preheater (*open triangle*), exiting section 1 of the heating side of the regenerative heater (*grey circle*), exiting section 4 of the heating side of the regenerative heater (*grey square*), exiting section 4 of the cooling side of the regenerative heater (*black square*), and exiting section 1 of the cooling side of the regenerative heater (*black circle*)

17. Thermophiles may be present in low numbers in raw milk. To confirm that the culture used to inoculate the milk is the same strain as the strain that has grown in the pasteurizer system, 5–10 colonies should be typed using a method such as random amplification of polymorphic DNA (RAPD) [12].
18. Thermophile counts should increase over time in section 4 of the heating side of the regenerative heater as well as in the cooling side. Figure 5 illustrates the results expected for thermophile counts in the milk over time.
19. To remove coupons from the modified Robbins device during the run, the system must be stopped and the coupons replaced with sterile coupons. Therefore, it is recommended that the coupons be removed only at the end of a run.
20. Alternatively, the biofilm can be visualized using epifluorescent microscopy by staining with an epifluorescent stain such as the LIVE/DEAD stain [13].
21. A sterile stainless steel frame should be used so that a defined area can be swabbed each time. The area can be swabbed according to the method described in ISO 18593:2004 [11].
22. If the first water rinse is milky in color, the caustic wash should be repeated.

References

1. Murphy PM, Lynch D, Kelly PM (1999) Growth of thermophilic spore forming bacilli in milk during the manufacture of low heat powders. Int J Dairy Technol 52:45–50
2. Scott SA, Brooks JD, Rakonjac J et al (2007) The formation of thermophilic spores during the manufacture of whole milk powder. Int J Dairy Technol 60:109–117
3. Flint SH, Palmer J, Bloemen K et al (2001) The growth of *Bacillus stearothermophilus* on stainless steel. J Appl Microbiol 90: 151–157
4. Seale RB, Flint SH, McQuillan AJ et al (2008) Recovery of spores from thermophilic dairy bacilli and effects of their surface characteristics on attachment to different surfaces. Appl Environ Microbiol 74:731–737
5. Burgess SA, Brooks JD, Rakonjac J et al (2009) The formation of spores in biofilms of *Anoxybacillus flavithermus*. J Appl Microbiol 107:1012–1018
6. Buckingham-Meyer K, Goeres DM, Hamilton MA (2007) Comparative evaluation of biofilm disinfectant efficacy tests. J Microbiol Methods 70:236–244
7. Luppens SBI, Reij MW, van der Heijden RWL et al (2002) Development of a standard test to assess the resistance of *Staphylococcus aureus* biofilm cells to disinfectants. Appl Environ Microbiol 68:4194–4200
8. McCoy WF, Bryers JD, Robbins J et al (1981) Observations of fouling biofilm formation. Can J Microbiol 27:910–917
9. British Standards Institution (1997) Chemical disinfectants and antiseptics. Quantitative suspension test for the evaluation of bactericidal activity of chemical disinfectants and antiseptics used in food, industrial, domestic, and institutional areas. Test method and requirements (phase 2, step 1). British Standard BS EN 1276:1997
10. Lindsay D, von Holy A (1999) Different responses of planktonic and attached *Bacillus subtilis* and *Pseudomonas fluorescens* to sanitizer treatment. J Food Prot 62:368–379
11. International Organization for Standardization (2004) ISO Standard 18593:2004. Microbiology of food and animal feeding stuffs—horizontal methods for sampling techniques from surfaces using contact plates and swabs
12. Ronimus RS, Parker LE, Turner N et al (2003) A RAPD-based comparison of thermophilic bacilli from milk powders. Int J Food Microbiol 85:45–61
13. Lindsay D, Brözel VS, Mostert JF et al (2002) Differential efficacy of a chlorine dioxide-containing sanitizer against single species and binary biofilms of a dairy-associated *Bacillus cereus* and a *Pseudomonas fluorescens* isolate. J Appl Microbiol 92:352–361

INDEX

Gianfranco Donelli (ed.), *Microbial Biofilms: Methods and Protocols*, Methods in Molecular Biology, vol. 1147, DOI 10.1007/978-1-4939-0467-9, © Springer Science+Business Media New York 2014

MIX
Papier aus verantwortungsvollen Quellen
Paper from responsible sources
FSC® C105338

If you have any concerns about our products,
you can contact us on
ProductSafety@springernature.com

In case Publisher is established outside the EU,
the EU authorized representative is:
Springer Nature Customer Service Center GmbH
Europaplatz 3, 69115 Heidelberg, Germany

Printed by Libri Plureos GmbH
in Hamburg, Germany